NONLINEAR CHANNEL MODELS AND THEIR SIMULATIONS

NONLINEAR CHANNEL MODELS AND THEIR SIMULATIONS

Yecai Guo

Nanjing University of Information Science and Technology, China
Wuxi University, China

清華大學出版社
TSINGHUA UNIVERSITY PRESS

World Scientific

Published by

World Scientific Publishing Co. Pte. Ltd.

5 Toh Tuck Link, Singapore 596224

USA office: 27 Warren Street, Suite 401-402, Hackensack, NJ 07601

UK office: 57 Shelton Street, Covent Garden, London WC2H 9HE

Library of Congress Cataloging-in-Publication Data

Names: Guo, Yecai, author.

Title: Nonlinear channel models and their simulations / Yecai Guo, Nanjing University of Information Science and Technology, China, Wuxi University, China.

Description: Singapore ; Hackensack, NJ : World Scientific Publishing Co. Pte. Ltd., [2022] | Includes bibliographical references and index.

Identifiers: LCCN 2021061681 | ISBN 9789811249440 (hardcover) | ISBN 9789811249457 (ebook for institutions) | ISBN 9789811249464 (ebook for individuals)

Subjects: LCSH: Signal processing--Digital techniques. | Radio frequency modulation--Transmitters and transmission--Simulation methods. | Nonlinear systems.

Classification: LCC TK5102.9 .G879 2022 | DDC 621.382/2--dc23/eng/20220222

LC record available at https://lccn.loc.gov/2021061681

British Library Cataloguing-in-Publication Data

A catalogue record for this book is available from the British Library.

Nonlinear Channel Models and Their Simulations

For any available supplementary material, please visit https://www.worldscientific.com/worldscibooks/10.1142/12647#t=suppl

Desk Editors: Balasubramanian Shanmugam/Steven Patt

Typeset by Stallion Press
Email: enquiries@stallionpress.com

Summary of Contents

The key technology in the field of satellite communication is satellite channel modeling and equalization, and the research on this technology is helpful in improving the feasibility and quality of satellite communication system services. In this book, the satellite communication channel is taken as the research object, and the nonlinearity of the satellite communication channel is taken as the foothold, so research on the establishment of a nonlinear channel model and the design of equalization technology is carried out.

This book is suitable for engineers, researchers, and postgraduates, who are engaged in information and communication engineering, control science and engineering, intelligent science and technology, and other disciplines!

Preface

With the growth of information globalization, the development of Internet, data multimedia communication, and video and audio services, and the growing need for communication individualization, mobility, and full coverage, satellite communication is taking advantage of the advantages of long communication distance, which has nothing to do with the distance communication cost, large coverage area without being limited by the geographical conditions, and wide communication frequency band, as well as large transmission capacity that is suitable for the transmission of a variety of services to develop in the direction of its unique advantages. However, the feasibility and quality of services provided by satellite communication system are largely affected by the channel characteristics between mobile terminals and communication satellites. In order to study these effects, it is an effective and feasible solution to adopt a channel model that can reflect the real characteristics of a satellite channel. It is no exaggeration to say that only a channel model close to the actual characteristics of a satellite channel is not established until the mechanism of fading, distortion of satellite signal in channel transmission is revealed and the scheme to eliminate fading, distortion of satellite signal is further studied. It is difficult to design a scheme that can effectively eliminate the fading, distortion of the satellite signals, or ill-considering the satellite communication environment.

This book is a collection of the achievements of the author and his team in the research of nonlinear channel model establishment and equalization technology based on the nonlinearity of satellite communication channel for many years. It mainly deals with the research contents of the satellite channel modeling and equalization theory, technology, and application based on the established satellite channel model. It has formed the theoretical and technological achievements system from satellite channel modeling to satellite channel equalization, which embodies the advanced, practical, and systematic content and is clear and readable. It helps readers understand and master this topic. The book consists of seven chapters. In Chapter 1, based on the analysis of the influencing factors of satellite communication quality, the research status of satellite channel modeling and equalization technology at home and abroad is summarized, and the research content from the perspective of system theory is put forward. Chapter 2 briefly analyzes the basic composition, basic parameters, layered propagation characteristics of satellite channel, classical satellite channel model and second-order statistical characteristic parameters, and satellite channel model simulation method, and so on. In Chapter 3, the satellite channel states in atmospheric environments and ground environments are analyzed, and the multi-state Markov chain model of satellite channel is established and simulated. In Chapter 4, aiming at the nonlinearity caused by traveling ware tube (TWT) amplifier and group delay in satellite, the nonlinear channel model both under the background of the Gaussian noise and alpha stable distributed noise and based on the combined effect of TWT amplifier and group delay is created and simulated. In Chapters 5 and 6, using the established nonlinear channel model and integrating various novel technologies into the equalization technology, the nonlinear channel equalization algorithms are proposed, and the performance of the proposed algorithms is verified. In Chapter 7, taking satellite MIMO channel and molecular MIMO channel as the research object, the multi-state Markov chain model, machine learning model, and simulation method of MIMO channel are analyzed.

The achievements of this book have been supported by the National Natural Science Foundation of China (61673222), the Natural Science Foundation of Jiangsu Province (BK2009410), the Major Natural Science Research Project of Colleges and Universities in Jiangsu Province (13KJA510001), the Natural Science Foundation of Colleges and Universities in Jiangsu Province (08KJB510010), and the second phase “information and Communication Engineering” of the construction of advantageous disciplines in Colleges and Universities in Jiangsu Province. Some of these achievements have been successively published in important academic conferences and journals at home and abroad. At the same time, this book has also absorbed the achievements of some excellent works, academic papers, and many relevant authors. For the achievements cited in this book, some have indicated the source in the book. The achievements of this book reflect the latest trends and frontier trends in the research of key technologies of satellite communication.

The co-researchers with the author of this book are Zhang Xiuzai, Han Yingge, Rao Wei, Yang Chao, Liao Juan, Ding Rui, Ji Tongying, Hu Lingling, Huang Wei, Sun Jing, Ding Xuejie, Wang Wei, Xu Wencai, Li Baoge, Zhao Weijuan, Bai He, Yuan Tao, Zhang Tao, Xu Ran, Yao Chaoran, and so on. The publication of this book has been strongly supported by the Tsinghua University Press. Here, I would like to express my sincere thanks to the researchers who participated in the research of the achievements of this book, the authors of the cited achievements of this book, and Tsinghua University Press!

Due to the limited level and experience of the author, the depth and breadth of the research in this book need to be further improved. Inappropriate places are inevitable. Please criticize and correct!

Guo Yecai
August 2021

About the Author

Guo Yecai is a Professor and doctoral advisor. He graduated from Anqing Normal University in 1986, majoring in physics. In 2003, he obtained the doctor's degree in underwater acoustic engineering from Northwestern Polytechnical University. He is one of the winners of the One Hundred Excellent Doctoral Dissertations in China, and the academic and technical leader of Anhui Province in China. Guo Yecai has presided over or undertaken more than 20 scientific research projects such as the special fund for National Excellent Doctoral Dissertation authors, the National Natural Science Foundation as well as national and provincial teaching and research projects. He has won seven provincial science and technology achievement awards and teaching achievement awards, published a Nationally Planned Textbook and four provincial key textbooks. More than 30 invention patents have been authorized by him.

Contents

Chapter 1

Introduction

Summary

Based on the analysis of the factors affecting the quality of satellite communication, the main achievements in the establishment of satellite channel models (single-state model, multi-state model, Ka-band model, and model simulation method) and the equalization methods (adaptive equalization and blind adaptive equalization) are summarized in this chapter. In view of the shortcomings of the existing research models and methods, from the perspective of system theory, a systematic theoretical framework and technical system were formed by a main line from the establishment of the satellite channel model to the research of the satellite channel equalization method.

Satellite communication can simply be viewed as the communication between radio communication stations on the earth (including the ground and the lower atmosphere) using satellite as relay. The satellite communication system consists of two parts: satellite and earth station. The wide range of satellite communication makes it easy to realize trans-oceanic and intercontinental communication. It is possible to communicate between any two points, so long as they are within the range of satellite radio waves. At the same time, it can be received in many places and can realize broadcast economically and multiple access communication (multiple access characteristics). The circuit setting is very flexible, which can disperse too concentrated telephone traffic at any time. The same channel can be used in different directions or in different intervals (multiple access connection). In satellite communication, the satellite acts as a relay station in the air, which can amplify the electromagnetic waves emitted by the earth station and then transmit them back to another earth station. The satellite data transmission capacity and quality are not only affected of the satellite equipment itself (such as traveling wave tube amplifier [TWT amplifier or TWTA]) but also affected by the satellite–ground link environment and the ground environment (called as the physical channel), such as atmospheric gases, the ionosphere and troposphere, cloud, rain, euphorbia crystals, eclipses and storms, and objects on the surface of the earth and electromagnetic environment interference, clutter interference of earth station equipment, intermodulation interference, adjacent channel interference, cross-polarization interference, artificial malicious interference and the universe noise, atmospheric noise, solar noise, electric noise, ground noise, and so on. These factors make the satellite channel produce many kinds of effects: the fading effect (mainly by atmospheric gases, rain, fog, clouds, euphorbia crystals, objects on the surface of the earth, and other factors), the multipath effect (mainly by atmospheric gases, the surface of the earth and the objects on it, and other factors), the shadow effect (mainly caused by many factors such as trees and buildings) and channel noises (various interference factors and all kinds of noise), the Doppler effect (caused by the relation between the satellite and the ground receiving station), and the time-varying effect (caused by the rapid fluctuation of the amplitude and phase of the receiving signals, it is difficult to accurately and promptly change the prediction mode when forecasting). The channel also presents time-varying characteristics [1], nonlinear effect (caused by the nonlinear TWTA, nonlinear physical channel, and nonlinear receiver in the satellite [2]), non-stationarity (caused by the fluctuation of the physical environment and meteorological factors [3]), and propagation delay effect (caused by factors such as troposphere and ionosphere).

These characteristics of the satellite channel are also affected by the satellite communication frequency band. The most suitable frequency band for satellite communication is 1–10 GHz, which is called as the microwave band. In order to improve the data capacity and quality of satellite transmission, the satellite communication frequency bands are extended to above 10 GHz (such as 12 GHz, 14 GHz, 20 GHz, and 30 GHz). The signal modulation mode is developing from constant modulus envelope modulation (such as multi-level phase shift keying, M-PSK) to non-constant modulus envelope modulation (such as quadrature amplitude modulation, QAM). However, at high frequencies, the effects of attenuation, shadowing, multipath, time variation, and non-linearity of the satellite channels are more serious, which makes the nonlinear distortion, inter-symbol interference (ISI), and inter-channel interference (ICI) of high-order modulated signals very prominent and further affects the quality of the satellite transmission data. These have become the key technical problems to be solved in the field of satellite communication.

In order to solve these problems, a satellite channel model which approximates the actual physical channel should be established according to the complexity of the satellite data transmission environment, and the correctness of the model is verified by comparing the actual received data with the simulation experiment. Secondly, according to the characteristics of the satellite channel, the compensation theory (also known as the equalization theory) and the method of influence of the satellite channel on data transmission should be established to improve the channel capacity and quality of the data received by the ground receiving system of the satellite.

1.1 Satellite Channel Modeling Research

In the study of the transmission characteristics of the satellite channel, there are mainly three models used to describe the transmission characteristics of the satellite channel: empirical model, geometric model, and statistical model. The empirical model is used to process the measured data and establish the corresponding simple formula with the mathematical fitting method. According to the spatial distribution of obstacles, the geometric model describes the propagation characteristics of the received signals by using scattering theory or ray tracking method. The statistical model simulates the statistical characteristics of the received signals in different environments through the combination of different probability distribution functions [4]. The empirical model can describe the sensitivity of the important parameters, but it cannot explain the nature of the signal propagation process. The geometric model can predict the action of single or multiple scattering sources and explain the fading phenomenon of the signal, but its analysis process is very complicated.

Compared with the empirical model and the geometric model, the statistical model simplifies the actual transmission of the signal, and the analysis process is simpler and the physical meaning is intuitive. Therefore, the transmission characteristics of the satellite channels are usually studied by the statistical model, which can be divided into single state and multi-state according to their description of channel states. The channel single-state model is mainly used to describe the stable and single channel environment. In the channel multi-state model, different states correspond to different probability density distributions, which are often used to describe the non-stationary and constantly changing dynamic environment.

1.1.1 *Channel single-state model*

In 1968, Clarke proposed a statistical model to describe flat small-scale fading, which is based on the scattering principle and assumes that the signal received by the receiving terminal is the superposition of multiple plane waves and it is called the Clarke model [5]. In 1977, Suzuki combined small-scale Rayleigh fading with large-scale shadowing fading to form a Suzuki model that obeys Rayleigh log-normal distribution [6], which assumes that there is no main path which is obviously different from other paths in the channel environment and mainly used for the typical urban environment where there is no direct viewing component. In 1985, C. Loo proposed to use the two-path model, which is called as the Loo model, to model the L-band transmission channel [7]. He believed that the received signal consisted of the main path and the slave path, in which the main path was affected by the shadowing effect, while the slave path was composed of pure multipath components. The Loo model combines the Rayleigh process with the lognormal process. Since the slave path is not affected by shadowing fading, it is also called the partial shadowing model, which is mainly applicable to open areas such as rural areas and suburbs. In 1994, Corazza proposed the full-shadowing channel model [8], which considered comprehensively the Rice process and the lognormal process, and the empirical formula of Corazza model parameters for L-band satellite channel was given by testing the actual satellite link to describe a variety of satellite mobile channel environments.

The Suzuki model, the Loo model, and the Corazza model are three classical satellite mobile channel models. Later, many channel models that were proposed are derivatives of these three models. In 1998, on the basis of the Suzuki model, Patzold proposed an extended Suzuki model [9]. In this proposed model, Patzold introduced the direct view component and adopted the correlated color Gaussian noise modeling. In the Loo model, the lognormal distribution describing the shadowing effect of the main path was modified into Nakagami distribution to establish the Abdi model [10]. Based on the Corazza model, the independent shadowing model was established based on the assumption that the direct view component and the multipath component were independently influenced by shadowing fading [11]. On the basis of the Corazza model, in 2002, a statistical model was proposed to describe channel transmission characteristics in non-urban areas [12]. In this statistical model, K-distribution was used to replace lognormal distribution to describe channel fading characteristics caused by the shadowing effect. In 2003, the Lognormal–Rician–Rayleigh model (LR^2 model) was proposed [13], which can be expressed as other classical satellite mobile channel statistical models by setting parameters of the model. The interrelationships among several satellite channel models are shown in Figure 1.1.

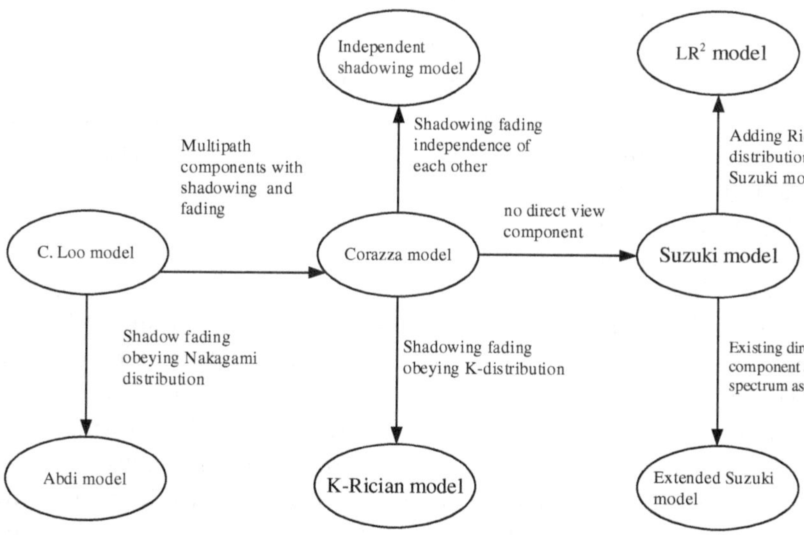

Figure 1.1 Relationship diagram among satellite channel single-state models.

1.1.2 *Channel multi-state model*

When the mobile terminals are moving in a wide range of environments, the channel single-state model cannot accurately describe the actual changing characteristics of the satellite channel. At this time, the channel characteristics can be described by the channel multi-state model.

Lutz proposed a channel two-state model [14] in 1991, which divides the channel environment into "good" and "bad" states according to whether there is a direct view component in the received signal. In 2001, Fonton proposed the channel model of the land mobile satellite (LMS) in a wide area environment [15], which divides the channel into three states according to the degree of signal shadowing and describes the different states of the channel via setting different parameters of the Loo model. In 2004, Lin proposed a two-level and multi-state Markov model, which is divided into different levels according to transmission loss, and each level is divided into several states [16]. On this basis, a six-state Markov chain model for LMS channel, published in [17] in 2005, divided the received signal into two states according to the shadowing fading degree of the received signal, and each state contained three different substates. In 2006, the extended Lutz model proposed the three-state Markov chain R^M model [18].

Before the establishment of the channel multi-state Markov chain model, it is necessary to make artificial assumptions about the number of channel states and the influencing factors of each state, which are usually based on experience and lack of theoretical support. Principal component analysis (PCA) and fuzzy clustering method were combined for the first time to model mobile satellite channels in [19–22], which does not make any assumptions about the number of channel states and the main influencing factors of the signals. The specific method is given as follows: firstly, the factors affecting the transmission characteristics of the channel are quantified, and the PCA method is used to process the quantified data to extract the main factors affecting the quality of satellite communication and select the corresponding probability distribution for description; secondly, the number of the channel states and state transition matrix are initialized, and the fuzzy clustering method is used to perform data iteration on the initial data to get the updated state number and state transition probability matrix; thirdly, state transition is realized according to Markov chain. The number of channel states and the fading distribution of each state are determined scientifically and accurately, and the error caused by artificial hypothesis is avoided. Simultaneous interpretation of the same part of the multi-state model is mostly based on the results of many experiments, so it has high credibility, while different parts of the channel show the channel characteristics under different transmission environments. The more the number of states, the more detailed and accurate the description of the channel characteristics is. However, as the number of channel states increases, the channel modeling parameters will increase, and the simulation will become more complex.

1.1.3 *Ka-band satellite channel statistical characteristics*

The congestion of the satellite channels has forced system designers to consider Ka-band applications [23–25]. In the high-frequency band, the influence of different meteorological conditions and transmission media on satellite data transmission is more obvious. In 1998, Loo analyzed and expanded the measurement data of the mobile satellite channel in the L band, and both the envelope and phase of the received signals of the Ka-band satellite affected by the weather were subject to Gaussian distribution [26]. In 2001, considering the influence of different weather conditions on the transmission characteristics of the satellite channel, W. Li established the model of the Ka-band LMS channel on the basis of the Loo model [27]. Then, the influence of rain attenuation, atmospheric absorption, cloud attenuation, scattering, and atmospheric scintillation on the channel fading in Ka-band satellite communication was studied [28–32], and a statistical model was established to describe the Ka-band satellite channel. In 2014, Kourogiorgas proposed a statistical dynamic model of the rain attenuation [33] to describe the signal attenuation in low earth orbit (LEO) satellite links caused by rainfall.

With the continuous development of satellite communication technology, the working frequency of satellite communication is getting higher and higher, and the influence of various fading factors in atmospheric environments on the signals in the link will be more and more obvious.

1.1.4 *Research on satellite channel simulation research*

In the research of the satellite communication system, it is impossible to test the satellite channel in real time due to the limitation of the conditions, so the channel emulator is usually used to describe the fading characteristics of the channel under different environments. The characteristic simulation of the satellite channel can be realized by means of software and hardware, both of which can be used to simulate the actual transmission characteristics of the satellite channel well. The flexibility of software simulation is high, and it is convenient to simulate and compare the performance of various statistical models. However, in the actual hardware debugging, the real-time performance of software simulation is poor due to the large amount of computation. The channel simulator implemented by hardware can enable researchers to test the performance of the communication prototype in real time, which can greatly improve the efficiency of research and development. However, due to its complex design and high cost, it also makes the hardware implementation more difficult.

At present, the software simulation of the satellite channel can be realized based on the MATLAB platform [34,35], and the realization methods of the channel model mainly include sum of sinusoid (SOS) and shaping filter method [36,37]. The sine wave superposition method is relatively small in computation and convenient in practice, but its simulation performance has some limitations. The shaping filter algorithm requires a bandwidth much lower than its sampling rate, so it is more complicated to be implemented. In terms of hardware implementation, an LMS channel simulator designed by Japan Communications Research Laboratory can simulate multipath and shadowing fading effects in channel transmission characteristics [38]. In the same year, Loo realized the simulation of Landsat mobile fading channel simulator by using the white Gaussian random process [39]. In 2000, the University of Electronic Science and Technology of China also completed the design of the principle prototype of the satellite mobile channel simulator [40].

At present, only a few foreign companies have the physical products of the satellite channel simulator. The SLE900 satellite channel link simulator, developed by DBM, USA, can simulate the channel characteristics of the satellite channels, such as time delay, frequency offset, Rayleigh fading, Rice fading, and Gaussian white noise. The SCATT8906 device of the GLOCOM in the United States divides the communication environment into ocean and land and uses the Rayleigh and Rice fading model to simulate the influencing factors such as multipath fading and Doppler shift. The MSCS-1 channel simulator, developed by the Australian Telecommunications Research Institute, adopts the Markov chain process to adjust the time-varying parameters of the channel model, so as to realize the simulation of time-varying channels. Most of these channel simulators are used to simulate several important factors in the ground mobile channel while ignoring the influence of other factors on the channel. Therefore, it is necessary to further study the satellite channels, so as to realize the channel characteristic description of the influence of various environmental factors.

1.2 Research on Satellite Channel Equalization

At present, the LMSs are taken as the research object, and there are two main methods to reduce or eliminate the impact of the satellite channel on the transmission data capacity and quality: the transmitter-based method [41–45] and the receiver-based method [46–60]. Among them, the transmitter-based method is very complicated and not suitable for the high-speed data communication system. The receiver-based method is realized by equalization technology, which refers to the design of an inverse filter (also known as equalizer) in the receiver, and is opposite to the characteristics of multipath, time-varying, fading, nonlinear, and non-stationary channel, so as to eliminate the inter-symbol interference (ISI) caused by the satellite channel and improve the channel capacity and transmission data quality. In the equalization technology, the traditional adaptive equalizer (non-blind), which needs the transmitter to send the known, periodic, and fixed-length training sequence, needs to occupy the bandwidth resources with limited rich means, which lead to reducing the channel capacity and increasing the complexity of the receiver design. However, the blind adaptive equalizer, which does not need the transmitter to send the training sequence, can overcome the shortcomings of the traditional adaptive equalizer. Therefore, it is necessary to design a blind equalizer in the receiver, which is opposite to the channel characteristics.

The main performance indicators of the equalizer are convergence speed, mean square error (MSE) or inter-symbol interference (ISI) or bit error rate (BER), computational complexity, and so on.

The main research achievements of satellite channel equalization technology can be summarized as follows:

(1) *ISI canceller*: It is suitable for equalizing 16QAM signal, but there is no report about BER performance, and it is only suitable for linear channels [51, 59, 60].
(2) *Volterra series equalizer*: It can reduce the nonlinear ISI between QPSK (quadrature phase shift keying) signal and QAM signal and is suitable for low-order nonlinear satellite channels [51, 53]. However, high computational complexity is the main reason to limit its application.
(3) *Maximum likelihood sequence estimation (MLSE) equalizer*: In the nonlinear satellite channels, for BPSK (Binary Phase Shift Keying) signals, the maximum likelihood sequence estimation equalizer is presented in [50], which is extended to the multi-stage structure, and consists of many filter bands. For a practical system, the number of filter bands is too many to be realized. However, the equalizer combining MLSE and decoding with complex structure is difficult to implement in engineering when using high-order modulation signal, and the likelihood receiver is highly sensitive to noise [58].
(4) *Coded modulation equalizer*: There are mainly equalizers based on grid modulation coding and turbo modulation coding. The former does not consider the nonlinear ISI, while the latter's computational complexity increases with the increase of the length of the channel impulse response and the order of the modulated signal. In order to reduce the computational complexity, it is necessary to reduce the computational amount of the equalization algorithm or use the interference canceller [59, 60]. An equalizer combining interference canceller with turbo coding is presented in [61, 62], which can reduce the computational complexity of 16QAM signals to a certain extent, but the improvement of the BER is not obvious.
(5) *Neural network equalizer*: The feedforward neural network (FNN) blind equalizer of the satellite general mobile channel is suitable for equalizing QAM signals, but its convergence speed

is slow. The bilinear cyclic neural network (BLRNN) equalizer is simpler and more effective than the neural network equalizer based on the Volterra series [63–65]. However, the constellations of 16QAM output still have a certain rotation and poor phase compensation ability to lead to large MSE.

(6) *Blind adaptive equalizer*: In 2008, Copjan and Krahulec proposed a constant modulus blind equalization algorithm based on the linear constraint, which can effectively recover the multi-carrier code-division multiple access (MC-CDMA) signal using the LMS channel and is based on the assumption that the channel noise is Gaussian white noise [66]. In 2009, Copjan and Krahulec proposed to use high-order amplitude phase keying (PSK) signal to improve the channel capacity of the satellite broadcasting system, but it is very sensitive to the nonlinear distortion of the channel.

To sum up, the existing problems of satellite channel modeling and equalization technology are summarized as follows:

(1) *Major problems existing in satellite channel modeling*: In the aspect of physical environment, the impact of the terrain and feature environments such as the urban, suburban, rural, and open land on the channel statistical characteristics is mainly based on the mobile terminal, but the spatial environments of the satellite–ground link are not considered. In the aspect of meteorological factors, the influence of rain attenuation and cloud attenuation is mainly considered, but other meteorological factors are rarely considered. In the aspect of channel state classification, the channel environments are mainly divided into two states, which are based on the measured data and experience and lack of clear physical meaning and theoretical support. When considering the channel stationarity, most satellite channels are assumed to be stationary, but in fact they are non-stationary.
(2) *Major problems in satellite channel equalization*: From the perspective of transformation domain, channel equalization techniques are mainly time-domain equalization, which has a large amount of computation, whereas the frequency domain equalization technology with lower computation loads is rarely

considered, but there is no report on frequency-domain equalization of the satellite channel. From the perspective of equalization methods, the traditional adaptive equalization techniques are mainly used, which waste the limited bandwidth resources and reduce the channel capacity. However, there are few studies on blind equalization of the satellite channels. From the perspective of mathematical tools, Volterra series, maximum likelihood sequence estimation and neural network, and so on are adopted, and the equalizers based on these mathematical tools have a complicated structure, a large amount of computation, and thus difficult to be implemented in engineering. In terms of considering channel noise, the channel noise is assumed to be Gaussian noise, the atmospheric noise is not Gaussian noise but the alpha distribution noise with a long tail.

1.3 Main Contents of the Book

According to the above analyses, in order to meet the needs of the development of the new generation satellite communication in the direction of high-frequency band, high speed and the use of high-order modulation signals and in view of the fact that the satellite data transmission environment is extremely complex and the satellite channel bandwidth resources are very limited, the main research contents of this book are as follows.

1.3.1 *Research on nonlinear channel modeling methods*

A practical satellite channel model is established by considering the effects of the atmospheric space physical environments, meteorological factors, the statistical characteristics of various interferences and noises, and so on, on the capacity and quality of satellite data transmission.

1.3.2 *Research on nonlinear channel equalization algorithm*

Based on modern signal processing technology and the established satellite channel model, the new theory and technology of the satellite channel blind equalization are developed to form a relatively complete theoretical framework and technical system from satellite channel modeling to satellite channel blind equalization.

References

[1] Qing L. and Yaohai D. Achievement and forecast of meteorological satellite technology in China. *Aerospace Shanghai*, 2008, 24(1): 1–10 (in Chinese).
[2] Aijun Z. Analysis on specification of FY-3 meteorological satellite data transmission. *Journal of Applied Meteorological Science*, 2006, 17(4): 494–501 (in Chinese).
[3] Aijun Z. and Jianping X. Development of meteorological satellite data transmission. *Meteorological Science and Technology*, 2005, 33(4): 295–299 (in Chinese).
[4] Patzold M., Li Y., and Laue F. A study of a land mobile satellite channel model with asymmetrical Doppler power spectrum and lognormally distributed line-of-sight component. *IEEE Transactions on Vehicular Technology*, 1998, 47(1): 297–310.
[5] Clarke R. H. A statistical theory of mobile-radio reception. *Bell Labs Technical Journal*, 1968, 47(6): 957–1000.
[6] Suzuki H. A statistical model for urban radio propagation. *IEEE Transactions on Communications*, 1977, 25(7): 673–680.
[7] Loo C. A statistical model for a land mobile satellite link. *IEEE Transactions on Vehicular Technology*, 1985, 34(3): 122–127.
[8] Corazza G. E. and Vatalaro F. A statistical model for land mobile satellite channels and its application to nongeostationary orbit systems. *IEEE Transactions on Vehicular Technology*, 1994, 43(3): 738–742.
[9] Patzold M., Killat U., and Laue F. An extended Suzuki model for land mobile satellite channels and its statistical properties. *IEEE Transactions on Vehicular Technology*, 1998, 47(2): 617–630.
[10] Abdi A., Lau W. C., Alouini M. S., *et al.* A new simple model for land mobile satellite channels: First-and second-order statistics. *IEEE Transactions on Wireless Communications*, 2003, 2(3): 519–528.
[11] Huang S. H., Kim K. J., Ahn J. Y. *et al.* A channel model for nongeostationary orbiting satellite system. *Vehicular Technology Conference, 1997, IEEE 47th.*, 1: 41–45.
[12] Kun X. and Shiqi W. A statistical mobile satellite channel model for non-urban areas. *Electronic Journals*, 2002, 30(7): 999–1002 (in Chinese).

[13] Xing L. and Shiqi W. A LR^2 model for mobile satellite channels and its system performance analysis. *Chinese Journal of Radio Science*, 2003, 18(3): 305–310 (in Chinese).
[14] Lutz E., Cygan D., Dippold M. *et al.* The land mobile satellite communication channel-recording, statistics, and channel model. *IEEE Transactions on Vehicular Technology*, 1991, 40(2): 375–386.
[15] Fontan F. P., Vázquez-Castro M., Cabado C. E. *et al.* Statistical modeling of the LMS channel. *IEEE Transactions on Vehicular Technology*, 2001, 50(6): 1549–1567.
[16] Lin H. P. and Yang C. Multi-state Markov model for satellite propagation channels. *Antennas and Propagation Society International Symposium, 1999, IEEE*, 3: 1486–1489.
[17] Dongya S., Jian R., Yihuai Y. *et al.* The six-state Markov model for land mobile satellite channels. *Microwave, Antenna, Propagation and EMC Technologies for Wireless Communications, 2005. MAPE 2005. IEEE International Symposium*, 2005, 2: 1619–1622.
[18] Jian R., Dongya S. and Jing S. R^M channel model for land mobile satellite. *Journal of Yunnan University* (Natural Science Edition), 2006, 28(6): 487–491 (in Chinese).
[19] Yecai G. and Tao Y. Channel state division of land satellite mobile and its multi-states Markov model based on fuzzy clustering. *Journal of Computational Information Systems*, 2014, 10(4): 1349–1356.
[20] Xiuzai Z., Guo Y. C., Yibo Z., *et al.* A simulation method for meteorological satellite Markov time-varying channel. Chinese Patent: ZL201510227901.9, 2017 September 5 (in Chinese).
[21] Tao Y. The research of satellite channel modeling and simulation method. Graduate Thesis, Nanjing: Nanjing University of Information Science and Technology, 2014 (in Chinese).
[22] Guo Y. C., Xiuzai Z., Tao Y., *et al.* A simulation method for meteorological satellite Markov time-varying channel. Chinese Patent: CN201310310998.0, 2013 November 20 (in Chinese).
[23] Gargione F., Iida T., Valdoni F. *et al.* Services, technologies, and systems at Ka band and beyond-a survey. *IEEE Journal on Selected Areas in Communications*, 1999, 17(2): 133–144.
[24] Evans J. V. Proposed US global satellite operating at Ka-band. *IEEE Aerospace Conference*, 1998, 4(3): 525–537.
[25] Karasawa Y. and Maekawa Y. Ka-band earth-space propagation research in Japan. *Proceedings of the IEEE*, 1997, 85(6): 821–842.
[26] Loo C. and Butterworth J. S. Land mobile satellite channel measurements and modeling. *Proceedings of the IEEE*, 1998, 86(7): 1442–1463.
[27] Li W., Law C. L., Dubey V. K. *et al.* Ka-band land mobile satellite channel model incorporating weather effects. *IEEE Communications Letters*, 2001, 5(5): 194–196.

[28] Aihua W. and Weixiong L. Modelling of Ka-band satellite communication channel and system performance simulation. *Journal of Communication*, 2001, 22(9): 61–69 (in Chinese).
[29] Yunjiang L., Guoce H., Shuchun Z., *et al.* An analysis of satellite channel capacity for performance under conditions of rain attenuation. *Journal of Air Force Engineering University*, 2004, 5(4): 50–59 (in Chinese).
[30] Chang S. and Po Z. A satellite channel model based on discrete rain attenuation states. *Command Control and Simulation*, 2007, 29(4): 48–51 (in Chinese).
[31] He H., Dongjin W., and Falin L. Modeling of Ka-band mobile satellite channel and Ber performance analysis. *Journal of University of Science and Technology of China*, 2005, 35(3): 346–353 (in Chinese).
[32] Hongxia W., Jianfang Y., and Chengsheng P. Simulation of Ka-band satellite communication channel characteristics and system performance. *Fire Control & Command Control*, 2008, 32(6): 121–124 (in Chinese).
[33] Kourogiorgas C. and Panagopoulos A. D. A rain-attenuation stochastic dynamic model for LEO satellite systems above 10 GHz. *IEEE Transactions on Vehicular Technology*, 2015, 64(2): 829–834.
[34] Prabhu G. S. and Shankar P. M. Simulation of flat fading using MATLAB for classroom instruction. *IEEE Transactions on Education*, 2002, 45(1): 19–25.
[35] Yanqing F., Kai X., Jia R., *et al.* Satellite communication simulation system based on MATLAB SIMULINK. *Communication Technology*, 2010, 43(3): 57–58 (in Chinese).
[36] Lu M., Lo T., and Litva J. A physical spatio-temporal model of multipath propagation channels. *Vehicular Technology Conference, 1997, IEEE 47th*. IEEE, 1997, 2: 810–814.
[37] Gutiérrez C. A. and Patzold M. The design of sum-of-cisoids Rayleigh fading channel simulators assuming non-isotropic scattering conditions. *IEEE Transactions on Wireless Communications*, 2010, 9(4): 1308–1314.
[38] Wakana H. A propagation model for land mobile satellite communications. *Antennas and Propagation Society International Symposium, 1991. AP-S. Digest*. IEEE, 1991: 1526–1529.
[39] Loo C. and Secord N. Computer models for fading channels with applications to digital transmission. *IEEE Transactions on Vehicular Technology*, 1991, 40(4): 700–707.
[40] Jianhao H., Tao L., and Shiqi W. Design of real-time low orbit satellite mobile communication channel simulator. *Journal of UEST of China*, 1998, 27(4): 351–356 (in Chinese).

[41] D'Andrea A. N., Lottici V., and Reggiannini R. RF power amplifier linearization through amplitude and phase predistortion. *IEEE Transactions on Communications*, 1996, 44(11): 1477–1484.

[42] Karam G. and Sari H. Data predistortion techniques using intersymbol interpolation. *IEEE Transactions on Communications*, 1990, 38(10): 1716–1723.

[43] D'Andrea A. N., Lottici V., and Reggiannini R. Nonlinear predistortion of OFDM signals over frequency-selective fading channels. *IEEE Transactions on Communications*, 2001, 49(5): 837–843.

[44] Di Benedetto M. G. and Mandarini P. A new analog predistortion criterion with application to high efficiency digital radio links. *IEEE Transactions on Communications*, 1995, 43(12): 2966–2974.

[45] Marsalek R., Jardin P., and Baudoin G. From post-distortion to predistortion for power amplifiers linearization. *IEEE Communications Letters*, 2003, 7(7): 308–310.

[46] Brajal A., Chouly A., and Fazel K. TCM schemes with predistortion techniques on nonlinear channels. *Communications, 1992. ICC'92, Conference Record, SUPERCOMM/ICC'92, Discovering a New World of Communications, IEEE International Conference*, 1992, 633–637.

[47] Benedetto S., Biglieri E. Nonlinear equalization of digital satellite channels. *IEEE Journal on Selected Areas in Communications*, 1983, 1(1): 57–62.

[48] Park D. C. and Jeong T. K. J. Complex-bilinear recurrent neural network for equalization of a digital satellite channel. *IEEE Transactions on Neural Networks*, 2002, 13(3): 711–725.

[49] Mesiya M., McLane P., and Campbell L. Maximum likelihood sequence estimation of binary sequences transmitted over bandlimited nonlinear channels. *IEEE Transactions on Communications*, 1977, 25(7): 633–643.

[50] Gutierrez A. and Ryan W. E. Performance of Volterra and MLSD receivers for nonlinear band-limited satellite systems. *IEEE Transactions on Communications*, 2000, 48(7): 1171–1177.

[51] Serfaty S., LoCicero J. L., and Atkin G. E. Cancellation of nonlinearities in bandpass QAM systems. *IEEE Transactions on Communications*, 1990, 38(10): 1835–1843.

[52] Serfaty S. Performance of combined trellis coded modulation and nonlinear cancellation. *IEEE Transactions on Communications*, 1993, 41(1): 7–10.

[53] De Gaudenzi R. and Luise M. Analysis and design of an all-digital demodulator for trellis coded 16QAM transmission over a

nonlinear satellite channel. *IEEE Transactions on Communications*, 1995, 43(234): 659–668.
[54] Tehan A. and Scanlan J. O. A simulation study of trellis-coded modulation for a satellite link. *IEEE Transactions on Communications*, 1997, 45(11): 1371–1374.
[55] Bai E. W., Li Q., and Zhang Z. Blind source separation/channel equalization of nonlinear channels with binary inputs. *IEEE Transactions on Signal Processing*, 2005, 53(7): 2315–2323.
[56] Patra J. C., Pal R. N., Baliarsingh R. *et al.* Nonlinear channel equalization for QAM signal constellation using artificial neural networks. *IEEE Transactions on Systems, Man, and Cybernetics, Part B (Cybernetics)*, 1999, 29(2): 262–271.
[57] Chang P. R. and Wang B. C. Adaptive decision feedback equalization for digital satellite channels using multilayer neural networks. *IEEE Journal on Selected Areas in Communications*, 1995, 13(2): 316–324.
[58] Benedetto S., Biglieri E., and Daffara R. Modeling and performance evaluation of nonlinear satellite links-a Volterra series approach. *IEEE Transactions on Aerospace and Electronic Systems*, 1979, (4): 494–507.
[59] Vogelbruch F. and Haar S. Improved soft ISI cancellation for turbo equalization using full soft output channel decoder's information. *Global Telecommunications Conference, 2003. GLOBECOM'03.* IEEE, 2003, 3: 1736–1740.
[60] Wang X. and Poor H. V. Iterative (turbo) soft interference cancellation and decoding for coded CDMA. *IEEE Transactions on Communications*, 1999, 47(7): 1046–1061.
[61] Heo S. W., Gelfand S. B., and Krogmeier J. V. Equalization combined with trellis coded and turbo trellis coded modulation in the nonlinear satellite channel. *MILCOM 2000. 21st Century Military Communications Conference Proceedings IEEE*, 2000, 1: 184–188.
[62] Marmarelis V. Z. and Zhao X. Volterra models and three-layer perceptrons. *IEEE Transactions on Neural Networks*, 1997, 8(6): 1421–1433.
[63] Bouchired S., Ibnkahla M., Roviras D., *et al.* Equalization of satellite mobile communication channels using combined self-organizing maps and RBF networks. *Acoustics, Speech and Signal Processing, 1998. Proceedings of the 1998 IEEE International Conference on IEEE*, 1998, 6: 3377–3379.
[64] Paquier W. and Ibnkahla M. Self-organizing maps for rapidly fading nonlinear channel equalization. *Neural Networks Proceedings, 1998. IEEE World Congress on Computational Intelligence. The 1998 IEEE International Joint Conference on IEEE*, 1998, 2: 865–869.
[65] Xiaoqin Z., Huakui W., Liyi Z., *et al.* A blind equalization algorithm based on bilinear recurrent neural network. *Intelligent Control and*

Automation, WCICA 2004. Fifth World Congress on IEEE, 2004, 3: 1982–1984.

[66] Copjan L., Krahulec J., and Marchevsky S. Blind receiver based on the LCCMA algorithm for MC-CDMA signals detection over the 3-states channel model. *Radioelektronika, 2008 18th International Conference*, IEEE, 2008, 1–5.

Chapter 2

The Theoretical Basis for the Establishment of the Satellite Channel Model

Summary

This chapter first describes the basic components of satellite communication systems (space satellites, intermediate links, ground control, and reception) and the basic parameters (elevation angle, azimuth, channel distance, communication frequency band, satellite–ground link distance, etc.). The characteristics of satellite channel layering (outer space, thermal layer and middle layer, as well as stratosphere and troposphere) are analyzed. Typical channel transmission models and classical satellite channel models (C Loo model, Suzuki model, Corazza model, and Lutz model) are analyzed and the statistical characteristics of the model performance as well as the simulation method of the classical satellite channel model are given.

The satellite communication system is a very complicated system, which includes on-board systems, ground transmitting and receiving systems, and intermediate link components, and so on. In order to model the channel of the intermediate link, which does not involve the structure and principle of the satellite communication system and on-board devices, it is necessary to understand the basic composition of the satellite communication system for subsequent channel analysis.

2.1 Basic Components of a Satellite Communication System

Figure 2.1 shows the schematic diagram of a typical satellite communication system. The entire satellite communication system consists of three major parts: satellite, intermediate link (channel), and ground control terminals and receiving terminals [1].

Space satellites are the main part of satellite communications and mainly responsible for detection and forwarding, and it can be divided into earth exploration satellites, meteorological satellites, mobile satellites, radio navigation satellites, and maritime satellites according to their uses. According to the satellite orbit, satellite communication systems are also classified into LEO (low earth orbit), MEO (middle earth orbit), and GEO (geostationary earth orbit), which are distributed in different orbits and have different detection ranges and working frequencies. For example, the orbit of the Iridium satellite in the LEO system is 500–2000 km away from the ground, so the transmission delay and power consumption are relatively small, but the coverage of each satellite is also relatively small. It takes about 66 satellites to cover the whole surface of the earth.

The orbit height of the GEO satellite communication system is 35800 km away from the ground, covering a large area. Only three satellites are needed to cover the whole earth (except the two poles). However, the transmission path is long, the transmission delay is large, and there is a considerable technical risk.

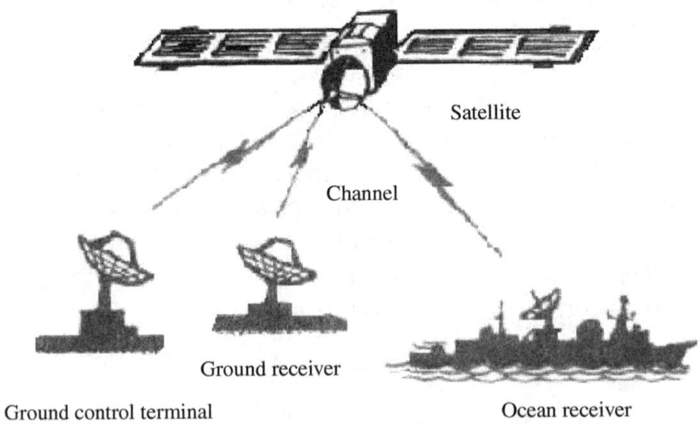

Figure 2.1 A typical satellite communication system.

The intermediate link part (channel), called as the physical channel, mainly realizes the wireless transmission from the satellite to the ground receiver. and the channel environment directly affects the communication quality. A variety of substances in the intermediate link, such as air, rain, and clouds, will cause all kinds of losses including free space loss, ionospheric scintillation loss, rain attenuation loss, gas absorption, and tropospheric loss and further lead to short-term satellite communication interruption and seriously affect the stability of the communication system. The link length of the satellite channel is directly related to the orbit height of the satellite. The link length of the GEO satellite communication system is the longest, whereas that of the LEO is the shortest. The main attenuation factors of different satellite channels are very different. In the GEO channel, because the free space accounts for more than 95% of the link, the cumulative attenuation is the main attenuation source of the link. However, the link of the LEO is relatively short, and the attenuation mainly comes from the absorption of atmospheric materials and the attenuation of multipath and shadowing.

The ground control and receiving part, including the control terminal and the receiving terminal, are mainly responsible for the ground part of the satellite communication, whose control terminal performs the necessary satellite management and control by tracking and telemetry and ensures the satellite safety in orbit, whereas the ground receiving part is composed of the earth surface terminals, mainly including fixed terminals, motor-driven terminals, and mobile terminals. Fixed terminals access satellites at fixed locations to provide different types of services and remain relatively inactive when communicating with the satellites. The motor-driven terminals are mobile and fixed at a certain location when it communicates with the satellite, and the communication connection can be established after the antennas are deployed. The mobile terminals communicate with the satellite in motion and can be further classified into land mobile terminals, aeronautical mobile terminals, and marine mobile terminals according to the position of the terminals on the surface of the earth or the adjacent surface. The mobile terminals can generate a Doppler effect and inter-symbol interference (ISI) when communicating with the satellite.

In the above three parts, the space satellite part, the ground control, and receiving part mainly study the electrical and mechanical characteristics of the equipment to provide feasible and robust hardware and software support system for satellite communication. The main research contents of the intermediate link are in the channel propagation characteristics to provide the best solution for satellite communication technology, which has become one of the researching hotspots of satellite communication technology. It is also the research focus of this book.

2.2 Basic Parameters of the Satellite Communication Link

The basic parameters of the satellite communication link mainly include the elevation angle, azimuth angle, channel distance, communication frequency band, and distance between the satellite and ground links.

Before the specific analysis, the input parameters, required to be calculated, are introduced including the longitude l_E and latitude L_E of the receiving terminal as well as the longitude l_S and the latitude L_E of the satellite. The units of these parameters are degree. The height H of the earth station is relative to the sea level in kilometers. Equatorial radius is denoted as r_E, orbital radius is denoted as r_S, orbital height is denoted as $h = r_S – r_E$, the earth's eccentricity is e_E, and the longitude difference between the earth station and the satellite is defined as $B = l_E - l_S$.

2.2.1 *Elevation from earth station to satellite*

The elevation angle θ is a very important parameter that determines the length of the inclined path through the earth's atmosphere and plays an important role in calculating parameters such as gas attenuation, spatial scintillation, atmospheric absorption, and rainfall attenuation. In general, the larger the elevation angle is, the longer the path is, and the more severe the atmospheric attenuation is. The elevation angle θ is written as shown in (2.2.1) [2, 3].

$$\theta = \arccos\left(\frac{r_E + h}{d}\sqrt{1 - \cos^2(B)\cos^2(L_E)}\right) \tag{2.2.1}$$

2.2.2 *Azimuth of earth station*

Before the azimuth φ_z of the satellite relative to the earth station is calculated, the angle A_i is first defined as:

$$A_i = \arc\sin\left(\frac{\sin(|B|)}{\sin\beta}\right) \tag{2.2.2}$$

where $|B|$ is the absolute value of the longitude difference, $\beta = \arc\cos[\cos(B)\cos(L_E)]$, and the azimuth φ_z is determined by the angle A_i. According to the different relative positions of the earth station and the earth's surface substellar (SS) point, there are four situations, as shown in Table 2.1.

Azimuth φ_z can be obtained directly in two special cases:

(1) If the earth station and the sub-satellite point are the same longitude, then the earth station is in the northern hemisphere and the azimuth is 180°. If the earth station is in the southern hemisphere, the azimuth is 0°.
(2) If the earth station is above the equator, then the earth is at the west side of the sub-satellite point and the azimuth is 90°. If the earth station is at the east side of the sub-satellite point, the azimuth is 270°.

2.2.3 *Link distance between satellite and ground*

It refers to the distance between the earth station and the satellite. The calculation of this distance requires the radius of the earth at the latitude and longitude of the earth station, i.e.,

$$R = \sqrt{l^2 + z^2} \tag{2.2.3}$$

Table 2.1 Azimuth φ_z division calculation table.

Standing on the earth station, look at the point below the satellite	$\varphi_z/(°)$
SS point is northwest of the earth station	A_i
SS point is northeast of the earth station	$360 - A_i$
SS point is southeast of the earth station	$180 - A_i$
SS point is southwest of the earth station	$180 + A_i$

where

$$l = \left(\frac{r_E}{\sqrt{1 - e_E^2 \sin^2(L_E)}} + H \right) \cos(L_E) \tag{2.2.4}$$

$$z = \left(\frac{r_E(1 - e_E^2)}{\sqrt{1 - e_E^2 \sin^2(L_E)}} + H \right) \sin(L_E) \tag{2.2.5}$$

Besides, the definition angle can be described as:

$$\psi_E = \arctan\left(\frac{z}{l}\right) \tag{2.2.6}$$

Then, the distance can be written as:

$$d = \sqrt{R^2 + r_S^2 - 2Rr_S \cos(\psi_E) \cos(B)} \tag{2.2.7}$$

The distance d can be used to determine several important parameters of the satellite link, including free-space link loss, which is directly dependent on the path length from the earth station antenna to the satellite antenna.

2.2.4 *Working frequency*

Satellite operating frequencies have a significant impact on the design and performance of satellite communication links. For example, in calculating free-space loss, wavelength or frequency is the main parameter affecting the atmosphere loss and link path loss. These attenuations caused by different influencing factors are related to the satellite operating frequency. For example, the rain attenuation effect must be considered in Ka-band communication and should not be considered in the lower frequency bands. Therefore, the choice of frequency bands is of great significance for satellite communication research.

The frequency of the radio waves is usually defined in two different ways. The first is the alphabetic representation, which divides the spectrum of 1–300 GHz into eight representative frequency bands, as shown in Figure 2.2.

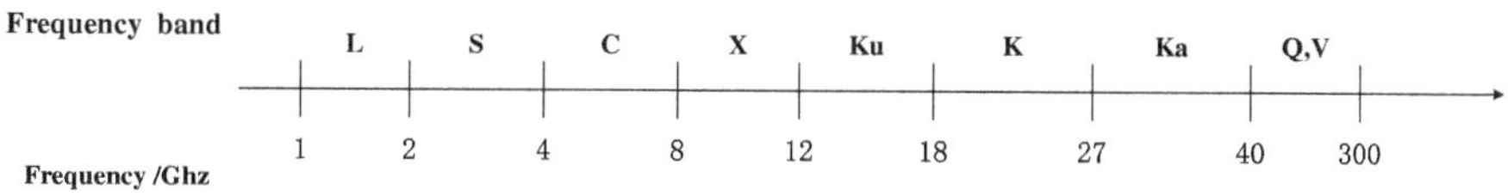

Figure 2.2 Alphabetic representation of the frequency band.

The second one is represented by the ten-band ladder method of the wavelength, which divides the frequency band from 3 Hz to 3000 GHz into 12 segments, as shown in Table 2.2.

2.2.5 *Key parameters in the communication link*

In the communication system link, there are several key parameters shown in Figure 2.3 [4]. P_r and P_t are the receiving power and the transmitter power, and their unit is $\mathrm{W} \cdot G_r$, G_t denote the receiving antenna gain and the transmitting antenna gain, r denotes the distance between the transmitter and the receiver, and its unit is m.

Effective omnidirectional radiation power is an important parameter for calculating RF power.

$$\mathrm{EIRP} = P_t G_t \tag{2.2.8}$$

Its decibel form is written as:

$$\mathrm{EIRP} = 20\log \mathrm{EIRP} = 20\log P_t + 20\log G_t \tag{2.2.9}$$

where EIRP is an unique quality factor index of the transmitter in wireless communication link.

2.2.6 *Power flux density*

Power flux density $(\mathrm{PFD})_r$ is defined as the power density at the distance r from the transmitting antenna and its unit is W/m^2.

$$(\mathrm{PFD})_r = \frac{P_t G_t}{4\pi r^2} \tag{2.2.10}$$

According to Equations (2.2.8) and (2.2.10) can be expressed as:

$$(\mathrm{PFD})_r = \frac{\mathrm{EIRP}}{4\pi r^2} \tag{2.2.11}$$

Table 2.2 Wavelength labeling band method.

Segment Number	Band Name	Frequency Range	Band Name	Wavelength Range in Free Space
1	Extremely low frequency	3–30 Hz	Extremely long wave	100000–10000 km
2	Ultra low frequency	30–300 Hz	Ultra long wave	10000–1000 km
3	Super low frequency	300–3000 Hz	Super long wave	1000–100 km
4	Very low frequency(VLF)	3–30 K	Very long wave	100–10 m
5	Low frequency (LF)	30–300 K	Long wave	10000–1000 m
6	Intermediate frequency (MF)	300–3000 K	Medium wave	1000–100 m
7	High frequency (HF)	3–30 M	Short wave	100–10 m
8	Very high frequency (VHF)	30–300 M	Meter wave	10–1 m
9	Ultra high frequency (UHF)	300–3000 M		1–0.1 m
10	Super high frequency(SHF)	3–30 G		100–10 mm
11	Extremely high frequency(EHF)	30–300 G		10–1 mm
12	Too high frequency	300–3000 G		1000–100 μm

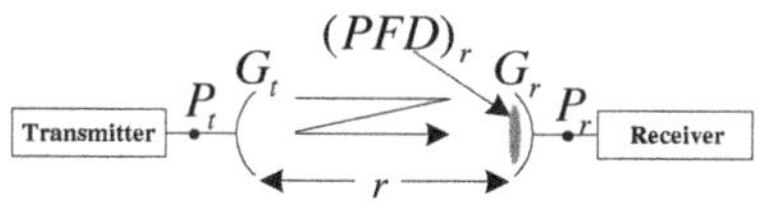

Figure 2.3 Basic communication link.

Power flux density is an important parameter for assessing the power requirements and interference levels of the satellite communication networks.

2.3 Layered Propagation Characteristics of the Satellite Channel

The analysis of satellite channel propagation characteristics is the basis of channel modeling. The channel propagation characteristics are related to many factors such as satellite height, carrier frequency, ground environment, and weather conditions, which are very complicated. The influence degree of various factors on the channel characteristics is very different. Some of them will bring signal amplitude fading, while others will change the signal phase. These factors can lead to signal distortion and seriously affect the communication quality. Therefore, it is of great significance to study the various factors in the satellite link to the special effects of the propagation characteristics of the satellite channel [5].

The main factors affecting the propagation characteristics of satellite channels include gas absorption, cloud occlusion, polarization effects, flicker effects, multipath, and shadowing effects, which may change the propagation direction and amplitude of wireless signal and result in the decline of communication quality. Only by fully studying the channel propagation characteristics and adopting appropriate modulation, coding, multiple access, and power control technology, the communication quality can be effectively improved and the "ideal" communication can be completed. A typical satellite channel downlink [6] is shown in Figure 2.4.

Figure 2.4 shows that the signals from the satellite to the ground receiving station go through the outer space, the dissipation layer, the thermal layer, the intermediate layer, the stratosphere, and the troposphere, which contains different substances and different propagation

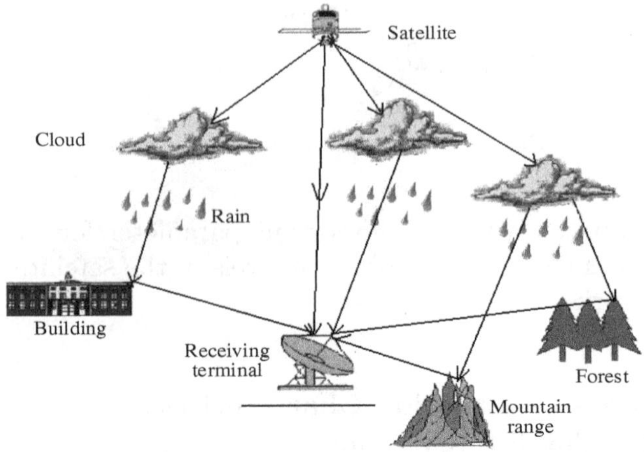

Figure 2.4 A typical satellite channel link diagram.

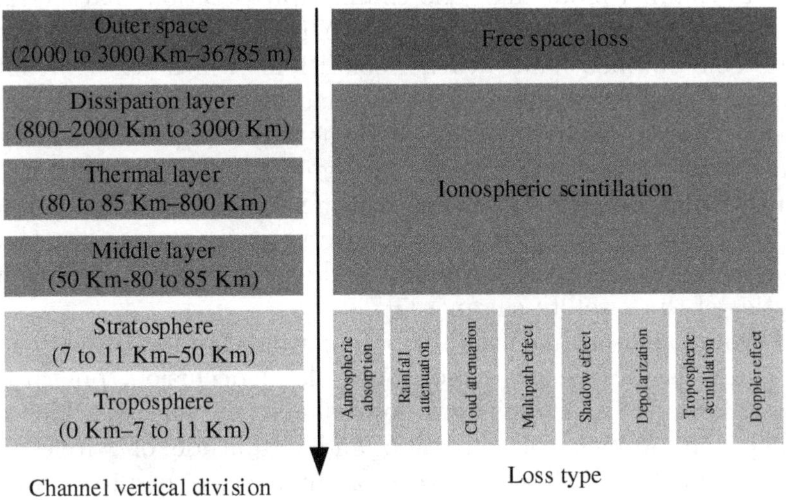

Figure 2.5 Propagation loss stratification.

environments and different types of losses and attenuations. According to the spatial order of the signal propagation in the satellite downlink, the main attenuation and interference types of each layer based on vertical layer are analyzed in turn and the corresponding quantitative calculation methods are given. The attenuation and interference are spatially distributed, as shown in Figure 2.5.

2.3.1 *Outer space*

The satellite transmits signals to the ground station at a height of tens of thousands of kilometers, and the signals first pass through the outer space, which contain very low outer space density substances in the form of plasma hydrogen. Although the density of the contained material in outer space is low, the path of free space is long and accounts for more than 95% of the total path length, so free space loss L_f is the most important loss [7]. Its calculation process is as follows:

Let the receiver antenna gain be G_r, the distance from the transmitter to the receiver is r, the transmitter power is P_t, the transmitting antenna gain is G_t, the receiving antenna power is P_r, and the free space loss is l_f. So P_r can be calculated by:

$$P_r = (PFD)_r A_e = \frac{P_t G_t}{4\pi r^2} A_e \tag{2.3.1}$$

where A_e is the effective area of the receiving antenna and

$$A_e = \frac{G_r \lambda^2}{4\pi} \tag{2.3.2}$$

Then, Equation (2.3.1) can be rewritten as:

$$P_r = \frac{P_t G_t}{4\pi r^2} \cdot \frac{G_r \lambda^2}{4\pi} = \left[\frac{P_t G_t}{4\pi r^2}\right] \cdot G_r \cdot \left[\frac{\lambda^2}{4\pi}\right] = P_t G_t G_r \left[\left(\frac{\lambda}{4\pi r}\right)^2\right] \tag{2.3.3}$$

Free-space loss is calculated by:

$$l_f = \left(\frac{4\pi r}{\lambda}\right)^2 \tag{2.3.4}$$

Its decibel form is defined as:

$$L_f = 20 \log \left(\frac{2\pi r}{\lambda}\right) \tag{2.3.5}$$

Since $c = \lambda f$, then Equation (2.3.5) can be written as:

$$L_f = 20 \lg f + 20 \lg r + 32.44 \tag{2.3.6}$$

Table 2.3 shows the orbital link losses for several typical geosynchronous orbit (GSO) or non-geostationary orbit (NGSO) satellites.

Table 2.3 Typical satellite link free-space path loss.

r/km	f/GHz	L_f/dB	r/km	f/GHz	L_f/dB
35900(GSO)	6	199	1000(NGSO)	12	174
	12	205		24	181
	20	209		2	138
	30	213		6	148
	44	216	100(NGSO)	12	154
1000(NGSO)	2	158		24	160
	6	168			

As can be seen from Table 2.3, the GSO orbital path loss is close to 200 dB, and the NGSO path loss is also about 150 dB. Therefore, in the communication link design, the influence of the path loss must be considered.

2.3.2 *Dissipation layer, thermal layer, and intermediate layer*

After the satellite signal propagates through the outer space, it will reach the dissipation layer, the thermal layer, and the intermediate layer. The gas in these layers is partially or fully ionized, which enables the radio wave to change the propagation speed, to cause refraction, reflection, and scattering, and to produce ionosphere scintillation, which can lead to rapid changes in the amplitude and phase of radio waves. The main reason is that the irregular distribution of electron density in the ionosphere leads to wave oscillation and sharp attenuation. In the satellite–ground link, scintillation effect can be observed from 30 MHz to 7 GHz. In VHF (very high frequency) band (30 MHz–300 MHz), the performance of ionospheric scintillation is particularly obvious. In some weather conditions, the flicker is very serious and the reliability of communication is seriously reduced. Ionospheric scintillation is closely related to season, radio frequency, geometry of observation points, and solar activity, especially in some special places and seasons. Ionospheric scintillation is mainly forward scattering and diffraction. The amplitude flicker index S_4 is usually used to quantitatively describe this flicker effect [8, 9], and S_4 is the ratio of the standard deviation to the mean of the signal intensity

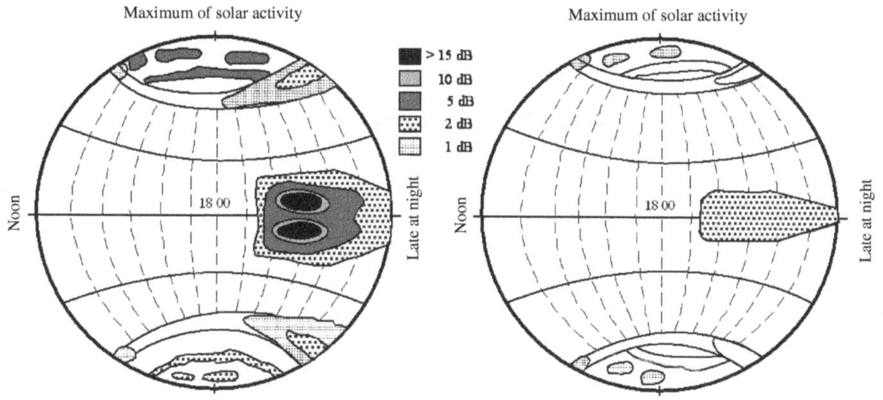

Figure 2.6 Ionospheric scintillation region distribution [9].

S per minute.

$$S_4^2 = \frac{<S^2> - [<S>]^2}{[<S>]^2} \tag{2.3.7}$$

where the time average of the signal amplitude is expressed as $< \cdot >$ and S_4 (in dB) is the ionospheric scintillation intensity.

The flicker intensity is related to the peak-to-peak fluctuation of the signal intensity. Statistics show that the value of S_4 obeys the Nakagami distribution. When the value of S_4 is close to 1, it approximates the Rayleigh distribution. Under certain special conditions, radio wave focusing caused by ionospheric irregularities may result in the values of S_4 to exceed 1.0 and reach 1.5.

When the fluctuation value P_{p-p} of the observed peak-to-peak power is known, S_4 is given by:

$$S_4 = P_{p-p}^{0.794} \tag{2.3.8}$$

In the International Telecommunications Union-Radio Communications Sector (ITU-R) report, an empirical value based on a range of flicker coefficients was given, where a P_{p-p} value of 10 dB corresponded to a scintillation coefficient of 0.5 dB, while the flicker factor of 1.0 corresponded to a value of above 27 dB. The intensity of the ionospheric scintillation intensity is the highest in the high latitudes and near the equatorial region. Figure 2.6 shows the L-band (1 GHz–2 GHz) scintillation region under the maximum and minimum solar activity conditions and gives a change process of day and night flashing. It can be found from Figure 2.6 that the flashing of the $\pm 20^0$ area near the equator at midnight is the most serious.

The ionospheric scintillation changes very quickly with a frequency in [0.1,1] Hz. Common ionospheric scintillation events begin after sunrise and last from 30 minutes to several hours. However, in the high years of solar activity near the equator, scintillation events occur almost every evening and after sunset, and the peak-to-peak value is above 10 dB. The ionospheric scintillation in high latitudes is relatively small and rarely exceeds 10 dB even in the high years of solar activity.

2.3.3 *Stratosphere and troposphere*

2.3.3.1 *Meteorological loss*

The earth's surface is surrounded by the atmosphere, which contains water vapor, dust, and so on. The temperature will change and the atmosphere will flow. These phenomena will produce polarization effects such as rainfall attenuation, atmospheric absorption, scintillation to the flow, rain, and hail.

2.3.3.1.1 Atmospheric absorption loss

The radio signals pass through the stratosphere and troposphere from top to bottom. These two layers concentrate almost all the atmosphere and water vapor. Among them, ozone in the stratosphere as well as oxygen and water vapor in the troposphere will cause absorption and loss of radio wave propagation. At present, there are two mature and accurate calculation methods [10], that is, the Bailey method and line-by-line summation method of water vapor and oxygen spectra, and the latter method is used in this book.

The meteorological parameters required by the line-by-line summation method are as follows: dry air pressure p_d(in hPa), where the unit is shown in the parentheses, similarly written hereinafter, water vapor partial pressure p_w(in hPa) and temperature T(in K), and their relationship is written as:

$$p_e = p_d + p_w \tag{2.3.9}$$

where p_e(in hPa) is a wide range of the available air pressures.

$$p_w = \frac{\rho\pi}{216.7} \tag{2.3.10}$$

where ρ(ing/m^3) is the water vapor density.

$$\rho = \frac{RH}{5.752}\theta_c^6 10^{(10-9.834\theta)} \tag{2.3.11}$$

where RH is the relative humidity, $\theta_c = \frac{300}{T_c+273.15°}$, and T_c (in °C) is the temperature. According to the commonly used meteorological parameters, Equation (2.3.9) can be rewritten as:

$$p_d = -\frac{\rho\pi}{216.7} + p_e \tag{2.3.12}$$

Ideally, the above meteorological parameters should be based on the actual local measurement values. In the absence of local measurements, the standard reference atmospheric pressure is given in ITU's supplementary recommendations, which also has good practicability.

Gas decay rate is defined and given by:

$$\gamma = \gamma_d + \gamma_w = 0.1820 f N''(f) \tag{2.3.13}$$

where γ_w and γ_d denote the value of the attenuation rate caused by water vapor and dry air, in units of dB/km, f is frequency in units of GHz, and $N''(f)$ is the imaginary part of the negative refractive index associated with frequency f, which can be expressed as:

$$N''(f) = \sum_i s_i F_i + N_d''(f) + N_w''(f) \tag{2.3.14}$$

where $N_w''(f)$ and $N_d''(f)$ denote wet and dry continuous spectrum, s_i is the intensity value of the ith spectral line, and F_i is the line shape factor.

The intensity of spectral line can be expressed as:

$$s_i = \begin{cases} a_1 \times 10^{-7} p_d \theta^3 \exp[a_2(1-\theta)] & \text{oxygen} \\ b_1 \times 10^{-1} p_w \theta^{3.5} \exp[b_2(1-\theta)] & \text{water vapor} \end{cases} \tag{2.3.15}$$

where, a_1,a_2,b_1, and b_2 are absorption parameters of the spectral line for oxygen and water vapor, provided by ITI-R P.676.

The line shape factor is given by:

$$F_i = \frac{f}{f_i}\left[\frac{\Delta f - \delta(f_i - f)}{(f_i - f)^2 + \Delta f^2} + \frac{\Delta f - \delta(f_i + f)}{(f_i + f)^2 + \Delta f^2}\right] \tag{2.3.16}$$

where f_i is the ith line spectrum frequency and the line spectrum width Δf is given by:

$$\Delta f = \begin{cases} a_3 \times 10^{-4}(p_d\theta^{(0.8-a_4)} + 1.1p_w\theta) & \text{oxygen} \\ b_3 \times 10^{-4}(p_d\theta^{b_4} + b_5 p_w\theta^{b_6}) & \text{water vapor} \end{cases} \tag{2.3.17}$$

The line correction factor δ of oxygen is given by:

$$\delta = \begin{cases} (a_5 + a_6\theta) \times 10^{-4}(p_d + p_w)\,\theta^{0.8} & \text{oxygen} \\ 0 & \text{water vapor} \end{cases} \tag{2.3.18}$$

The continuous spectrum $N_d''(f)$ of dry air determined by the non-resonant Debye spectrum of oxygen (below 10 GHz) and the nitrogen attenuation (above 100 GHz) caused by pressure is represented as:

$$N_d''(f) = f p_d \theta^2 \left[\frac{6.14 \times 10^{-5}}{d\left[1 + \left(\frac{f}{d}\right)^2\right]} + \frac{1.4 \times 10^{-12} p_d \theta^{1.5}}{1 + 1.9 \times 10^{-5} f^{1.5}}\right] \times \left[\frac{6.14 \times 10^{-5}}{d\left[1 + \left(\frac{f}{d}\right)^2\right]}\right] \tag{2.3.19}$$

where d is the width coefficient of Debye spectrum and

$$d = 5.6 \times 10^{-4}(p_d + p_w)\theta^{0.8} \tag{2.3.20}$$

Humid air continuous spectrum $N_w''(f)$ is represented as:

$$N_w''(f) = f(3.570\theta^{7.5} p_w + 0.113 p_d) p_w \theta^3 \times 10^{-7} \tag{2.3.21}$$

The atmosphere can be divided into multiple layers along the path, and the temperature, pressure, and humidity of each layer are definite values, so the attenuation rate of the inclined paths of each layer of the atmosphere can be calculated and then the attenuation rate of the entire path can be integrated to obtain the total attenuation along the inclined path. The attenuation of the ground station with elevation $\theta \geq 0$ and height above sea level h is defined as:

$$A(h,\ \varphi) = \int_h^{\infty} \frac{\gamma(H)}{\sin \Phi} \mathrm{d}H \tag{2.3.22}$$

where the angle Φ between the inclined path and the ground is written as:

$$\Phi = \arccos\left(\frac{c}{(r+H)\times n(H)}\right) \tag{2.3.23}$$

$$c = (r+h)\times n(h)\times \cos\theta \tag{2.3.24}$$

where the atmospheric RF refractive index $n(H)$ can be calculated from the water vapor pressure, along the path pressure, and temperature. When $\Phi = 0$, the Equation (2.3.22) is infinite. In practical engineering applications, let $\mu^4 = H - h$, then such problems can be solved. If $\Phi < 0$, then there is a minimum height $h_{\min}$.

In practical applications, numerical calculations are often used instead of the integral calculation of Equation (2.3.22). The geometric model of the numerical calculation is shown in Figure 2.7. According to Figure 2.7, the atmosphere is divided into multiple levels, a_1 and a_2 are the length of the path "1" and the path "2" corresponding to layer thickness δ_1 and δ_2, n_1 and n_2 are the refractive index, and α_1, α_2 and β_1, β_2 are the incident angle and the exit angle of the first layer and the second layer, respectively. r_1, r_2, and r_3 are the radius from the center of the earth to each layer and

$$\begin{aligned} a_1 &= -r_1\cos\beta_1 + \frac{1}{2}\sqrt{4r_1^2\cos^2\beta_1 + 8r_1\delta_1 + 4\delta_1^2} \\ a_2 &= -r_2\cos\beta_2 + \frac{1}{2}\sqrt{4r_2^2\cos^2\beta_2 + 8r_2\delta_2 + 4\delta_2^2} \end{aligned} \tag{2.3.25}$$

Angle α_1 and α_2 are obtained by:

$$\begin{aligned} \alpha_1 &= \pi - \arccos\left(\frac{-a_1^2 - 2r_1\delta_1 - \delta_1^2}{2a_1r_1 + 2a_1\delta_1}\right) \\ \alpha_2 &= \pi - \arccos\left(\frac{-a_2^2 - 2r_2\delta_2 - \delta_2^2}{2a_2r_2 + 2a_2\delta_2}\right) \end{aligned} \tag{2.3.26}$$

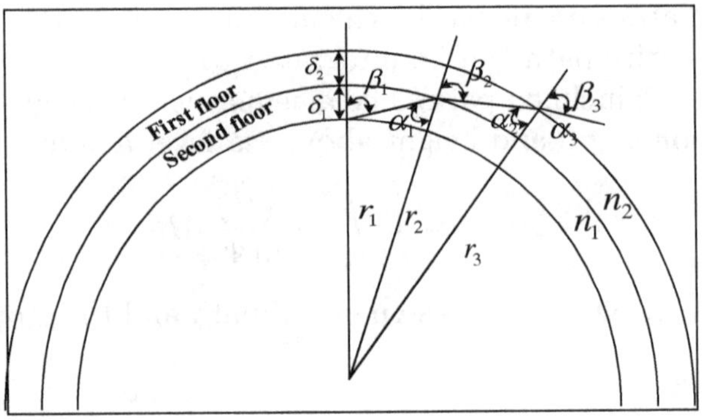

Figure 2.7 Geometric relationship of the gas attenuation layer.

where β_1 is the incidence angle of the location of the ground station and θ is the remainder of elevation. Angles β_2 and β_3 are given by:

$$\begin{aligned} \beta_2 &= \arcsin\left(\frac{n_1}{n_2}\sin\alpha_1\right) \\ \beta_3 &= \arcsin\left(\frac{n_2}{n_3}\sin\alpha_2\right) \end{aligned} \tag{2.3.27}$$

The gas attenuation of the total bias path is defined as:

$$A_{\text{gas}} = \sum_{i=1}^{k} a_i \gamma_i \tag{2.3.28}$$

where γ_i is the attenuation rate, a_i is the length of the ith layer path, and the unit of A_{gas} is dB.

2.3.3.1.2 Rain attenuation

Studies have shown that when the frequency is higher than 10 GHz, rainfall attenuation has become one of the main attenuations of radio waves passing through the troposphere. Many scholars have done a lot of long-term rainfall attenuation experimental measurement and modeling, but the measurement data will always be limited by the specific location and link parameters and its practicality is not strong. To this end, ITU-R presented various predictive models to predict the expected rainfall attenuation for a given communication system [11, 12].

In the ITU-R rainfall attenuation model, the relationship between the attenuation rate γ and the rainfall rate R is given by using the parameters a and b as $\gamma = aR^b$, where the unit of γ and R is dB/km and mm/h, respectively. The organization ITU-R also announced that the modeling method can estimate annual path attenuation statistics up to 55 GHz at a given location. In order to calculate the rainfall attenuation, the input parameters of the model, such as ground receiving station altitude h_s (in km), working frequency f (in GHz), the elevation angle of the ground θ (in degree), the polarization dip angle τ (in degree) relative to the ground station plane, and the ground station latitude φ (in degree), must be obtained. The measurement rate is 0.01% of the annual average rainfall rate $R_{0.01}$. The specific calculation steps are as follows:

Step 1: Determine the rainfall height h_R of the ground receiving station

The rainfall height (above sea level) of the ground receiving station is calculated by:

$$h_R = h_o + 0.36 \tag{2.3.29}$$

where the unit of h_R is km. h_o is the height of the annual average 0° isotherm, that is, the height of the upper atmosphere where the rain is in the transition state between water and ice.

Step 2: Calculate the length of the inclined path and the horizontal projection

The slope path length L_s and horizontal projection L_G can be calculated based on the elevation and height of the ground receiving station and the local rainfall height. The length L_s of the slope path is defined as the link length between satellites and ground affected by the rainy clusters, as shown in Figure 2.8.

L_s is calculated by:

$$L_s = \begin{cases} \dfrac{h_R - h_S}{\sin\theta} & \theta \geq 5^\circ \\[2ex] \dfrac{2(h_R - h_S)}{\sqrt{\sin^2\theta + \frac{2(h_R - h_S)}{R_E}} + \sin\theta} & \theta < 5^\circ \end{cases} \tag{2.3.30}$$

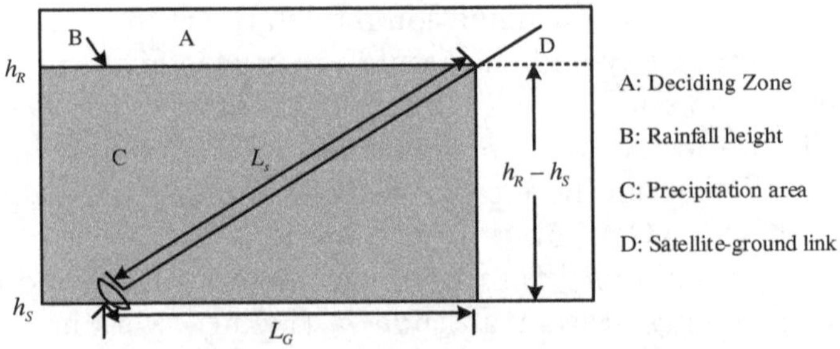

Figure 2.8 Schematic diagram of the inclined path of rain.

where h_S is the altitude of the ground receiving station, θ is the elevation angle, and R_E is the equivalent radius of the earth. $L_s < 0$, when the rainfall height is less than the receiving station height.

The horizontal projection is calculated as:

$$L_G = L_S \cos\theta \tag{2.3.31}$$

Step 3: Determine the average annual rainfall rate of 0.01%
The rainfall $R_{0.01}$ (1 min cumulative time) is obtained, where the probability of measuring the ground position exceeds 0.01% of the annual average. If there is no long-term statistical data in the local area, the rainfall at the 0.01% interruption rate from the rainfall intensity map, published by the World Meteorological Organization, can be used. The rainfall intensity map provides rainfall intensity contours that are exceeded by an average of 0.01% over the ocean and on land.

Step 4: Calculate the decay rate
The decay rate is calculated by:

$$\gamma_R = kR_{0.01}^{\alpha} \tag{2.3.32}$$

where γ_R represents the decay rate. k and α denote the dependent variables and are functions of frequency, elevation, and polarization dip and can be calculated from

k_H,k_V,α_H, and α_V by:

$$k = [k_H + k_V + (k_H - k_V)\cos^2\theta\cos 2\tau]/2$$
$$\alpha = [k_H\alpha_H + k_V\alpha_V + (k_H\alpha_H - k_V\alpha_V)\cos^2\theta\cos 2\tau]/2k \tag{2.3.33}$$

where τ is the polarization angle of linear polarization transmission in the horizontal plane. For circular polarization transmission, $\tau = 45°$.

Step 5: Calculate the horizontal reduction factor

The horizontal reduction factor $R_{0.01}$ is calculated by using the rainfall rate as:

$$r_{0.01} = \frac{1}{1 + 0.78\sqrt{\frac{L_G\gamma_R}{f}} - 0.38(1 - e^{-2L_G})} \tag{2.3.34}$$

where L_G is the horizontal projection determined in Step 2.

Step 6: Calculate the vertical adjustment factor

The vertical adjustment factor of 0.01% time is calculated by:

$$v_{0.01} = \frac{1}{1 + \sqrt{\sin\theta}\left[31(1 - e^{-\theta/(1+\chi)})\frac{\sqrt{L_R\gamma_R}}{f^2} - 0.45\right]} \tag{2.3.35}$$

where

$$\chi = \begin{cases} 36 - |\varphi| & |\varphi| < 36° \\ 0 & |\varphi| \geq 36° \end{cases} \tag{2.3.36}$$

$$L_R = \begin{cases} \dfrac{L_G r_{0.01}}{\cos\theta} & \zeta > \theta \\ \dfrac{h_R - h_S}{\sin\theta} & \zeta \leq \theta \end{cases} \tag{2.3.37}$$

where $\zeta = \arctan\left(\dfrac{h_R - h_s}{L_G r_{0.01}}\right)$.

Step 7: Determine the equivalent path length

The equivalent path length L_E (in km) is calculated by:

$$L_E = L_R v_{0.01} \tag{2.3.38}$$

Step 8: Calculate the attenuation over the annual average of 0.01%

The attenuation prediction exceeding the annual average of 0.01% is calculated by:

$$A_{0.01} = \gamma_R L_E \tag{2.3.39}$$

2.3.3.1.3 Cloud and fog attenuation

The attenuation of clouds and fog has a great impact on the design of satellite communication systems operating at frequencies above 20 GHz. Clouds and fog are suspended waters, whose diameters generally do not exceed 0.01 cm, and are the main factors causing the depolarization attenuation of signals. The attenuation of the satellite signals increases with the increase of frequency and the decrease of elevation.

The communication systems with cloud particle sizes below 100 μm and carrier frequencies below 300 GHz were proposed by ITU-R and a calculation method of the attenuation of clouds and fog was introduced [13]. The input parameters in this calculation method are frequency f (in GHz), elevation angle θ (in degree), ground temperature T, and total liquid water content L(in kg/m^2). In addition, there is the reciprocal of the temperature constant Φ, which belongs to an intermediate parameter. Its expression is given as:

$$\Phi = \frac{300}{T} \tag{2.3.40}$$

The calculation steps of cloud and fog attenuation are as follows:

Step 1: Calculate the relaxation frequency

The second-order and first-order relaxation frequencies of the double Debye model of the water medium constant are calculated by:

$$\begin{aligned} f_s &= 590 - 1500(\Phi - 1) \\ f_p &= 20.09 - 142(\Phi - 1) + 294(\Phi - 1)^2 \end{aligned} \tag{2.3.41}$$

Step 2: Calculate the complex permittivity

Based on the first- and second-order relaxation frequencies, the real and imaginary parts of the water complex permittivity are given by:

$$\begin{aligned}\varepsilon'(f) &= \frac{\varepsilon_0-\varepsilon_1}{\left[1+\left(\frac{f}{f_p}\right)^2\right]}+\frac{\varepsilon_1-\varepsilon_2}{\left[1+\left(\frac{f}{f_s}\right)^2\right]}+\varepsilon_2\\ \varepsilon''(f) &= \frac{f\cdot(\varepsilon_0-\varepsilon_1)}{f_p\cdot\left[1+\left(\frac{f}{f_p}\right)^2\right]}+\frac{f\cdot(\varepsilon_1-\varepsilon_2)}{f_s\cdot\left[1+\left(\frac{f}{f_s}\right)^2\right]}\end{aligned} \tag{2.3.42}$$

where $\varepsilon_0 = 77.6 + 103.3(\Phi - 1)$, $\varepsilon_1 = 5.48$, and $\varepsilon_2 = 3.51$.

Step 3: Calculate the attenuation rate coefficient

The calculation equation of the attenuation rate coefficient is given as:

$$K_1 = \frac{0.819f}{\varepsilon''(1+\eta^2)} \tag{2.3.43}$$

where $\eta = \frac{2+\varepsilon'}{\varepsilon''}$. The decay rate coefficient represents the "dot" attenuation at a particular frequency and water vapor concentration.

Step 4: Calculate the liquid water column content

The liquid water column content $L(\text{in kg/m}^2)$, when there is no statistical data on liquid water content in the cloud, can be calculated by four global maps of liquid water content contours in clouds provided by ITU-R P.840.5. Contour lines are 20%, 10%, 5%, and 1% over the average annual time, respectively. These maps are derived from two-year data with a spatial resolution of 1.5° in longitude and latitude.

The positions of different grid points can be obtained by a bilinear difference operation on the values of the nearest four grid points. Figure 2.9 shows the normalized total content of liquid water in the cloud over 0.01% of the time of one year, and its horizontal axis represents latitude and vertical axis represents longitude.

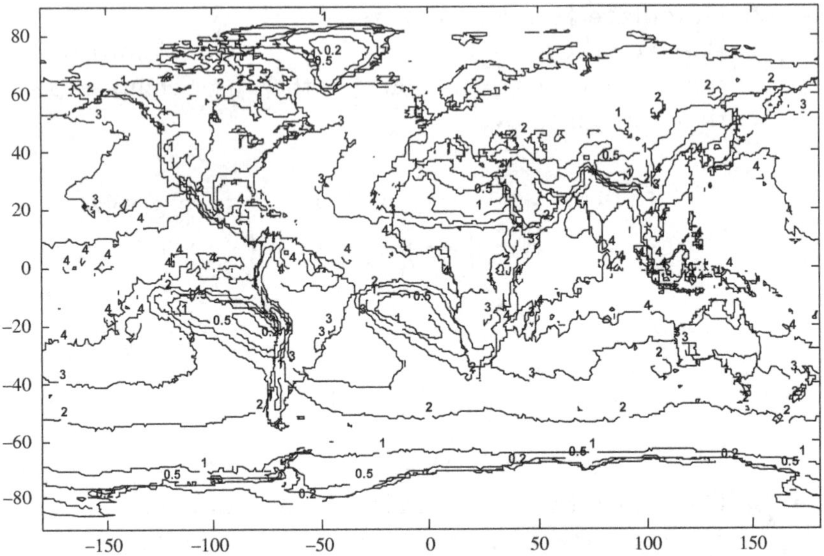

Figure 2.9 Normalized total content of liquid water in the cloud over 0.01% of the time of one year.

Step 5: Calculate the total attenuation

The total attenuation of the cloud is calculated by:

$$A_c = \frac{LK_1}{\sin\theta} \tag{2.3.44}$$

where $10^\circ \leq \theta \leq 90^\circ$. When the elevation angle is less than 10°, it is not suitable to be calculated by the relational expression $1/\sin\theta$, because it assumes that the range of the cloud is close to infinity. When the elevation angle is close to 0°, a physical limit should be imposed on the path length.

2.3.3.1.4 Tropospheric scintillation

The tropospheric scintillation reflects the rapid fluctuation of the signal level in the satellite system, which is caused by turbulent fluctuations in the small range of refractive index under certain temperature, humidity, and pressure conditions. Tropospheric scintillation causes a change in the phase and amplitude of the electric field. For systems with low elevation (below 10°) and low margin (below 3 dB) above 10 GHz, the performance degradation caused by a flicker is greater than that of the rainfall attenuation. Therefore, the prediction and modeling of the tropospheric scintillation is very important.

ITU-R provided a method for quickly calculating tropospheric scintillation based on local meteorological conditions [9], which may calculate the standard deviation of the instantaneous signal strength. The input parameters of the model are as follows: frequency f and $4\,\text{GHz} \leq f \leq 20\,\text{GHz}$, path elevation $\theta \geq 4°$, antenna physical aperture D, average relative temperature T of the ground station surface over a period of more than one month, humidity H, as well as antenna efficiency η and in the unknown case $\eta = 0.5$.

The calculation steps of the tropospheric scintillation are as follows:

Step 1: Calculate saturated water vapor pressure

In average relative temperature, the saturated water vapor pressure p_{ws} (kPa) is calculated by:

$$p_{w_s} = 6.1121\text{exp}\left(\frac{17.502T}{T + 240.97}\right) \tag{2.3.45}$$

Step 2: Calculate the wet-term of the radio refractive index

The wet-term N_{wet} of the refractive radio index is given by:

$$N_{\text{wet}} = \frac{3732e_s H}{(273 + T)^2} \tag{2.3.46}$$

Step 3: Calculate the standard deviation of the signal amplitude

The standard deviation σ_{ref} of the signal amplitude is calculated as:

$$\sigma_{\text{ref}} = 3.6 \times 10^{-3} + N_{\text{wet}} \times 10^{-4} \tag{2.3.47}$$

Step 4: Calculate the effective path length

The effective path length is defined as:

$$L = \frac{2h_L}{\sqrt{\sin^2\theta + 2.35 \times 10^{-4}} + \sin\theta} \tag{2.3.48}$$

where the height h_L of the stratosphere is usually taken as $h_L = 1000\,\text{m}$.

Step 5: Calculate the equivalent antenna aperture
The relationship among equivalent antenna aperture D_{eff} (in m), geometric aperture D, and antenna efficiency η is written as:

$$D_{\text{eff}} = \sqrt{\eta} D \tag{2.3.49}$$

Step 6: Calculate the antenna averaging factor
The average factor of the antenna is given by:

$$g(x) = \sqrt{3.86(x^2+1)^{11/12} \sin\left[\frac{11}{6}\operatorname{arc\,tan}\left(\frac{1}{x}\right)\right] - 7.08x^{5/6}} \tag{2.3.50}$$

where

$$x = 1.22\frac{D_{\text{eff}}^2 f}{L} \tag{2.3.51}$$

Step 7: Calculate the standard deviation of the signal during considering the time period and the transmission path
The signal standard deviation is given by:

$$\sigma = \sigma_{\text{ref}} f^{7/12} \frac{g(x)}{(\sin\theta)^{1.2}} \tag{2.3.52}$$

Step 8: Calculate the time percentage factor
The time percentage factor is calculated by:

$$a(v) = -0.061(\log v)^3 + 0.072(\log v)^2 - 1.71(\log v) + 3.0 \tag{2.3.53}$$

where v refers to the percentage of interruption, $0.01 < v \leq 50$.

Step 9: Calculate the depth of scintillation fading
The formula for calculating the scintillation fading depth $A_s(v)$ is written as:

$$A_s(v) = a(v) \cdot \sigma \tag{2.3.54}$$

where the unit of $A_s(v)$ is dB.

2.3.3.1.5 Depolarization effect

In the satellite communication system, linear polarization or circular polarization is usually used to generate radio waves. These two polarization states have their own advantages, but both of them use biorthogonal polarization technology to double the transmission link capacity on the basis of frequency multiplexing. Depolarization effect refers to the change of polarization characteristics of radio waves when they propagate through the atmosphere. The main reasons for causing this effect are the influence of rainfall, ice particles, and multipath effect in the transmission path. In the design of satellite–ground link, cross-polarization discrimination (XPD) and isolation are usually used to measure the polarization effect [14].

The input parameters in the calculation method of rain and ice depolarization effects in ITU-R include elevation angle θ of the ground station, polarization dip angle τ relative to horizontal plane, operating frequency f, and rainfall attenuation A_v over time ratio in signal transmission path v. The calculation procedure is as follows:

Step 1: Calculate the term related to frequency

$$C_f = 30 \lg f \tag{2.3.55}$$

where GHz $8 \leq f \leq 35$.

Step 2: Calculate the term related to rainfall attenuation

$$C_A = V(f) \log A_v(dB) \tag{2.3.56}$$

where

$$V(f) = \begin{cases} 12.8 f^{0.19} & 8 \leq f \leq 20\,\text{GHz} \\ 22.6 & 20 \leq f \leq 35\,\text{GHz} \end{cases} \tag{2.3.57}$$

Step 3: Calculate the polarization improvement factor

$$C_\tau = -10 \log[1 - 0.484(1 + \cos(4\tau))] \tag{2.3.58}$$

when $\tau = 45°$, $C_\tau = 0$. When $\tau = 0°$ or $\tau = 90°$, the maximum value is 15 dB.

Step 4: Calculate the term related to elevation

$$C_\theta = -40\log(\cos\theta), \theta \leq 60° \tag{2.3.59}$$

Step 5: Calculate the term related to tilt angle

$$C_\sigma = 0.00522\sigma^2 \tag{2.3.60}$$

where σ refers to the effective standard deviation of the raindrop tilt angle distribution. When $\upsilon = 1\%$, $\upsilon = 0.1\%$, $\upsilon = 0.01\%$, and $\upsilon = 0.001\%$ of the time percentage factor, σ is $0°$, $5°$, $10°$ and $15°$, respectively.

Step 6: Calculate the XPD of rain with time percentage less than p%

$$\text{XPD}_{\text{rain}} = C_f - C_A + C_\tau + C_\theta + C_\sigma \tag{2.3.61}$$

Step 7: Calculate the item related to the crystal

$$C_{\text{ice}} = \frac{\text{XPD}_{\text{rain}} \times (0.3 + 0.1 \lg_{10} p)}{2} \tag{2.3.62}$$

Step 8: Calculate the effects of rain and ice with less than the percentage υ of time

$$\text{XPD}_p = \text{XPD}_{\text{rain}} - C_{\text{ice}} \tag{2.3.63}$$

Step 9: Calculate XPD below 8 GHz frequency

For frequencies in the $4\,\text{GHz} \le f < 8\,\text{GHz}$ range, the equation is given by:

$$\text{XPD}_2 = \text{XPD}_1 - 20 \lg \left(\frac{f_2}{f_1}\right) \tag{2.3.64}$$

where $f_1 = 8$, XPD_1 can be calculated according to Step 8 and f_2 is the working frequency in the $4\text{GHz} \le f < 8\text{GHz}$ range.

2.3.3.2 *Non-meteorological loss*

2.3.3.2.1 Multipath effect

As shown in Figure 2.10(a), in the satellite communication propagation environment, the signals, which pass through trees, buildings, mountains, etc., cause signals to reflect, scatter, refract, etc., and reach the receiving terminal as multipath signals, and each path

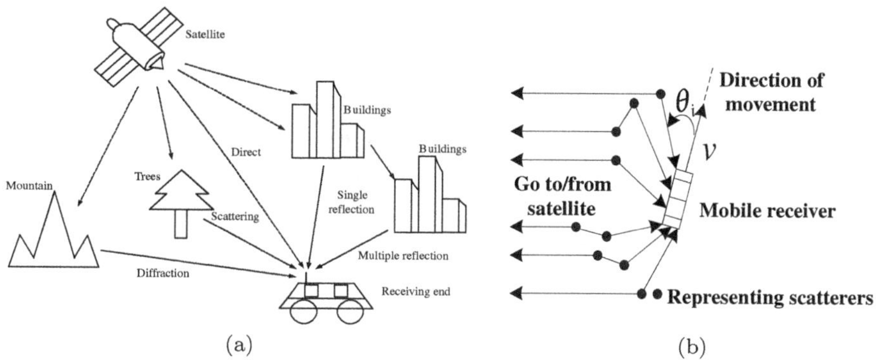

Figure 2.10 Multipath environment diagram: (a) Multipath propagation and (b) multipath signal model.

length is different, the arrival time and the phase are also different. It can be known from the wave propagation theory that waves of different phases are superimposed, and the in-phase waves are strengthened and the anti-phase waves weaken each other, resulting in multipath attenuation.

The propagation diagram shown in Figure 2.10(a) can be simplified as shown in Figure 2.10(b) [9]. The mobile speed of the receiving terminal is v, and θ_i is the angle between the direction of the ith reflected signal from the first scatterer and the mobile receiving end. When the wireless signal is a narrowband signal, the complex envelope $s'(t)$ of the signals can be expressed as:

$$s'(t) = \mathrm{Re}[s(t)e^{j2\pi f_c t}] \tag{2.3.65}$$

where $\mathrm{Re}[\cdot]$ represents the operation of taking real part, $s(t)$ refers to the transmitted signal, and f_c is the carrier frequency.

Let the length of the ith path be x_i and its reflection coefficient be a_i, then the received signal is written as:

$$\begin{aligned} y'(t) &= \sum_i a_i s'\left(t - \frac{x_i}{c}\right) = \sum_i a_i \mathrm{Re}\left\{s\left(t - \frac{x_i}{c}\right)\exp\left[j2\pi f_c\left(t - \frac{x_i}{c}\right)\right]\right\} \\ &= \mathrm{Re}\left\{\sum_i a_i s\left(t - \frac{x_i}{c}\right)\exp\left[j2\pi\left(f_c t - \frac{x_i}{\lambda}\right)\right]\right\} \end{aligned} \tag{2.3.66}$$

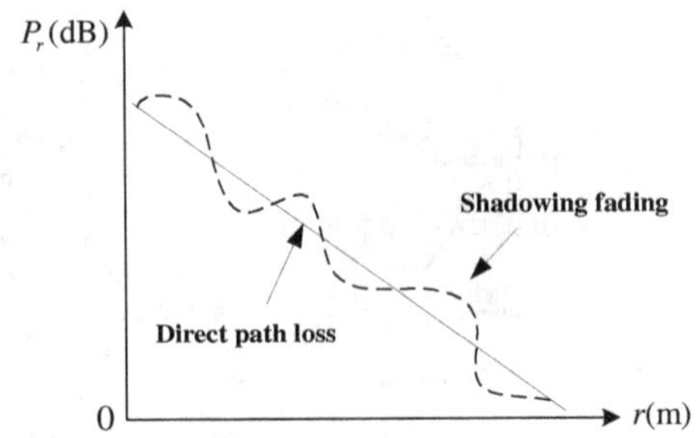

Figure 2.11 A schematic diagram of the shadowing attenuation component.

where c is the speed of light, wavelength $\lambda = c/f_c$. The Equation (2.3.66) can be written as:

$$y'(t) = \mathrm{Re}[y(t)e^{j2\pi f_c t}] \tag{2.3.67}$$

The complex envelope of the received signal is the sum of the multipath components with different delay, phase shift, and attenuation, i.e.,

$$y(t) = \sum_i a_i e^{-j2\pi \frac{x_i}{\lambda}} s\left(t - \frac{x_i}{\lambda}\right) = \sum_i a_i e^{-j2\pi f_c \tau_i} s(t - \tau_i) \tag{2.3.68}$$

where the delay $\tau_i = x_i/c$. The Equation (2.3.68) is the complex envelope model of multipath loss.

2.3.3.2.2 Camouflage shadowing effect

Satellite mobile communications are highly mobile and flexible, and terrestrial communication locations cannot be pre-selected. Therefore, when the radio wave propagation path encounters obstacles such as buildings, trees, and rolling hills, the level of the received signal will decrease. This phenomenon is called as the shadowing effect, and the signal fading caused by it is called as shadowing fading [15].

The camouflage attenuation component of the received signal of the terminal varies with the average power of the region by several wavelengths, as shown in Figure 2.11.

In Figure 2.11,

$$\bar{P}_r = 10\log(P_t G_t G_r G(r)), \tag{2.3.69}$$

where $G(r)$ is the path loss factor. P_t,G_t, and G_r are defined as in Section 2.2.

The shadowing effect is represented by a Gaussian random variable x with a mean of zero and a variance of σ^2, i.e.,

$$f(x) = \frac{1}{\sqrt{2\pi\sigma^2}} e^{-\frac{x^2}{2\sigma^2}} \tag{2.3.70}$$

The total received power is defined as:

$$P_r = P_t G_t G_r G(\mathrm{r}) \times 10^{x/10} \times \alpha^2 = \bar{P}_r \times 10^{x/10} \times \alpha^2 \tag{2.3.71}$$

where the first and second terms represent the received power terms in statistical significance due to shadowing attenuation, which may be defined as the local average power P_{SF} of the shadowing attenuation, i.e.,

$$P_{SF} = \bar{P}_r 10^{x/10} \tag{2.3.72}$$

Its decibel form is written as:

$$P_{SF} = \bar{P}_r + x \tag{2.3.73}$$

The probability density function of P_{SF} can be expressed as Gaussian distribution with a mean of $\bar{P}_{SF}$, i.e.,

$$f(P_{SF}) = \frac{1}{\sqrt{2\pi\sigma^2}} e^{-\frac{(P_{SF} - \bar{P}_{SF})^2}{2\sigma^2}} \tag{2.3.74}$$

The long-period camouflage attenuation caused by the topographical changes in the study area is strongly correlated with the local environment. Experiments have shown that when the range of motion of a mobile terminal is several hundred meters, the range of P_{SF} is 40 dB or more, and the value σ varies between 6 dB and 10 dB.

Now consider the statistics of the link shadowing attenuation of the mobile receiving terminal, the threshold of the power P_{SF} received by the mobile terminal is set to P_0(in dB). Then, the probability of the signal power, received by the terminal and is greater than or equal to the threshold, is written as:

$$P(P_{SF} \geq P_0) = \int_{P_0}^{\infty} \frac{1}{\sqrt{2\pi\sigma^2}} e^{-\frac{(P_{SF} - \bar{P}_{SF})}{2\sigma^2}} dx \tag{2.3.75}$$

2.3.3.2.3 Doppler effect

The relative movement between the receiving terminal and the satellite results in a significant frequency shift among multipath signals, which is called as Doppler shift and leads to inter-symbol interference and affects the quality of communication in satellite communications.

The Doppler effect is highly dependent on the relative moving speed, the direction of motion, and the incidence angle of multipath signals [16]. Let the mobile receiving terminal move in the schematic shown in Figure 2.12.

In Figure 2.12, the mobile terminal's active range (the distance between X and Y) is denoted as d, the moving speed is v, the satellite is at the point S, the angle between the received multipath signals and the ground is α_i and α_j, respectively, and the radio wave of the point S is received by the mobile terminal at the point X and point Y. The distance difference between them is defined as:

$$\Delta l = d \cos \alpha_i = v \Delta t \cos \alpha_i \tag{2.3.76}$$

where Δt denotes the time difference between the terminal to the point X and the terminal to the point Y. In actual satellite communication, the propagation path of the signal is long, and it can be assumed that the incidence angles of the point X and the point Y are equal, that is, $\alpha_i = \alpha_i = \alpha$, and the phase difference caused by the different two-path is written as:

$$\Delta \varphi = \frac{2\pi \Delta l}{\lambda} = \frac{2\pi v \Delta t}{\lambda} \cos \alpha \tag{2.3.77}$$

where λ is the wavelength.

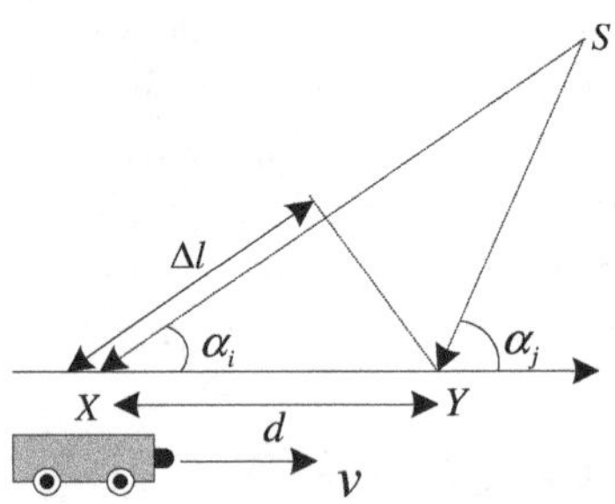

Figure 2.12 Doppler effect diagram.

According to Equation (2.3.77), the Doppler shift is given by:

$$f_d = \frac{1}{2\pi}\frac{\Delta\varphi}{\Delta t} = \frac{v}{\lambda}\cos\alpha = f_m \cos\alpha \tag{2.3.78}$$

where f_m is the maximum Doppler shift.

From Equations (2.3.76)–(2.3.78), if the angle between the receiving terminal and the incident wave is less than 90°, the Doppler shift is positive (the receiving frequency rises), otherwise it is the opposite. In practical communication technology applications, the received signals are a sum of multipath components that arrive at the receiver in different directions, which cause Doppler spread and broaden the signal bandwidth.

2.4 Classic Satellite Channel Model

According to the subjective experience of researchers, the traditional channel modeling methods think that there are only one or two factors that affect the channel propagation characteristics in some cases, and the factors considered in modeling are relatively single, which cannot describe all propagation characteristics of a link, but it has strong pertinence and describes the studied factors clearly.

The classical satellite channel models established by traditional methods mainly include lognormal model, Rayleigh model, and Suzuki model. This section first introduces the common probability distribution function and then introduces the classical satellite channel model, which lays the necessary foundation for the subsequent channel modeling research.

2.4.1 *Common probability distribution functions*

In the study of satellite channel transmission characteristics, four kinds of probability distribution functions are used, which are Gaussian distribution function, Rician distribution function, lognormal distribution function, and Nakagami distribution function. These can be described as:

2.4.1.1 *Gaussian distribution*

Gaussian distribution is also called normal distribution. In the satellite communication, the influence of atmospheric environmental

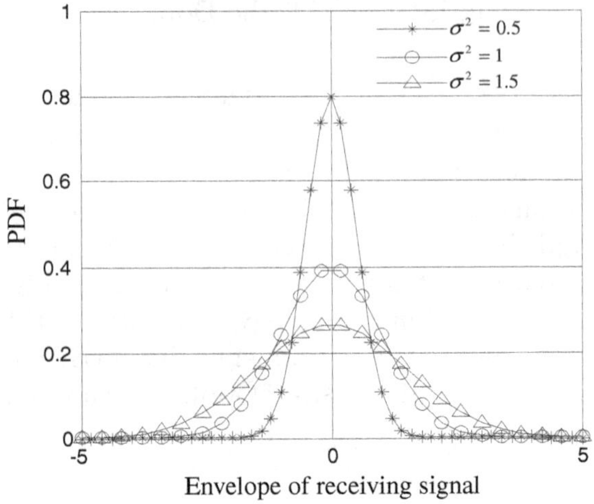

Figure 2.13 PDF curves of Gaussian distribution.

factors (rainfall, snowfall, cloud, etc.) on signal transmission can be described by Gaussian distribution, which can describe the distribution of signal envelope and phase [17, 18].

The probability density function (PDF) of Gaussian distribution can be expressed as:

$$f_{\text{Gus}}(r) = \frac{1}{\sqrt{2\pi}\sigma}\exp\left(\frac{-(r-m_G)^2}{2\sigma_G^2}\right) \tag{2.4.1}$$

where m_G and σ_G^2 are the mean and variance of Gaussian distribution, respectively. The mean determines the position of the distribution, the variance determines the magnitude of the distribution, and the probability density function curve is symmetric. Figure 2.13 shows the curves of the Gaussian probability density function with the mean $m_G = 0$ and variance σ_G^2 taking different values.

2.4.1.2 *Rice/Rayleigh distribution*

In the process of radio wave propagation, it will encounter various obstacles such as buildings, trees, hills, and so on, which will produce reflected wave, scattered wave, and diffraction wave, to make the received signal be the superposition of multipath signals. Due to the different transmission time in each path, the phase is different when it reaches the receiver. When the signals of different phases are superimposed, they are often enhanced in phase, weakened in opposition, and sometimes even offset each other. In this way, the amplitude of the received signal will reduce sharply, which leads to multipath fading. If there are multipath component and direct component in the received signals, the signal envelope obeys Rice distribution [18]. If there is only reflected wave and no direct wave in the received signals, the envelope of the received signal obeys Rayleigh distribution [19].

Let the unmodulated carrier signal be given as:

$$s(t) = \cos(\omega_c t + \varphi_c) \tag{2.4.2}$$

The Doppler shift of the nth path is denoted as $f_{dn} = v \cos \alpha_n / \lambda$. If the received signal is a superposition of multipath signals and a direct view signal, the amplitude $r(t)$ is given by:

$$r(t) = c_0 \cos(\omega_c t + \varphi_c + \Delta\omega_0 t) + \sum_{n=1}^{N} c_n \cos(\omega_c t + \varphi_c + \phi_n + \Delta\omega_n t) \tag{2.4.3}$$

where the first term is the line of sight (LOS) component, and the latter term is the superposition term of the multipath signals. $\Delta\omega_n = 2\pi f_{dn}$, ϕ_n is the phase deviation of the nth path. $\theta_1 = \varphi_c + \Delta\omega_0 t$, $\theta_2 = \phi_c + \varphi_n + D\omega_n t$, then Equation (2.4.3) can be written as:

$$\begin{aligned} r(t) &= c_0 \cos(\omega_c t + \theta_1) + \sum_{n=1}^{N} c_n \cos(\omega_c t + \theta_2) \\ &= c_0 \cos \omega_c t \cos \theta_1 - c_0 \sin \omega_c t \sin \theta_1 + \sum_{n=1}^{N} c_n \cos \omega_c t \cos \theta_2 \\ &\quad - \sum_{n=1}^{N} c_n \sin \omega_c t \sin \theta_2 \qquad (2.4.4) \\ &= \left(c_0 \cos \theta_1 + \sum_{n=1}^{N} c_n \cos \theta_2 \right) \cos \omega_c t \\ &\quad - \left(c_0 \sin \theta_1 + \sum_{n=1}^{N} c_n \sin \theta_2 \right) \sin \omega_c t \end{aligned}$$

where $I(t) = c_0 \cos\theta_1 + \sum_{n=1}^{N} c_n \cos\theta_2$, $Q(t) = c_0 \sin\theta_1 + \sum_{n=1}^{N} c_n \sin\theta_2$, then

$$r(t) = I(t)\cos\omega_c t - Q(t)\sin\omega_c t \tag{2.4.5}$$

If $T_c(t) = \sum_{n=1}^{N} c_n \cos\theta_2, T_s(t) = \sum_{n=1}^{N} c_n \sin\theta_2$, it is known by the central limit theorem that $T_c(t)$ and $T_s(t)$ are all Gaussian random process. For a given time t_0, there are the following relationships.

$$I = I(t_0) = c_0 \cos\theta_1 + T_c(t_0) = c_0 \cos\theta_1 + T_c \tag{2.4.6}$$

$$Q = Q(t_0) = c_0 \sin\theta_1 + T_s(t_0) = c_0 \sin\theta_1 + T_s \tag{2.4.7}$$

where the PDFs of T_c and T_s are written as:

$$f_{T_c}(t_c) = \frac{1}{\sigma_c\sqrt{2\pi}} \exp\left(-\frac{t_c^2}{2\sigma_c^2}\right) \tag{2.4.8}$$

$$f_{T_s}(t_s) = \frac{1}{\sigma_s\sqrt{2\pi}} \exp\left(-\frac{t_s^2}{2\sigma_s^2}\right) \tag{2.4.9}$$

where $E[T_c] = E[T_s] = 0$, $D[T_c] = D[T_s] = \sigma_c^2 = \sigma_s^2 = \sigma^2$.

Because θ_n is uniform distribution in $[0, 2\pi]$, the covariance of T_c and T_s is written as

$$\mathrm{Cov}(T_c, T_s) = E[T_c T_s] = \sum_{n=1}^{N}\sum_{m=1}^{N} c_n c_m E\left[\frac{\sin 2\theta_n}{2}\right] = 0 \tag{2.4.10}$$

In addition, the PDFs of I and Q are obtained by:

$$\begin{aligned} f_I(i) &= f_{T_c}(i - c_0\cos\theta_1) f|(i - c_0\cos\theta_1)'| \\ &= \frac{1}{\sigma\sqrt{2\pi}} \exp\left[-\frac{(i - c_0\cos\theta_1)^2}{2\sigma^2}\right] \end{aligned} \tag{2.4.11}$$

$$\begin{aligned} f_Q(q) &= f_{T_c}(q - c_0\sin\theta_1) f|(q - c_0\sin\theta_1)'| \\ &= \frac{1}{\sigma\sqrt{2\pi}} \exp\left[-\frac{(q - c_0\sin\theta_1)^2}{2\sigma^2}\right] \end{aligned} \tag{2.4.12}$$

Similarly, the mean and the covariance of I and Q are written as:

$$E[I] = c_0 \cos\theta_1, E[Q] = c_0 \sin\theta_1 \tag{2.4.13}$$

$$\mathrm{Cov}(I,Q) = E[(I - E[I])(Q - E[Q])] = E[T_s T_c] = 0 \tag{2.4.14}$$

Since I and Q are independent of each other, then

$$\begin{aligned} f_{IQ}(i,q) &= f_I(i) f_Q(q) \\ &= \frac{1}{2\pi\sigma^2} \exp\left(-\frac{(i^2+q^2) - 2c_0\sqrt{i^2+q^2}\cos\left(\theta_1 + \arctan\frac{q}{i}\right) + c_0^2}{2\sigma^2}\right) \end{aligned} \tag{2.4.15}$$

Let

$$r^2 = i^2 + q^2 \tag{2.4.16}$$

$$\theta = \theta_1 + \mathrm{arc}\tan\frac{q}{i}$$

$$i = r\cos\theta, \quad q = r\sin\theta$$

Then, the Jacobian determinant is written as:

$$J = \frac{\partial(i,q)}{\partial(r,\theta)} = \begin{vmatrix} \cos\theta & -r\sin\theta \\ \sin\theta & r\cos\theta \end{vmatrix} = r \tag{2.4.17}$$

Therefore, the joint probability density function of both phase θ and envelope r of the received signal is given by:

$$\begin{aligned} f_r(r) &= \int_{-\pi}^{\pi} f_{r\theta}(r,\theta) d\theta \\ &= \frac{r}{2\pi\sigma^2} \exp\left(-\frac{r^2 + c_0^2}{2\sigma^2}\right) \int_{-\pi}^{\pi} \exp\left(\frac{c_0 r}{\sigma^2}\cos\theta\right) d\theta \end{aligned} \tag{2.4.18}$$

As is known to all, the first type of zero-order Bessel function is known as:

$$I_0(x) = \frac{1}{2\pi}\int_{-\pi}^{\pi} \exp(x\cos\theta) d\theta \tag{2.4.19}$$

So, Equation (2.4.18) can be rewritten as:

$$f_r(r) = \frac{r}{2\pi\sigma^2} \exp\left(-\frac{r^2 + c_0^2}{2\sigma^2}\right) I_0\left(\frac{c_0 r}{\sigma^2}\right) \tag{2.4.20}$$

The magnitude of the multipath fading can also be expressed by the Rice factor K, which represents the ratio of the power of the direct view component to that of the multipath component i.e., $K = z^2/(2\sigma^2)$. The larger the Rice factor is, the lower the power of the multipath component in the received signals relative to the direct view component power. When $K = 0$, the direct component of the total received signals is smaller.

When the direct view component z is zero, the Rice distribution becomes a Rayleigh distribution, which is a type of common probability distribution to describe the characteristics of flat fading channels. When there is no direct view component in the received signals, only the superposition of multipath components, the envelope of the received signal obeys the Rayleigh distribution.

The PDF of the Rayleigh distribution is written as:

$$f_{Ray}(r) = \frac{r}{\sigma^2} \exp\left(-\frac{r^2}{2\sigma^2}\right) \tag{2.4.21}$$

where σ^2 is the variance of the Rayleigh distribution and $r \geq 0$.

The n-order origin moment of the multipath signal r is written as:

$$E[r^n] = n\sigma^2 E[r^{n-2}], n \geq 2 \tag{2.4.22}$$

Figure 2.14 is a plot of the probability density function for the variance $\sigma^2 = 1$, direct component $z = 1$, and $z = 0$, respectively.

2.4.1.3 *Lognormal distribution*

When satellite signals propagate through obstacles such as leaves and thin clouds, radio waves create shadowing effect duo to absorption or scattering. The shadowing effect exists in both LOS (Line of Sight) propagation and multipath transmission. In this section, the shadowing effects under LOS propagation are discussed [20].

Let the unmodulated carrier signal be written as:

$$s(t) = A\cos(\omega_c t + \phi_c) \tag{2.4.23}$$

The Doppler shift of the LOS signal is denoted as $f_d = v\cos\alpha/\lambda$, and the shadow masking factor is denoted as $C(t) = \prod_{i=1}^{N} c_i(t)$. For

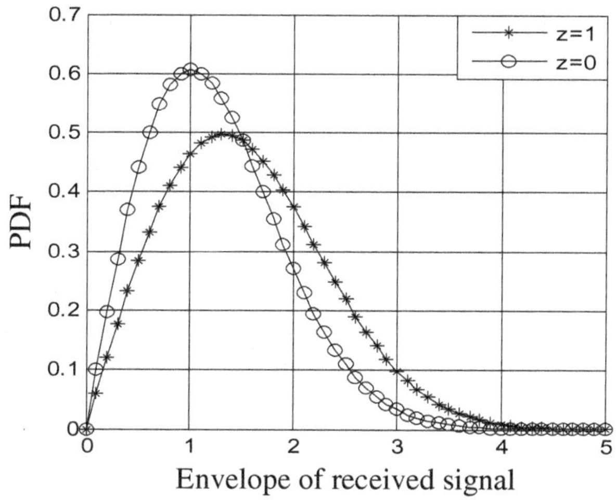

Figure 2.14 PDF curves of Rice distribution.

the carrier signal, it is the multiplicative factor, and the shadowing signal at the receiving end is represented as:

$$\begin{aligned} r(t) = C(t)s(t) &= \prod_{i=1}^{N} c_i(t) A \cos(\omega_c t + \varphi_c + 2\pi f_d t) \\ &= \prod_{i=1}^{N} c_i(t) A \cos\theta \cos\omega_c t - \prod_{i=1}^{N} c_i(t) A \sin\theta \sin\omega_c t \end{aligned} \tag{2.4.24}$$

where $\theta = \phi_c + 2\pi f_d t$. At a certain point in time t_0, $r = r(t_0)$. The envelope r of the received signal $r(t)$ is written as:

$$r = \sqrt{\left(\prod_{i=1}^{N} c_i(t_0) A \cos\theta\right)^2 + \left(\prod_{i=1}^{N} c_i(t_0) A \sin\theta\right)^2} = A \prod_{i=1}^{N} c_i(t_0) \tag{2.4.25}$$

Equation (2.4.25) may be modified as:

$$\ln r = \ln A + \sum_{i=1}^{N} \ln c_i(t_0) \tag{2.4.26}$$

Let $\ln r = y, \ln A = k, \sum_{i=1}^{N} \ln c_i(t_0) = x$, then $y = k + x$. Based on the central limit theorem, $x \sim N(m_X, \sigma_X^2)$, its PDF can be rewritten as:

$$f_X(x) = \frac{1}{\sigma_X\sqrt{2\pi}} \exp\left[-\frac{(x - m_X)^2}{2\sigma_X^2}\right] \tag{2.4.27}$$

So,

$$f_r(y) = f_X(y - k)|(y - k)'| = \frac{1}{\sigma_X\sqrt{2\pi}} \exp\left[\frac{(y - k - m_X)^2}{2\sigma_X^2}\right] \tag{2.4.28}$$

As $\ln r = y$, it can be derived as:

$$f_r(r) = \frac{1}{\sigma_X r\sqrt{2\pi}} \exp\left[-\frac{(\ln r - k - m_X)^2}{2\sigma_X^2}\right] \tag{2.4.29}$$

The mean of $\ln r$ is obtained by:

$$m_Y = E[\ln r] = E\left[\sum_{i=1}^{N} \ln c_i(t_0)\right] = k + m_X \tag{2.4.30}$$

While its variance is obtained by:

$$\sigma_Y^2 = D[\ln r] = D[\ln A] + D\left[\sum_{i=1}^{N} \ln c_i(t_0)\right] = 0 + \sigma_X = \sigma_X \tag{2.4.31}$$

Therefore, under the LOS propagation mode, the PDF of the envelope r of the signal is written as:

$$f_r(r) = \frac{1}{r\sigma_Y\sqrt{2\pi}} \exp\left[-\frac{(\ln r - m_Y)^2}{2\sigma_Y^2}\right] \tag{2.4.32}$$

Furthermore, let $z = r, d_0 = \sigma^2$, then

$$f_z(z) = \frac{1}{z\sqrt{2\pi d_0}} \exp\left[-\frac{(\ln z - m_Z)^2}{2d_0}\right] \tag{2.4.33}$$

In addition, the PDF curves of the lognormal distribution, when taking different parameter values m and σ, are shown in Figure 2.15.

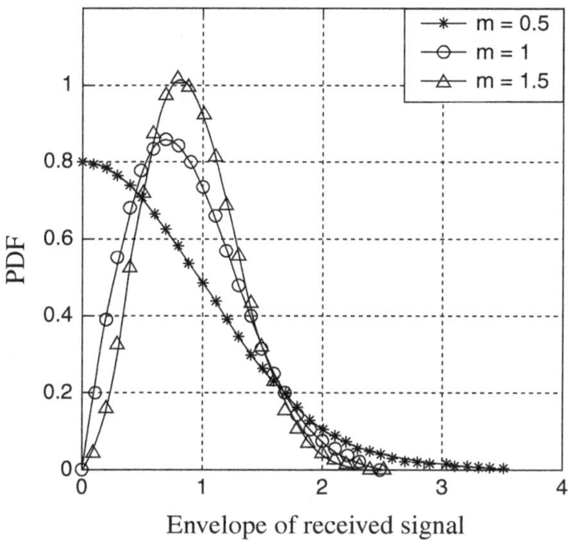

Figure 2.15 Probability density function curves of lognormal distribution.

2.4.1.4 *Nakagami distribution*

The Nakagami distribution is a kind of generalized Rayleigh distribution. In the actual wireless environment, the Nakagami distribution can well approximate the actual measured results, and by adjusting the parameters, channel simulations of different fading degrees can be realized, so it can be widely applied to the simulation of fading channels [21].

The probability density function of the Nakagami distribution is written as:

$$f(r) = \frac{2m^m r^{2m-1}}{\Omega^m \Gamma(m)} \exp\left(-\frac{mr^2}{\Omega}\right) \tag{2.4.34}$$

where $\Omega = E[r^2]$ is the average power of the signal, $\Gamma(\cdot)$ represents the Gamma function, and $m = \Omega^2/E[(r^2 - \Omega)^2]$ is the fading factor of the Nakagami distribution, which indicates the fading degree of the signal caused by channel multipath effects. The smaller the value of m is, the more serious the channel attenuation is. The PDF of the Nakagami distribution with different values of m is shown in Figure 2.16.

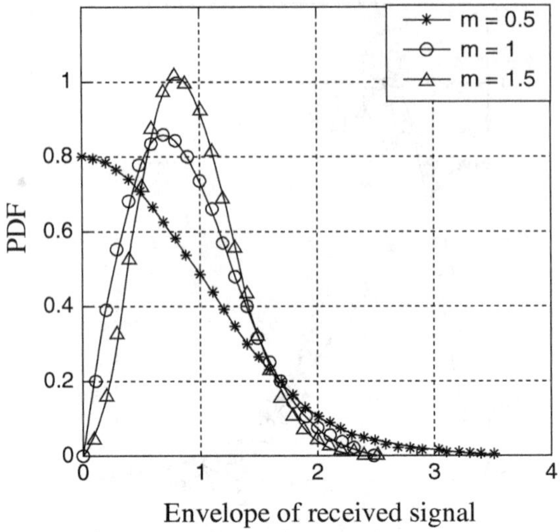

Figure 2.16 PDF curve of Nakagami distribution.

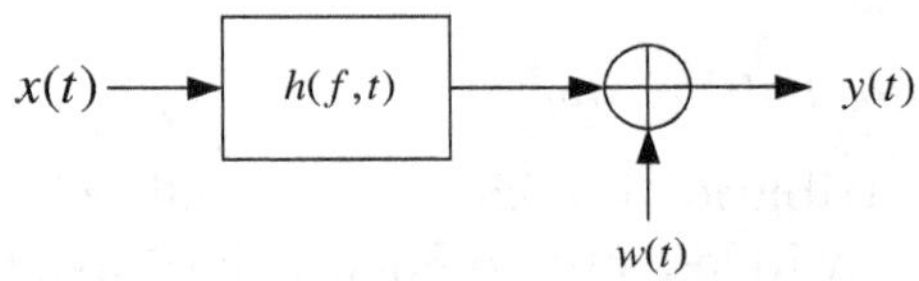

Figure 2.17 Classic channel model.

2.4.2 *Classic satellite channel modeling*

The main task of channel modeling and simulation is to reflect the signal changes in the real transmission environment as much as possible. In the actual transmission environment, there are many factors that affect the signal transmission. At this time, the channel model cannot simulate every factor accurately. However, channel modeling and simulation can describe the main factors affecting signal transmission relatively accurately. Through the simulation of these main factors, it provides a theoretical basis and technical support for the design of a satellite communication system. A typical channel model is shown in Figure 2.16.

In Figure 2.17, the modulation signal $x(t)$ is written as:

$$x(t) = \mathrm{Re}[x_1(t)e^{j2\pi f_c t}] \tag{2.4.35}$$

where $x_1(t)$ is the equivalent low-pass signal of the transmitted signal and f_c is the carrier frequency.

In the absence of noise, the received signal of the system can be expressed as:

$$y(t) = \int_{-\infty}^{\infty} h(f,t) X_1(f) e^{j2\pi ft} df \qquad (2.4.36)$$

where $X_1(f)$ is the Fourier transform of the equivalent low-pass signal and $h(f,t)$ is the transfer function of the communication system. For a flat fading channel, it can be regarded as a random quantity independent of frequency, which is equivalent to $h(0,t)$, and at this time, the received signal can be simplified as:

$$y(t) = h(0,t) \int_{-\infty}^{\infty} X_1(f) e^{j2\pi ft} df = h(0,t) x_1(t) \qquad (2.4.37)$$

Equation (2.4.37) shows that the received signal can be expressed as the product of the transmitted signal and the random variable $h(0,t)$ with time-varying characteristics, that is, the effect of the wireless channel on the transmitted signal is expressed as a product relationship, so it is a kind of multiplicative interference to the transmitted signal. $h(0,t)$ is a function of time and reflects changes of channel characteristics with time. As interference will cause random attenuation and distortion with time, it can only be described by random process. The modeling and simulation of channel is the study and implementation of $h(0, t)$.

In the satellite communication system, multipath fading and shadowing fading of signals are the main factors affecting transmission characteristics of the channel. These fading effects affect the envelope and phase of the received signal. However, in the actual research, the loss caused by the signal phase is smaller than the fading of the signal envelope. Therefore, in the study of satellite channel model, only the influence of signal amplitude is usually considered, and the influence of the signal phase is ignored.

2.4.2.1 *C. Loo model*

The C. Loo model is a statistical model of the channel established by a probability density function describing the multipath fading and shadowing fading [22]. In the C. Loo model, the received signal of the terminal is the combination of the direct view component and the multipath components, and the direct view component is affected by the shadowing fading, while the multipath components are not affected by the shadowing fading. Therefore, the C. Loo model is also called partial shadowing model and is mainly used to describe rural or suburban communication environment.

The terminal receiving signal $r(t)$ for the C. Loo model is denoted as:

$$r(t) = z(t)s(t) + d(t) \tag{2.4.38}$$

where $z(t)$ is the direct view component, $d(t)$ is the multipath component, and $s(t)$ is the shadowing fading.

When the amplitude of direct view component $z(t)$ remains the same, the received signal $r(t)$ can be regarded as a combination of the direct view component and the multipath components. At this time, the envelope of $r(t)$ is subject to the PDF of the Rice distribution, i.e.,

$$f(r|z) = \frac{r}{\sigma^2} \exp\left(-\frac{r^2 + z^2}{2\sigma^2}\right) I_0\left(\frac{rz}{\sigma^2}\right) \tag{2.4.39}$$

where σ^2 is the average power of the multipath components and $I_0(x)$ is the first type of modified zero-order Bessel function.

Since the direct view component is affected by the shadowing fading, the PDF of the envelope of $z(t)$ obeys the lognormal distribution and can be written as:

$$f(z) = \frac{1}{\sqrt{2\pi\sigma_Z}z} \cdot \exp\left(-\frac{(\ln z - m_Z)^2}{2\sigma_Z}\right) \tag{2.4.40}$$

where m_Z and σ_Z are the mean and variance of the direct view component obeying the lognormal distribution, respectively.

Therefore, from the Equations (2.4.39) to (2.4.40), the PDF of the received signal envelope for the C. Loo model is expressed as:

$$\begin{aligned} f_{Loo}(r) &= \int_0^\infty f(r|z)f(z)dz \\ &= \frac{r}{\sigma^2\sqrt{2\pi\sigma_z}} \int_0^\infty \frac{1}{z} \exp\left(-\frac{r^2 + z^2}{2\sigma^2} - \frac{(\ln z - m_z)^2}{2\sigma_z}\right) \\ &\quad \times I_0\left(\frac{rz}{\sigma^2}\right) dz \end{aligned} \tag{2.4.41}$$

Table 2.4 Parameters of the C. Loo model.

Environmental Conditions	σ^2	m_z	$\sqrt{\sigma_z}$
Light shadow	0.1585	−0.23	0.0115
Moderate shadow	0.1	−1.3	0.046
Heavy shadow	0.0398	−1.95	0.46

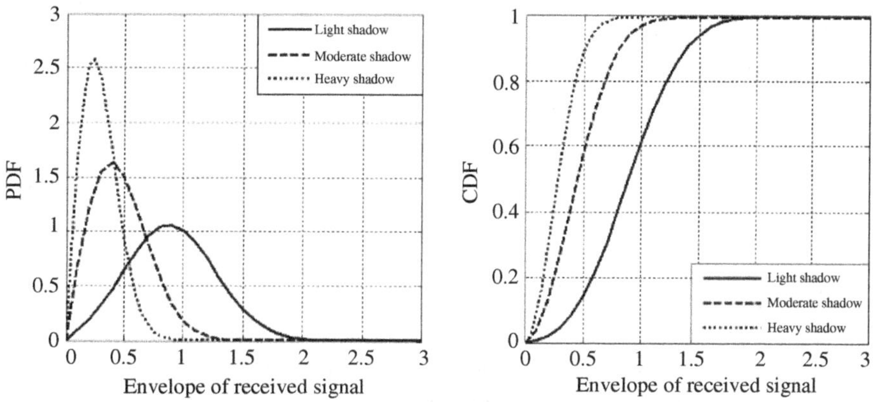

Figure 2.18 Statistical characteristics of the C. Loo model in three environments (PDF and CDF).

The C. Loo model, proposed in the literature [52], describes the parameters of the Ka-band satellite channel affected by three different degrees of shadowing, as shown in Table 2.4. The statistical characteristics of the C. Loo model under these three channel environments are shown in Figure 2.18.

2.4.2.2 *Suzuki model*

The Suzuki model is a statistical model established by both multipath fading effects and shadowing fading effects [23]. It is assumed that there is no direct view component between the mobile terminal and the satellite, and the received signal is a superposition of the reflected signals of the individual independent paths as well as its envelope obeys the Rayleigh distribution. At the same time, the power of the Rayleigh process changes significantly due to the effects of shadowing fading. In this case, the Suzuki model can more accurately describe the transmission characteristics of the channel under typical urban environments.

The received signal $r(t)$ of the Suzuki model can be represented by the product of the Rayleigh process $d(t)$ and the lognormal process $s(t)$, i.e.,

$$r(t) = d(t)s(t) \tag{2.4.42}$$

where $r(t)$ is the received signal, $d(t)$ is the multipath components of the signal, and $s(t)$ is the shadowing fading effect of the signal.

According to the theory of stochastic process, the PDF of the received signal envelope r is defined as:

$$f(r) = \int_{-\infty}^{\infty} \frac{1}{|s|} f_{\text{Ray}-LN}\left(\frac{r}{s}, s\right) ds \tag{2.4.43}$$

where $f_{\text{Ray}-LN}(x,y)$ is the joint PDF of both $d(t)$ and $s(t)$. If the multipath fading effect and the shadowing fading effect are independent of each other, the joint PDF is expressed as $f_{\text{Ray}-LN}(x,y) = f_{\text{Ray}}(x) \cdot f_{LN}(y)$. Therefore, the PDF of the received signal envelope for the Suzuki model can be expressed as:

$$f(r) = \int_{0}^{\infty} \frac{1}{s} f_{\text{Ray}}\left(\frac{r}{s}\right) f_{LN}(s) ds \tag{2.4.44}$$

where $s(t)$ describes the shadowing fading effect of the signal and obeys a lognormal distribution, and its probability density function is shown in Equation (2.4.30). From the Equations (2.4.42) to (2.4.44), the PDF of the received signal enveloper for the Suzuki model is obtained by:

$$\begin{aligned} f_{\text{Suz}}(r) &= \int_{0}^{\infty} \frac{1}{s} f_{\text{Ray}}\left(\frac{r}{s}\right) f_{LN}(s) ds \\ &= \frac{r}{\sqrt{2\pi}\sigma_0^2 \sigma_{LN}} \int_{0}^{\infty} \frac{1}{s^3} \exp\left(-\frac{r^2}{2s^2\sigma_0^2} - \frac{(\ln s - m_{LN})^2}{2\sigma_{LN}^2}\right) ds \end{aligned} \tag{2.4.45}$$

where σ_0^2 is the average power of the multipath components, m_{LN} is the mean of the lognormal distribution, and σ_{LN}^2 is the variance of the lognormal distribution.

Table 2.5 Parameters of the Suzuki model.

Environmental Conditions	σ_0	σ_{LN}	m_{LN}
Light shadowing	0.3273	0.0205	0.1882
Heavy shadowing	0.1847	0.1860	0.3515

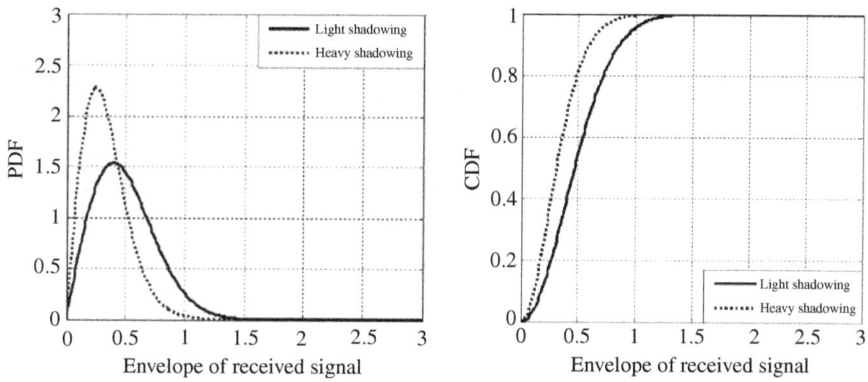

Figure 2.19 Statistical characteristics of light and heavy shadowing areas.

The Suzuki model can describe most fading channels affected by shadow in the urban environment [24], and its specific parameters are given in Table 2.5. Figure 2.19 shows the statistical characteristic curves under light and heavy shadowing areas.

2.4.2.3 *Corazza model*

The Corazza model consists of Rician process and lognormal process. In the Corazza model, the received signals of the terminal include direct view component and multipath components and is affected by shadowing fading during transmission. Therefore, the Corazza model is also called the total shadowing model, can the simulate satellite channels in different environments, including rural, suburban, and urban areas, based on setting parameters reasonably.

The received signal of the terminal for the Corazza model is written as:

$$r(t) = [z(t) + d(t)]s(t) \tag{2.4.46}$$

where $r(t)$ is the terminal receiving signal, $d(t)$ is the multipath components of the signal, $z(t)$ is the direct view component, and $s(t)$ denotes the shadowing fading effect of the signal. $z(t)$ and $d(t)$ reflect the multipath fading effect of the channel, and these two fading effects of the channel are independent of each other. According to the stochastic process theory, the probability density function of the terminal receiving signal envelope $\boldsymbol{r}$ in the Corazza model is defined as:

$$f(r) = \int_0^{\infty} \frac{1}{s} f_{\text{Rice}}\left(\frac{r}{s}\right) f_{LN}(s) ds \tag{2.4.47}$$

The shadow fading effect $s(t)$ is assumed to be constant, and according to the full probability formula, the PDF of the received signal envelope r is defined as:

$$f(r) = \int_0^{\infty} f_{\text{Rice}}(r|s) f_{LN}(s) ds \tag{2.4.48}$$

From Equations (2.4.47) and (2.4.48), the conditional PDF of Rice distribution is defined as:

$$f_{\text{Rice}}(r|s) = \frac{1}{s} f_{\text{Rice}}\left(\frac{r}{s}\right) \tag{2.4.49}$$

Since $z(t)$and $d(t)$ are the multipath fading effects describing the signals and obey the Rice distribution, so

$$f_{\text{Rice}}(r|s) = \frac{1}{s} f_{\text{Rice}}\left(\frac{r}{s}\right) = \frac{r}{s^2\sigma_0^2} \exp\left(-\frac{r^2}{2s^2\sigma_0^2} - \frac{z^2}{2\sigma_0^2}\right) I_0\left(\frac{rz}{s\sigma_0^2}\right) \tag{2.4.50}$$

where σ_0^2 is the variance of the Rice distribution.

The Rice factor K is used to describe the size of the multipath fading, i.e.,

$$K = \frac{z^2}{2\sigma_0^2} \tag{2.4.51}$$

Then,

$$K + 1 = \frac{z^2 + 2\sigma_0^2}{2\sigma_0^2} \tag{2.4.52}$$

In order to normalize the power of total received signal, let $z^2 + 2\sigma_0^2 = 1$, then

$$K + 1 = \frac{1}{2\sigma_0^2} \tag{2.4.53}$$

Substituting from Equation (2.4.51) to Equation (2.4.53) into Equation (2.4.50), then

$$f_{\text{Rice}}(r|s) = \frac{2r(K+1)}{s^2} \exp\left(-\frac{r^2(K+1)}{s^2} - K\right) I_0\left(\frac{2r\sqrt{K(K+1)}}{s}\right) \tag{2.4.54}$$

where $s(t)$ is subject to a lognormal distribution and is expressed in decibels as $s_{\text{dB}} = 20\lg s$, then its PDF is written as:

$$f_{LN}(s) = \frac{20}{\sqrt{2\pi}\sigma_{\text{dB}} s \ln(10)} \cdot \exp\left(-\frac{(20\lg s - m_{\text{dB}})^2}{2\sigma_{\text{dB}}^2}\right) \tag{2.4.55}$$

where m_{dB} is the mean of $20\lg s$, σ_{dB}^2 is the variance of $20\lg s$, and their units are both dB.

Therefore, according to Equations (2.4.54) and (2.4.55), the PDF of the receiving signal envelope r in the Corazza model is obtained by:

$$\begin{aligned} f(r) &= \int_0^\infty f_{\text{Rice}}(r|s) f_{\text{LN}}(s) ds \\ &= \frac{40r(K+1)}{\sqrt{2\pi}\sigma_{\text{dB}}} \exp(-K) \\ &\quad \times \int_0^\infty \frac{1}{s^3} \exp\left(-\frac{r^2(K+1)}{s^2} - \frac{(20\lg s - m_{\text{dB}})^2}{2\sigma_{\text{dB}}^2}\right) \\ &\quad \times I_0\left(\frac{2r\sqrt{K(K+1)}}{s}\right) ds \end{aligned} \tag{2.4.56}$$

where, when $K = 0$, the Corazza model is transformed into a pure multipath channel model describing the shadowing fading effect, that is, the Suzuki model is a special case of the Corazza model.

It is known from Equation (2.4.57) that the parameters determining the statistical properties of the signal in the Corazza model

Table 2.6 Coefficients of Corazza model parameters.

K	m_{dB}	σ_{dB}
$K_0 = 2.731$	$m_0 = -2.331$	$\sigma_0 = 4.5$
$K_1 = -0.1704$	$m_1 = 0.1142$	$\sigma_1 = -0.05$
$K_2 = 2.774 \times 10^{-3}$	$m_2 = -1.939 \times 10^{-3}$	
	$m_3 = 1.094 \times 10^{-5}$	

are the Rice factor K and the mean m_{dB} and the variance σ^2_{dB} of the shadowing fading. Reference [25] pointed out that the size of the three parameters is related to the elevation between the ground mobile terminal and the satellite, and the fitting formula of the measured L-band signal data is given by:

$$K(\theta) = K_0 + K_1\theta + K_2\theta^2 \tag{2.4.57}$$

$$m_{\mathrm{dB}}(\theta) = m_0 + m_1\theta + m_2\theta^2 + m_3\theta^3 \tag{2.4.58}$$

$$\sigma_{\mathrm{dB}}(\theta) = \sigma_0 + \sigma_1\theta \tag{2.4.59}$$

where θ is the elevation angle of the satellite and its value ranges from 20° to 80°. The specific coefficients in the parameter fitting equation are shown in Table 2.6. For a given satellite elevation angle, the corresponding parameters of the model can be calculated by the fitting equation. Figure 2.20 shows the statistical characteristics of the Corazza model at three different satellite elevation angles when the shadowing environment in the country is shaded.

2.4.2.4 *Lutz model*

The Lutz model is a two-state channel model based on a probability distribution function to describe a wide area environment. In the Lutz model, the satellite channel is divided into two states of "good" and "bad" according to the presence or absence of the direct view component in the actual communication process [26]. The switching between the two states in the Lutz model is determined by changes of the actual communication environment.

In the "good state" environment, the quality of the channel transmission signal is good, and the terminal receiving signal is a combination of the direct view component and the multipath components

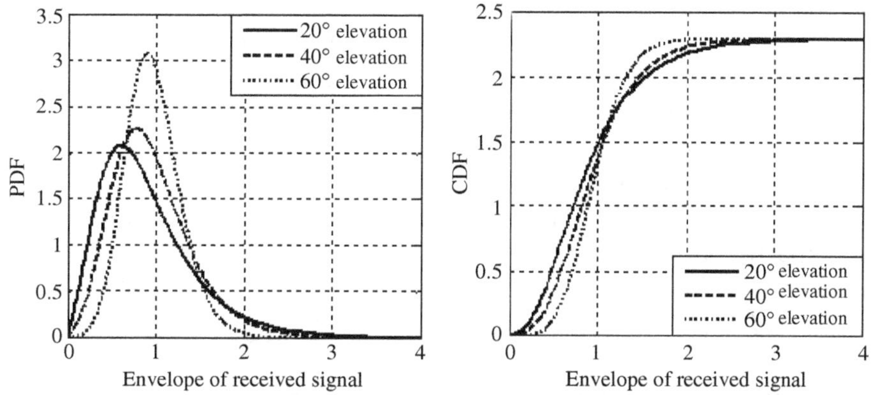

Figure 2.20 Statistical characteristics of different elevations in rural environments (PDF and CDF).

and is not affected by the shadowing fading. At this time, the envelope of the terminal receiving signal obeys the probability density function (PDF) of the Rice distribution. This PDF is defined as:

$$f_r(r) = \frac{r}{\sigma_g^2} \exp\left(-\frac{r^2+z^2}{2\sigma_g^2}\right) I_0\left(\frac{rz}{\sigma_g^2}\right) \tag{2.4.60}$$

where σ_g^2 is the variance of the Rice distribution under the "good state".

The Lutz model is used to model satellite channel from the perspective of the received signal power. Therefore, let $\mathrm{P_T}$ be the power of the terminal receiving signal, that is, $\mathrm{P_T} = r^2$, and its PDF is obtained by:

$$f_g(\mathrm{P_T}) = \frac{1}{2\sigma_g^2} \exp\left(-\frac{\mathrm{P_T}+z^2}{2\sigma_g^2}\right) I_0\left(\frac{\sqrt{\mathrm{P_T}}z}{\sigma_g^2}\right) \tag{2.4.61}$$

In order to normalize the Rice factor, let $z = 1$, then

$$K = \frac{1}{2\sigma_g^2} \tag{2.4.62}$$

Substituting Equation (2.4.62) into Equation (2.4.61), the PDF of the received signal power in the "good state" is obtained by:

$$f_g(\mathrm{P_T}) = K \exp[-K(\mathrm{P_T}+1)] I_0(2K\sqrt{\mathrm{P_T}}) \tag{2.4.63}$$

In the "bad state", the channel environment is relatively complicated, the direct view component of the transmitted signals is completely blocked, and the terminal receiving signals only have multipath components, which are also affected by shadowing fading. At this point, the envelope of the received signal is subject to a Rayleigh–lognormal distribution.

When the shadowing fading effect is constant, the envelope of the multipath components obeys the Rayleigh distribution, i.e.,

$$f_r(r|\text{shadow is constant}) = \frac{r}{\sigma_b^2}\exp\left(-\frac{r^2}{2\sigma_b^2}\right) \tag{2.4.64}$$

where σ_b^2 is the variance of the Rayleigh distribution under the "bad state".

Let the power of the multipath component be P_0, then $P_0 = 2\sigma_b^2$. When the shadowing fading is constant, it is equivalent to P_0, so the PDF of the received power is constant and written as:

$$f_P(\mathrm{P_T}|\mathrm{P_0}) = f_b(\sqrt{\mathrm{P_T}}|\mathrm{P_0})(\sqrt{\mathrm{P_T}})' = \frac{1}{\mathrm{P_0}}\exp\left(-\frac{\mathrm{P_T}}{\mathrm{P_0}}\right) \tag{2.4.65}$$

The multipath component is also affected by the shadowing fading effect, and its power P_0 obeys the lognormal distribution. The PDF of P_0 is given by:

$$f_{\mathrm{P_0}}(\mathrm{p_0}) = \frac{10}{\ln(10)\sqrt{2\pi}\sigma_b\mathrm{P_0}}\cdot\exp\left(-\frac{(10\lg\mathrm{P_0} - m_b)^2}{2\sigma_b^2}\right) \tag{2.4.66}$$

where m_b is the mean of $10\lg\mathrm{P_0}$ in the "bad state" and σ_b is the standard deviation of $10\lg\mathrm{P_0}$ in the "bad state".

Therefore, according to Equations (2.4.65) and (2.4.66), the PDF of the received signal power in the "bad state" is given by:

$$\begin{aligned} f_b(\mathrm{p}_T) &= \int_0^\infty f_{P_T}(\mathrm{P_T}|\mathrm{P_0})f_{P_0}(\mathrm{P_0})d\mathrm{P_0} \\ &= \frac{10}{\ln(10)\sqrt{2\pi}\sigma_b}\int_0^\infty \frac{1}{\mathrm{P_0^2}}\cdot\exp\left(-\frac{\mathrm{P}}{\mathrm{P_0}} - \frac{(10\lg\mathrm{P_0} - m_b)^2}{2\sigma_b^2}\right)d\mathrm{P_0} \end{aligned} \tag{2.4.67}$$

In Lutz model, the difference between "good state" and "bad state" is whether the received signal is affected by the shadowing fading effect. In order to reflect the state of the communication channel

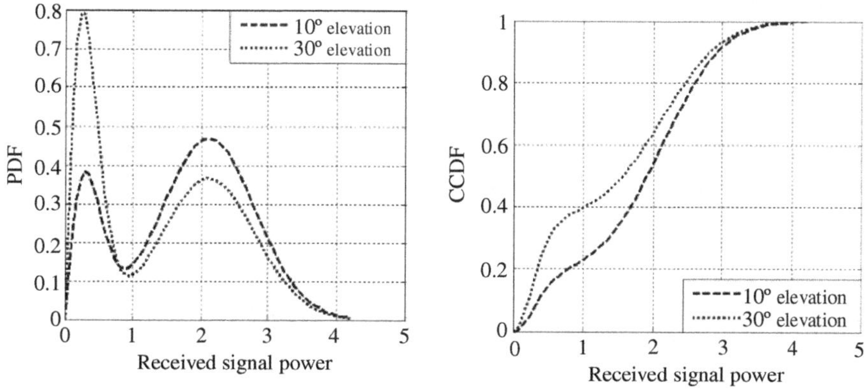

Figure 2.21 Statistical characteristics of the Lutz model (PDF and CCDF).

completely, let A represent the percentage of time of the signal being shaded and fading during the whole communication process. According to Equations (2.4.63) and (2.4.67), the PDF of the received signal power $\mathrm{P_T}$ in Lutz model is as follows:

$$f_{\mathrm{Lutz}}(\mathrm{P_T}) = (1 - A) f_g(\mathrm{P_T}) + A f_b(\mathrm{P_T}) \tag{2.4.68}$$

The Lutz model describes the two-state channel characteristics when the geographical environment changes. The typical values of L-band mobile satellite channel under different satellite elevation described by Lutz model were given by [26]. Figure 2.21 shows the statistical characteristics of the Lutz model when the urban and suburban environment changes at two elevations.

2.5 Statistical Characteristics of Satellite Channels

The main first-order parameters to measure channel index are probability statistic characteristics of envelope and phase, and the second-order statistical characteristics mainly include fading rate, level crossing rate (LCR), and average fading duration(AFD), which, respectively, describe the channel quality from different angles. Taking Rayleigh and Rice distribution as examples, the first-order statistical properties and second-order statistical properties are introduced, respectively.

2.5.1 *First-order statistical properties*

2.5.1.1 *Probability density function of the envelope*

The probability statistics characteristics of the envelope can directly reflect all kinds of interference in the process of signal propagation. The PDF of signal envelope obeys Rayleigh distribution and is defined as:

$$f_{\text{Rayleigh}}(r) = \frac{r}{\sigma^2} \exp\left[-\frac{r^2}{2\sigma^2}\right] \tag{2.5.1}$$

The PDF of the signal envelope obeys the Rice distribution signal and is written as:

$$f_{\text{Rician}}(r) = \frac{r}{2\pi\sigma^2} \exp\left(-\frac{r^2 + c_0^2}{2\sigma^2}\right) I_0\left(\frac{c_0 r}{\sigma^2}\right) \tag{2.5.2}$$

It can be seen that the Rice distribution and the Rayleigh distribution are mutually convertible. When $c_0 \to 0$ in the Rice distribution, the Rice distribution can be converted into a Rayleigh distribution. Therefore, the Rayleigh distribution is a special case of the Rice distribution, and the Rice distribution is an extension of the Rayleigh distribution [27].

2.5.1.2 *Probability density function of phase*

The PDF of the signal phase obeys the Rayleigh distribution and is denoted as:

$$f_{\text{Rayleigh}}(\phi) = \frac{1}{2\pi} \tag{2.5.3}$$

The PDF of the signal phase in the Rice distribution is denoted as:

$$f_{\text{Rician}}(\varphi) = \frac{\exp(-c_0/2\pi)}{2\pi}\left\{1 + \sqrt{\frac{\pi}{2}}\frac{c_0}{\sigma}\cos(\theta_s - \varphi)e^{\frac{c_0^2\cos^2(\theta_s-\varphi)}{2\sigma^2}}\right.$$
$$\left.\times\left[1 + erf\left(\frac{c_0\cos(\theta_s - \varphi)}{\sqrt{2}\sigma}\right)\right]\right\}$$

where $0 \leq \varphi < 2\pi$, $erf(\cdot)$ is the error function.

2.5.2 *Second-order statistical property*

2.5.2.1 *Fading rate*

The fading rate is a physical quantity that reflects the rate at which the channel fluctuates with time, space, and frequency. Figures 2.22(a) and 2.22(b) are the curves of the signal envelope level obeying the Rayleigh distribution with time. Obviously, these two curves are very different, and the change of the curves in Figure 2.22(a) is significantly faster than that in Figure 2.22(b). Their fading rates [28, 29] are different.

The channel fading rate is based on the time derivative of the channel. The time-varying channel impulse response is denoted as $\tilde{h}(t)$ and its derivative is $d\tilde{h}(t)/dt$. Because $E[d\tilde{h}(t)/dt] = 0$, it is impossible to describe the fluctuation of the channel with $d\tilde{h}(t)/dt$. While $E[|d\tilde{h}(t)/dt|^2] \neq 0$, the fluctuation degree of the channel can be described by $|d\tilde{h}(t)/dt|^2$, that is, the fading rate can be described.

For a generalized stationary random (WSS, Wide-Sense Stationary) channel with a fixed phase,

$$\Phi(t) = \arg[\tilde{h}(t)] \tag{2.5.4}$$

If the center frequency of the Doppler spectrum $S_{\tilde{h}(t)}(\omega)$ is not zero, the random phase is non-stationary and its mathematical expectation is a function of time, i.e.,

$$E[\Phi(t)] = \phi_0 + \varpi t \tag{2.5.5}$$

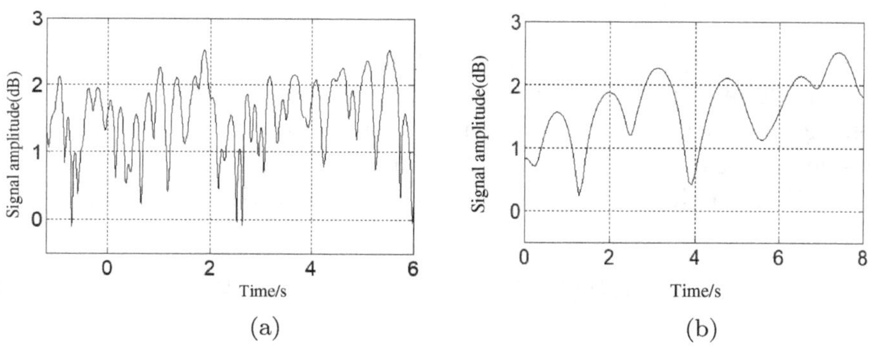

Figure 2.22 Rayleigh distribution at different fading rates: (a) Fast change over time and (b) slowly changing over time.

where ϖ is the center of the Doppler spectrum. In order to reduce the influence of the unsteady phase, the complex exponent $\exp(-j\varpi t)$ can be multiplied by $\tilde{h}(t)$ to obtain a new channel fluctuation size, i.e.,

$$\sigma_t^2 = E\left\{\left|\frac{d[\tilde{h}(t)\exp(-j\varpi t)]}{dt}\right|^2\right\} \tag{2.5.6}$$

where σ_t^2 is the variance of the fading rate.

Based on Equation (2.5.6), the variance of the fading rate in the frequency domain is written as:

$$\sigma_f^2 = E\left\{\left|\frac{d[\tilde{h}(f)\exp(-j2\pi\bar{\tau}f)]}{df}\right|^2\right\} \tag{2.5.7}$$

where $\bar{\tau}$ is the center of the delay spectrum. From the dual relationship, the variance of the spatial domain fading rate is obtained by:

$$\sigma_r^2 = E\left\{\left|\frac{d[\tilde{h}(r)\exp(-j\bar{k}r)]}{dr}\right|^2\right\} \tag{2.5.8}$$

where $\bar{k}$ is the spectral center of the beam.

2.5.2.2 *Level crossing rate*

Level crossing rate (LCR) is a statistical process of time. In channel modeling, it is defined as the average number of times that the envelope r of the signal is lower than a given level threshold in one second [30, 31]. The level crossing rate $N_t(r)$ is defined by both the joint PDF of the envelope r and its time derivative dr/dt as:

$$N_t(R) = \int_0^{\infty} \dot{r} f_{r\dot{r}}(R,\dot{r})d\dot{r} \quad R \geq 0 \tag{2.5.9}$$

where $f_{r\dot{r}}(R,\dot{r})$ is the joint PDF of both the envelope r and its time derivative when $r = R$.

Let's take the Rice distribution as an example to derive its time domain level crossover rate. The PDF, of the Rice distribution signal is shown in Equation (2.4.20). The PDF of the phase is given by:

$$f_r(\theta) = \frac{e^{-\frac{c_0^2}{2\sigma^2}}}{2\pi} \cdot \left\{ 1 + \sqrt{\frac{\pi}{2\sigma^2}} c_0 \cos(\theta - \theta_0) e^{\frac{c_0^2 \cos^2(\theta-\theta_0)}{\sqrt{2\sigma^2}}} \cdot \left[1 + erf\left(\frac{c_0 \cos(\theta - \theta_0)}{\sqrt{2\sigma^2}} \right) \right] \right\} \tag{2.5.10}$$

where c_0 and θ_0 are the amplitude and phase of the direct view component, $2\sigma^2$ is the signal power, $erf(\cdot)$ is the error function, and θ is in the range $0 \sim 2\pi$.

The time domain level crossing rate $f_{r\dot{r}}(r, \dot{r})$ of the Rice process is written as:

$$\begin{aligned} f_{r\dot{r}}(R, \dot{r}) &= \sqrt{\frac{1}{2\pi b_2}} \exp\left(-\frac{\dot{r}^2}{2b_2}\right) \cdot \frac{R}{b_0} \exp\left(-\frac{R^2 + c_0^2}{2b_0}\right) \cdot I_0\left(\frac{Rc_0}{b_0}\right) \\ &= f(\dot{r}) \cdot f_r(R) \end{aligned} \tag{2.5.11}$$

Substituting Equation (2.5.12) into Equation (2.5.10), then

$$N_t(R) = \sqrt{\frac{b_n}{2\pi}} \cdot f_r(R) \tag{2.5.12a}$$

where

$$b_n = \begin{cases} c_0^2 & n = 0 \\ (2\pi f_m)^n b_0 \int_0^{2\pi} f(\theta) \cos^2 \theta \, d\theta & n \neq 0 \end{cases} \tag{2.5.12b}$$

where f_m is the maximum of the Doppler shift and $f(\theta)$ is the PDF of the incident angle. From Equation (2.5.11), the level crossing rate of the Rice distribution is the product of its PDF and the constant $\sqrt{b_2/2\pi}$, where $\sqrt{b_2/2\pi}$ is determined by f_m and $f(\theta)$ together.

When the PDF of the incident angle is uniform distribution within $[0, 2\pi]$, i.e.,

$$f(\theta) = \frac{1}{2\pi} \tag{2.5.13}$$

Equation (2.5.12) is rewritten as:

$$b_n = \begin{cases} b_0(2\pi f_m)^n \cdot \dfrac{1 \cdot 2 \ldots (n-1)}{2 \cdot 4 \ldots . n}, & n \text{ is an even number} \\ 0, & n \text{ is an odd number} \end{cases} \tag{2.5.14}$$

From the above derivation, the level crossing rate of the Rice distribution is known as:

$$N_t(R) = \sqrt{2\pi(K+1)} f_m c_0 e^{-k-(k+1)c_0^2} \cdot I(2c_0\sqrt{k(k+1)}) \tag{2.5.15}$$

where K is the Rice factor.

2.5.2.3 *Average fading duration*

The average fading duration (AFD) is defined as the duration of each signal level below a given threshold, and its calculation method is similar to that of the level crossing rate. Let the given threshold be R and the fading signal be $r(t)$, then the average fading duration [32] is defined as:

$$\bar{t}(R) = \frac{F_r(R)}{N_t(R)} \tag{2.5.16}$$

where $N_t(R)$ is the level crossing ratio of the signal $r(t)$ and $F_r(R)$ is the cumulative probability distribution function of the signal $r(t)$, which represents the probability that the signal is below a given level R.

2.5.3 ***Doppler power spectrum***

If there is relative movement of the receiving terminal or satellite, a Doppler effect will occur, which causes a Doppler shift. After all the scattering components are superimposed, a continuous Doppler spectrum is formed, which is also known as the "Doppler power spectrum". The common power spectrums are the classical power spectrum and the Gaussian power spectrum [33, 34].

2.5.3.1 *Classic power spectrum*

Assume that there are incident waves in N directions, whose incident power in the range of $0 \sim 2\pi$ is continuous, $p(\alpha)d\alpha$ is the incident energy with an incident angle of $[\alpha, \alpha + d\alpha]$, $\bar{P}$ is the average received power of the omnidirectional antenna, and $G(\alpha)$ is the antenna gain in the α direction. If N tends to infinity, then $p(\alpha)d\alpha$ is a continuous distribution. The total incident energy in the range of $0 \sim 2\pi$ is defined as:

$$P_r = \int_0^{2\pi} AG(\alpha)p(\alpha)d\alpha \tag{2.5.17}$$

The carrier frequency of the transmitted signal is f_c, then the Doppler frequency is calculated by:

$$f(\alpha) = \frac{v}{\lambda} \cdot \cos\alpha + f_c = f_m \cdot \cos\alpha + f_c \tag{2.5.18}$$

where v is the relative moving speed, λ is the carrier wavelength, and f_m is the maximum of the Doppler shift. Since the cosine function is an even function, $f(\alpha)$ is also an even function, that is, there is $f(-\alpha) = f(\alpha)$.

Let the channel impulse response be written as:

$$h(t) = h_1(t)\cos(2\pi f_c t) + h_2(t)\sin(2\pi f_c t) \tag{2.5.19}$$

Let $S_{hh}(f)$ denote the received signal power spectrum, then

$$S_{hh}(f)|df| = A[p(\alpha)G(\alpha) + p(-\alpha)G(-\alpha)]|d\alpha| \tag{2.5.20}$$

At the same time, by derivation of Equation (2.5.18) with respect to frequency f, the following equation can be obtained.

$$|df| = f_m|-\sin\alpha||d\alpha| \tag{2.5.21}$$

where

$$\alpha = \cos^{-1}\left[\frac{f - f_c}{f_m}\right] \tag{2.5.22}$$

According to Equation (2.5.22),

$$\sin\alpha = \sqrt{1 - \left(\frac{f - f_c}{f_m}\right)^2} \tag{2.5.23}$$

Substituting Equations (2.5.21) and (2.5.23) into Equation (2.5.20), then

$$S_{hh}(f) = \begin{cases} \dfrac{A[p(\alpha)G(\alpha) + p(-\alpha)G(-\alpha)]}{f_m\sqrt{1-\left(\frac{f-f_c}{f_m}\right)^2}}, & |f - f_c| < f_m \\ 0, & |f - f_c| > f_m \end{cases} \tag{2.5.24}$$

The above equation shows that the power spectrum is centered on the carrier frequency and distributed in the range of the carrier frequency $[-f_m, f_m]$. Each arriving wave has a different frequency depending on the direction of arrival.

For a vertically polarized antenna with a wavelength $\lambda/4$, the gain $G(\alpha)$ is constant $2\sigma_0^2$ in all directions, and the incident energy is uniform distribution in $[0, 2\pi]$, i.e., $p(\alpha) = 1/2\pi$, then

$$S_{hh}(f) = \frac{\sigma_0^2}{\pi f_m\sqrt{1-\left(\frac{f-f_c}{f_m}\right)^2}} \quad |f - f_c| < f_m \tag{2.5.25}$$

when $f_c = 0$,

$$S_{hh}(f) = \frac{\sigma_0^2}{\pi f_m\sqrt{1-\left(\frac{f}{f_m}\right)^2}} \quad |f| < f_m \tag{2.5.26}$$

The correlation function can be obtained from $S_{hh}(f)$.

$$R_{hh}(\tau) = \int_{-\infty}^{\infty} S_{hh}(f)e^{j2\pi f\tau}df = \frac{2\sigma_0^2}{\pi f_m}\int_0^{f_m} \frac{\cos(2\pi f\tau)}{\sqrt{1-(f/f_m)}}df \tag{2.5.27}$$

Let $f = f_m \cos x$, then Equation (2.5.27) can be rewritten as:

$$R_{hh}(\tau) = \frac{2\sigma_0^2}{\pi}\int_0^{\pi/2} \cos 2\pi f_m\tau \cos x\, dx \tag{2.5.28}$$

Figure 2.23 shows the classical power spectrum and autocorrelation function curves of Rice distribution. In Figure 2.23, $f_m = 110\,\text{Hz}$, $\sigma_0^2 = 0.1$.

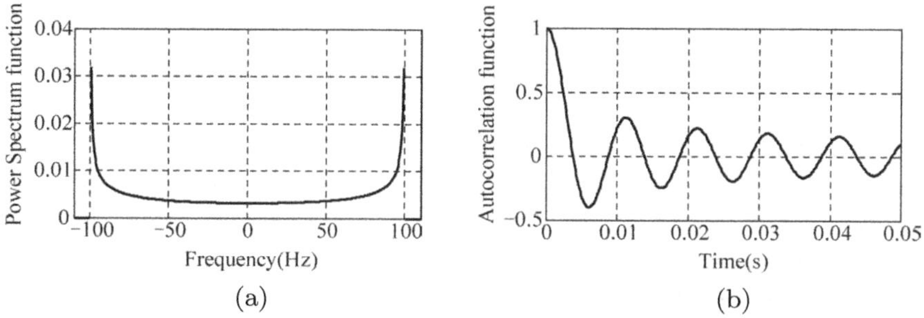

Figure 2.23 Classic power spectrum of Rice distribution: (a) Gaussian power spectrum and (b) autocorrelation function.

2.5.3.2 *Gaussian power spectrum*

Gaussian power spectrum [35] is calculated by:

$$S_{hh}(f) = \frac{\sigma_0^2}{f_{3\mathrm{d}}}\sqrt{\frac{\ln 2}{\pi}}e - \ln 2\left(\frac{f}{f_{3\mathrm{d}}}\right)^2 \tag{2.5.29}$$

where f_{3d} is the 3 dB cutoff frequency.

Similarly, the autocorrelation function of $h(t)$ can be obtained by the inverse Fourier transform of Equation (2.5.29).

$$R_{hh}(\tau) = \sigma_0^2 e^{-(\pi f_{3d}\tau\sqrt{\ln 2})^2} \tag{2.5.30}$$

2.6 Satellite Channel Model Simulation Method

The research task of the satellite channel implementation method is to design the probability statistics model into a random process that can be realized in a computer, DSP, or FPGA by designing relevant algorithms. The first- and second-order statistical characteristics of the random process should be the same as the original probability statistics model, and other performance indicators should also meet the design requirements. This section gives the implementation method of the common single state model and the calculation method of related parameters.

2.6.1 *Generation method of colored Gaussian noise*

All channel simulations are based on several real colored Gaussian processes. The difference between different channel models lies in the number of real Gaussian processes and different processing methods. For example, Rayleigh and Rice models are simulated by two real Gaussian processes, while Suzuki model [36] is simulated by three real Gaussian processes.

The above analyses show that real Gaussian process is the basis and premise of the channel simulation [37, 38]. The correlation of real Gaussian process determines the superior performance of the simulation model. How to generate a real Gaussian process to meet the requirements becomes the key of research. The generation process of the colored Gaussian process $\hat{\mu}_i(t)$ is shown in Figure 2.24.

From Figure 2.24,

$$\hat{\mu}_i(t) = \lim_{N_i \to \infty} \sum_{n=1}^{N_i} c_{i,n} \cos(2\pi f_{i,n} t + \theta_{i,n}) \tag{2.6.1}$$

where $c_{i,n}$ is the weight of the ith harmonic of the first real Gaussian random process, also known as the Doppler coefficient, and

$$c_{i,n} = 2\sqrt{\Delta f_i S_{\mu_i \mu_i}(f_{i,n})} \tag{2.6.2}$$

$$f_{i,n} = n\Delta f_1 \tag{2.6.3}$$

where $\theta_{i,n}$ is a Doppler phase shift and a random variable distributed within $[0, 2\pi]$. When $N_i \to \infty$, $\Delta f_i \to 0$, the frequency tends to be continuously distributed, then $\hat{\mu}_i(t) \to \mu_i(t)$, $S_{\mu_i \mu_i}(\cdot)$ is the power spectral density.

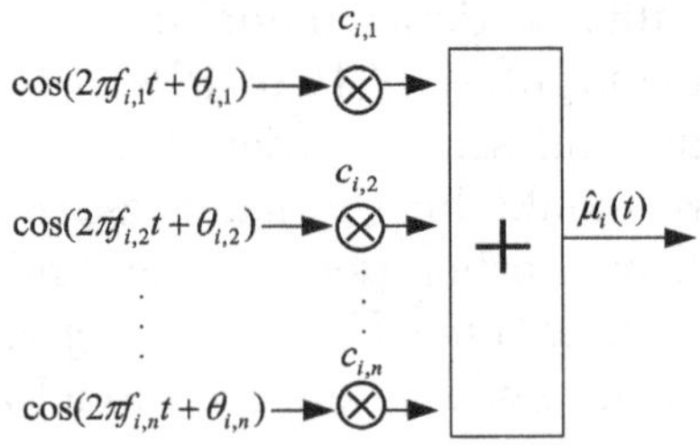

Figure 2.24 Generation block diagram of colored Gaussian process.

2.6.2 *Calculation method of Doppler coefficient and Doppler frequency*

There are many calculation methods for parameters $c_{i,n}$ and $f_{i,n}$ such as equidistance method, MSE method, equal area method, and Monte Carlo method, which have their own advantages and disadvantages [39, 40] and can be selected according to the needs of the system design. Based on the analyses of above several calculation methods, an improved Doppler coefficient and frequency calculation method is proposed in this section.

2.6.2.1 *Equidistance method*

One of the main features of the equidistance method is that adjacent discrete Doppler frequency pairs have equal distances. Discrete Doppler frequency is defined as:

$$f_{i,n} = \frac{\Delta f_i}{2}(2n-1) \tag{2.6.4}$$

where $n = 1, 2, \ldots, N_i$. The distance between the ith adjacent discrete Doppler frequency pairs of the deterministic process $\hat{\mu}_i(t)$ is defined as:

$$\Delta f_{i,n} = f_{i,n} - f_{i,n-1} \tag{2.6.5}$$

where $n = 2, 3, \ldots, N_i$.

In practical applications, since $|f| < f_{\max}$, the distance Δf_i between two adjacent discrete Doppler frequency pairs can be taken as $f_{\max}/N_i$, so

$$f_{i,n} = \frac{f_{\max}}{2N_i}(2n-1) \tag{2.6.6}$$

When calculating the Doppler coefficient $c_{i,n}$, the power spectral density $\hat{S}_{\mu_i\mu_i}(f)$ of the deterministic simulation model must be equal to the power spectral density $S_{\mu_i\mu_i}(f)$ of the reference model in the interval $L_{i,n} = \left[f_{i,n} - \frac{\Delta f_i}{2}, f_{i,n} + \frac{\Delta f_i}{2}\right)$, i.e.,

$$\int_{f\in L_{i,n}} \hat{S}_{\mu_i\mu_i}(f)df = \int_{f\in L_{i,n}} S_{\mu_i\mu_i}(f)df \tag{2.6.7}$$

The Doppler coefficient $c_{i,n}$ is obtained as:

$$c_{i,n} = \sqrt{\left\{\frac{4\sigma_0^2}{\pi}\left[\arcsin\left(\frac{n}{N_i}\right) - \arcsin\left(\frac{n-1}{N_i}\right)\right]\right\}} \tag{2.6.8}$$

2.6.2.2 *Equal area method*

The equal area method (MAE) means that the area of the Doppler power spectrum $S_{\mu_i\mu_i}(f)$ is equal to $\sigma_{\mu_i}^2/2N_i$ in any adjacent frequency range $f_{i,n-1} < f < f_{i,n}$ under the condition that the power spectral density function is constant, i.e.,

$$A_{\mu_i} = \int_{f_{i,n-1}}^{f_{i,n}} S_{\mu_i\mu_i}(f)df = \frac{\sigma_{\mu_i}^2}{2N_i}, n = 1, 2, \ldots, N_i \tag{2.6.9}$$

where, when $n = 0$, $f_{i,0} = 0$. In order to calculate the discrete Doppler frequency $f_{i,n-1}$, the following function is defined by:

$$F_{\mu_i}(f_{i,n}) = \int_{-\infty}^{f_{i,n}} S_{\mu_i\mu_i}(f)df \tag{2.6.10}$$

It is known from Equation (2.5.26) that $S_{\mu_i\mu_i}(f)$ is an even function, that is, $S_{\mu_i\mu_i}(f) = S_{\mu_i\mu_i}(-f)$, so

$$F_{\mu_i}(f_{i,n}) = \frac{\sigma_{\mu_i}^2}{2} + \sum_{v=1}^{n}\int_{f_{i,v-1}}^{f_{i,v}} S_{\mu_i\mu_i}(f)df = \frac{\sigma_{\mu_i}^2}{2}\left(1 + \frac{n}{N_i}\right) \tag{2.6.11}$$

Denote that the inverse function of F_{μ_i} as $F_{\mu_i}^{-1}$, then

$$f_{i,n} = F_{\mu_i}^{-1}\left[\frac{\sigma_{\mu_i}^2}{2}\left(1 + \frac{n}{N_i}\right)\right] \quad n = 1, 2, \ldots, N_i \tag{2.6.12}$$

In interval $L_{i,n} = [f_{i,n}, f_{i,n})$, the average power of $S_{\mu_i\mu_i}(f)$ is equal to $c_{i,n}^2/4$, then

$$c_{i,n} = \sqrt{4A_{\mu_i}} = \sigma_{\mu_i}\sqrt{\frac{2}{N_i}} \tag{2.6.13}$$

It can be seen from the above equation that the Doppler coefficients in the equal area method are equal.

2.6.2.3 *MSE method*

The data of the MSE method (MSEM) is obtained when the MSE value takes the minimum value [41]. Let the MSE of $R_{\mu\mu}(\tau)$ be defined as:

$$E_{R_{\mu\mu}} = \frac{1}{\tau_{\max}} \int_0^{\tau_{\max}} (R_{\mu\mu}(\tau) - \tilde{R}_{\mu\mu}(\tau))^2 d\tau \tag{2.6.14}$$

where $\tau_{\max}$ represents the maximum value of this interval τ and is calculated by $\tau_{\max} = T_i/4 = 1/2\Delta f_i$. $R_{\mu\mu}(\tau)$ is an arbitrary autocorrelation function of describing the theoretical reference model process $\mu(t)$ and $\tilde{R}_{\mu\mu}(\tau)$ is the autocorrelation function of the deterministic model.

In practical applications, the solution of discrete Doppler frequency shift $f_{i,n}$ is usually set as a set of relatively simple equidistance values, e.g.,

$$\left\{ f_{i,n} | f_{i,n-1} = \frac{\Delta f_i}{2}(2n-1), n = 1, 2, \ldots, N_i \right\} \tag{2.6.15}$$

Equation (2.6.14) is used to calculate the partial derivative of Doppler coefficient c_n, and let $\frac{\partial E_{R_{\mu\mu}}}{\partial c_n} = 0$, then

$$c_{i,n} = \sqrt{\frac{4}{\tau_{\max}} \int_0^{\tau_{\max}} R_{\mu\mu}(\tau)\cos(2\pi f_{i,n}\tau) d\tau}, \quad n = 1, 2, \ldots, N_i \tag{2.6.16}$$

When $\tau_{\max} = T_i/4 = 1/2\Delta f_i$ and $\Delta f_i \to 0$, the parameter $c_{i,n}$ is obtained by:

$$c_{i,n} = \lim_{\Delta \mathrm{f_i} \to 0} 2\sqrt{\Delta f_i S_{\mu\mu}(f_{i,n})} \tag{2.6.17}$$

Numerical analysis shows that when $\Delta f_i > 0$, the parameter $c_{i,n}$ is regained by:

$$c_{i,n} \approx 2\sqrt{\Delta f_i S_{\mu\mu}(f_{i,n})} \tag{2.6.18}$$

At this time,

$$\begin{aligned}
&\lim_{\mathrm{N_i}\to\infty} \tilde{R}_{\mu\mu}(\tau) \\
&= \lim_{\mathrm{N_i}\to\infty} \sum_{n=1}^{N_i} \frac{c_{i,n}^2}{2} \cos(2\pi f_{i,n}\tau) \\
&= \lim_{N_i\to\infty} 4 \sum_{n=1}^{N_i} \int_0^{\frac{1}{2\Delta f_i}} R_{\mu\mu}(\tau') \cos(2\pi f_{i,n}\tau') \cos(2\pi f_{i,n}\tau) d\tau' \Delta f_i \\
&= 2 \int_0^{\infty} S_{\mu\mu}(f) \cos(2\pi f_{i,n}\tau) df \\
&= R_{\mu\mu}(\tau)
\end{aligned} \tag{2.6.19}$$

That is, the design requirement that the MSE is the smallest is satisfied.

The calculation processes for determining the parameters $c_{i,n}$, $f_{i,n}$, and $\theta_{i,n}$, according to the equal area method, the equidistance method, and the minimum MSE method, are given above, but the discrete Doppler frequencies obtained by the three methods have the same distance characteristic between adjacent frequency pairs. However, they all have one common drawback, that is, the designed deterministic Gaussian process has a relatively small period, and the calculation process associated with Gaussian power spectrum is highly complicated [42, 43].

2.6.2.4 *Improved Doppler coefficient and frequency calculation method*

In view of the shortcomings of the traditional method for calculating the parameters, an improved method of the Doppler coefficient and frequency is proposed in this section. The starting point of this method is as follows:

$$J_0(z) = \frac{2}{\pi} \int_0^{\pi/2} \cos(z \sin\alpha) d\alpha \tag{2.6.20}$$

Its limit is expressed as:

$$J_0(z) = \lim_{N_i\to\infty} \frac{2}{\pi} \sum_{n=1}^{N_i} \cos(z \sin\alpha_n) \Delta\alpha \tag{2.6.21}$$

where $\alpha_n = (2n-1)\pi/4N_i$ and $\Delta\alpha = \pi/(2N_i)$.

The autocorrelation function of the classical power spectrum is gained by:

$$R_{\mu\mu}(\tau) = \sigma_0^2 J_0(2\pi f_m \tau) \tag{2.6.22}$$

After Equation (2.6.21) is substituted into Equation (2.6.22), $R_{\mu\mu}(\tau)$ is rewritten as:

$$R_{\mu\mu}(\tau) = \lim_{N_i \to \infty} \frac{\sigma_0^2}{N_i} \sum_{n=1}^{N_i} \cos\left\{2\pi f_{\max} \sin\left[\frac{\pi}{2N_i}\left(n - \frac{1}{2}\right)\right]\tau\right\} \tag{2.6.23}$$

In a random process $\tilde{\mu}(t)$ consisting of a finite number of oscillators, when $N_i \to \infty$, there is $R_{\mu\mu}(\tau) = \tilde{R}_{\mu\mu}(\tau)$, so

$$\tilde{R}_{\mu\mu}(\tau) = \frac{\sigma_0^2}{N_i} \sum_{n=1}^{N_i} \cos\left\{2\pi f_{\max} \sin\left[\frac{\pi}{2N_i}\left(n - \frac{1}{2}\right)\right]\tau\right\} \tag{2.6.24}$$

Compared with the previous autocorrelation function,

$$c_{i,n} = \sigma_0 \sqrt{\frac{2}{N_i}} \tag{2.6.25}$$

$$f_{i,n} = f_{\max} \sin\left(\frac{\pi}{2N_i}\left(n - \frac{1}{2}\right)\right) \tag{2.6.26}$$

From the above analyses, the improved method of calculating Doppler coefficient and frequency is relatively simple.

2.6.3 *Doppler phase calculation method*

In the calculation of the parameters $c_{i,n}$ and $f_{i,n}$, it is assumed that the Doppler phase $\theta_{i,n}$ is uniformly distributed in $[0, 2\pi)$, but this may result in a certain correlation among $c_{i,n}$, $f_{i,n}$, and $R_{\mu\mu}(\tau)$ in the effective range, which cannot satisfy the high-precision design. The demand for colored Gaussian noise is highly transient [41,44]. To solve this problem, a standard phase vector $\mathbf{\Theta}_i$ with N_i deterministic elements is now introduced, i.e.,

$$\mathbf{\Theta}_i = \left(2\pi \frac{1}{N_i + 1}, 2\pi \frac{2}{N_i + 1}, \ldots, 2\pi \frac{N_i}{N_i + 1}\right) \tag{2.6.27}$$

At this point, let the Doppler phase shift vector $\boldsymbol{\theta}_i = (\theta_{i,1}, \theta_{i,2}, \ldots, \theta_{i,N_i})$ be the combination of the standard phase shift

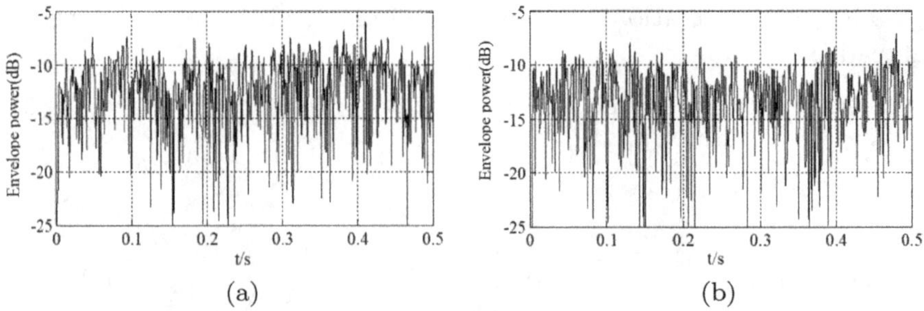

Figure 2.25 Comparison of two methods for generating phase shifts: (a) Random method and (b) combination method.

vector $\mathbf{\Theta}_i$, and there are $N_i!$ combination modes. After the above processing, the design requirements for the random value of $\theta_{i,n}$ in the $[0, 2\pi)$ can be truly satisfied.

Figure 2.25 shows the curves of two Doppler phase calculation methods. It can be seen from Figure 2.25 that the combined phase shift method can well simulate the channel model.

The above methods for the implementation of colored Gaussian noise process and the calculation method for Doppler coefficients, Doppler frequency, and Doppler phase calculation methods are studied. This will lay the foundation for the simulation implementation research of the subsequent channel model.

2.6.4 *Simulation implementation method of the classical channel model*

In the satellite channel model, common channel models include Rayleigh, Rice, Suzuki, and lognormal. Other channel models can be transformed by combining these common models. Therefore, in the channel model design, it is very necessary to fully study the simulation implementation method of the classical channel model. In this section, the simulation implementation method of the classical channel model is obtained by the combination of colored Gaussian noise [39, 45].

2.6.4.1 *Simulation implementation method of Rayleigh channel model*

Such a channel model is commonly used to describe multipath transmission of signals, which can be obtained by adding two colored

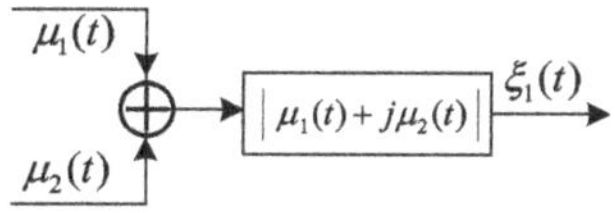

Figure 2.26 Simulation of Rayleigh channel model.

Gaussian noises, i.e.,

$$\xi_1(t) = |\mu_1(t) + j\mu_2(t)| \tag{2.6.28}$$

That is,

$$\xi_1(t) = \sqrt{\mu_1^2(t) + \mu_2^2(t)} \tag{2.6.29}$$

where $\mu_1(t)$ and $\mu_2(t)$ represent colored Gaussian processes and the simulation implementation process is shown in Figure 2.24. $\xi_1(t)$ is a random variable obeying the Rayleigh distribution, and the simulation implementation process is shown in Figure 2.26.

The mathematical expectation of $\xi_1(t)$ is obtained by:

$$E[\xi_1(t)] = \sigma_0\sqrt{\frac{\pi}{2}} \tag{2.6.30}$$

The variance is obtained by:

$$\mathrm{Var}[\xi_1(t)] = \sigma_0^2\left(2 - \frac{\pi}{2}\right) \tag{2.6.31}$$

2.6.4.2 *Simulation implementation method of Rice channel model*

Both the Rice channel model and the Rayleigh channel model can describe multipath channels. The most significant difference between them is that there is a direct view component in the Rice channel that reaches the receiving terminal through the direct view path, while the signals in the Rayleigh channel are all multipath signals, without direct view path. The Rice channel model is suitable for the suburban environment, whereas the typical scene of the Rayleigh channel model is the urban environment. The Rice channel model is

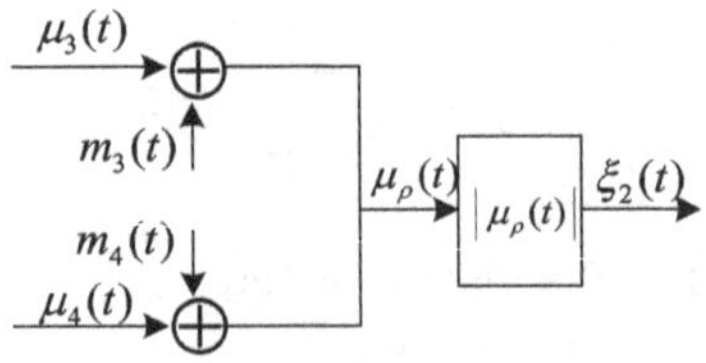

Figure 2.27 Implementation process of Rice channel model simulation.

denoted as:

$$\xi_2(t) = |\mu_\rho(t)| = |\mu(t) + m(t)| \tag{2.6.32}$$

$$\mu(t) = \mu_3(t) + j\mu_4(t) \tag{2.6.33}$$

$$m(t) = m_3(t) + jm_4(t) \tag{2.6.34}$$

where $\mu_3(t)$ and $\mu_4(t)$ are colored Gaussian process and the simulation implementation process is shown in Figure 2.24. Two direct view signals are $m_3(t)$ and $m_4(t)$, and $m_i(t) = \cos[2\pi f + \theta + (i+1)\pi/2]$ is the ith direct view signal and $i = 3, 4$. f and θ represent the frequency and phase of the direct view signal. The Rice channel model is obtained by constructing the complex random process and taking its modulus. The simulation implementation process is shown in Figure 2.27.

2.6.4.3 *Simulation implementation method of lognormal channel model*

In the propagation process of the satellite signals, when satellite signals pass though urban environments such as clouds and polluted areas, they will have a shadowing fading effect, which is called as a large-scale fading, and the entire propagation path of the satellite signals exists in a non-ideal propagation environment. The envelope of the satellite signals can be described by lognormal distribution. Its mathematical expression is written as:

$$\lambda(t) = e^{\mu_\lambda(t)} \tag{2.6.35}$$

$$\mu_\lambda(t) = \mu_5(t)\sigma_\mu + m_\mu \tag{2.6.36}$$

where $\mu_5(t)$ is a colored Gaussian process and the simulation implementation process is shown in Figure 2.24. σ_μ, m_μ are the variance

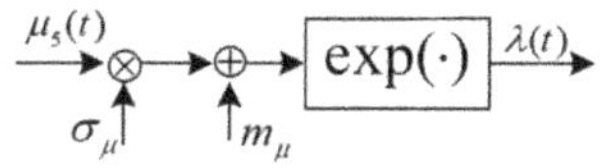

Figure 2.28 Simulation implementation of lognormal channel model.

and the mean, respectively. The simulation implementation process is shown in Figure 2.28.

Its mathematical expectation and variance are written as:

$$\mathrm{E}[\lambda(t)] = e^{m_\mu + \frac{\sigma_\mu}{2}} \tag{2.6.37}$$

$$\mathrm{Var}[\lambda(t)] = e^{2m_\mu + \sigma_\mu^2}(e^{\sigma_\mu^2} - 1) \tag{2.6.38}$$

2.6.4.4 *Simulation implementation method of Suzuki channel model*

When the weather conditions are poor (clouds and rain), the signal propagation through the satellite channel is affected by the shadowing effect and the multipath effect of the absence of direct signals, which can be described by the Suzuki model consisting of Rayleigh distribution and lognormal distribution [46, 47], which is given by:

$$f_r(r) = \int_0^\infty \frac{r}{\sigma_4^2} \exp\left(-\frac{r_0^2}{2\sigma_4^2}\right) \frac{1}{\sqrt{2\pi}\sigma_2\sigma_4} \cdot \exp\left(-\frac{(\ln\sigma_4 - m_{LN})^2}{2\sigma_2^2}\right) d\sigma_4, 0 \leq r < \infty \tag{2.6.39}$$

where σ_4 is the standard deviation of the Gaussian components in the Rayleigh distribution, while m_{LN} and σ_2 are the mean and standard deviation of the lognormal signal envelope σ_4 obeying the lognormal distribution, respectively. Equation (2.6.39) shows that the Suzuki channel model is obtained by integrating the standard deviation σ_4 of the Rayleigh signal envelope of Rayleigh distribution in the case of obeying the lognormal distribution, and it can realize the transformation from local characteristics to global characteristics.

The Rayleigh distribution signal envelope is multiplied with the lognormal distribution signal envelope to obtain the signal envelope

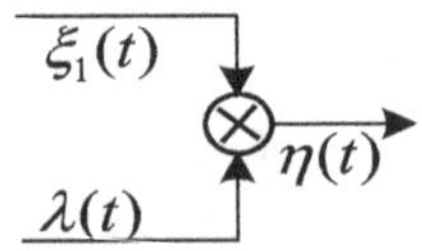

Figure 2.29 Simulation implementation of Suzuki model.

$\eta(t)$ of the Suzuki channel model, i.e.,

$$\eta(t) = \xi_1(t) \cdot \lambda(t) \tag{2.6.40}$$

where $\eta(t)$ obeys the distribution according to Equation (2.4.45). The simulation implementation process of the Suzuki channel model is shown in Figure 2.29.

The mathematical expectation and variance of $\eta(t)$ are calculated by:

$$\mathrm{E}[\eta(t)] = \sigma_0 \sqrt{\frac{\pi}{2}} e^{\mu_2 + \frac{\sigma_2^2}{2}} \tag{2.6.41}$$

$$\mathrm{Var}[\eta(t)] = \sigma_0^2 \cdot e^{2\mu_2 + \sigma_2^2} \cdot \left(2e^{\sigma_2^2} - \frac{\pi}{2}\right) \tag{2.6.42}$$

The simulation implementation method of these classical channel models is the basis and premise for establishing the simulation model of a more complex channel.

References

[1] Cheffena M. and Fontan F. P. Channel simulator for land mobile satellite channel along roadside trees. *IEEE Transactions on Antennas and Propagation*, 2011, 59(5): 1699–1706.

[2] Tao Y. The research of satellite channel modeling and simulation method. Nanjing University of Information Science and Technology, 2014.

[3] Tao Z. Design and Implementation of LEO satellite channel Emulator. Nanjing University of Information Science and Technology, 2014.

[4] Ghasemi A., Abedi A., and Ghasemi F. Radiowave propagation in satellite communication. *Propagation Engineering in Radio Links Design*. Springer New York, 2013.

[5] Lutz E. Modeling of the land mobile satellite communications channel. *Proceedings of the 2013 IEEE-APS Topical Conference on Antennas*

and Propagation in Wireless Communications. Piscataway, NJ: IEEE, 2013. pp. 199–202.

[6] Kopke A., Willig A., and Karl H. Chaotic maps as parsimonious bit error models of wireless channels. *Joint Conference of the IEEE Computer and Communications*, 2003, 1(1): 513–523.

[7] ITU-R Recommendation P.525-2. Calculation of free-space attenuation. International Telecommunication Union, August, 1944.

[8] Lu S., Liu T., and Zheng M. A wideband space time statistical model for characterization of satellite communication channel in dense multipath environment. *Vehicular Technology Conference*, 2010, 5(1): 1–5.

[9] ITU-R Recommendation P.531-12. Ionospheric propagation data and prediction methods required for the design of satellite services and systems. International Telecommunication Union, October, 2013.

[10] ITU-R Recommendation P.676-10. Attenuation by atmospheric gases. International Telecommunication Union, October, 2012.

[11] ITU-R Recommendation P.837-6. Characteristics of precipitation for propagation modeling. International Telecommunication Union, January, 2012.

[12] ITU-R Recommendation P.838-3. Specific attenuation model for rain for use in prediction method. International Telecommunication Union, March, 2005.

[13] ITU-R Recommendation P.840-6. Attenuation due to clouds and fog. International Telecommunication Union, September, 2013.

[14] ITU-R Recommendation P.618-11. Propagation data and prediction methods required for the design of earth-space telecommunication systems. International Telecommunication Union, September, 2013.

[15] Seyedi Y., Shirazi M., Moharrer A., *et al.* Use of shadowing moments to statistically model mobile satellite channels in urban environments. *IEEE Transactions on Wireless Communications*, 2013, 12(8): 3760–3769.

[16] Patzold M., Yingchun L., and Laue F. A study of a land mobile satellite channel model with asymmetrical Doppler power spectrum and lognormally distributed line-of-sight component. *IEEE Transactions on Vehicular Technology*, 1998, 47(1): 297–310.

[17] Jun X., Jong J. -H., and Ravishankar C. Channel modeling for a land mobile satellite system. *IEEE Global Telecommunication Conference*, 2007, pp. 4596–4600.

[18] Loo C. and Butterworth J. S. Land mobile satellite channel measurements and modeling. *Proceedings of the IEEE*, 1998, 86(7): 1442–1463.

[19] Alimohammad A., Fard S. F., Cockburn B. F., *et al.* Compact Rayleigh and Rician fading simulator based on random walk processes. *IET Communications*, 2009, 3(8): 1333–1342.

[20] Rafiq G. and Patzold M. The impact of shadowing on the capacity of mobile fading channels. *International Symposium on Wireless Communication Systems*, 2007, pp. 209–214.

[21] Wu T. M. and Tzeng S. Y. Sum-of-sinusoids-based simulator for Nakagami-m fading channels. *Vehicular Technology Conference*, 2003, 1(1): 158–162.

[22] Loo C. Digital transmission through a land mobile satellite channel. *IEEE Transactions on Communications*, 1990, 38(5): 693–697.

[23] Suzuki H. A statistical model for urban radio propagation. *IEEE Transactions on Communications*, 1977, 25(7): 673–680.

[24] Patzold M., Killat U., and Laue F. An extended Suzuki model for land mobile satellite channels and its statistical properties. *IEEE Transactions on Vehicular Technology*, 1998, 47(2): 617–630.

[25] Corazza G. E. and Vatalaro F. A statistical model for land mobile satellite channels and its application to nongeostationary orbit systems. *IEEE Transactions on Vehicular Technology*, 1994, 43(3): 738–742.

[26] Lutz E. Modelling of the land mobile satellite communications channel. *Antennas and Propagation in Wireless Communications*, Torino, 2013, pp. 199–202.

[27] Seyedi Y., Shirazi M., Moharrer A., *et al.* Use of shadowing moments to statistically model mobile satellite channels in urban environments. *IEEE Transactions on Wireless Communications*, 2013, 12(8): 3760–3769.

[28] Abdi A., Lau W. C., Alouini M. S., *et al.* A new simple model for land mobile satellite channels: first-and second-order statistics. *IEEE Transactions on Wireless Communications*, 2003, 2(3): 519–528.

[29] Li Y., Patzold M., Killat U., *et al.* An efficient deterministic simulation model for land mobile satellite channels. *Vehicular Technology Conference*, 1996, 2(2): 1028–1032.

[30] Li Y., Song Y., Chen J., *et al.* On the second-order statistics of a new fading model for land mobile satellite channels. *Wireless, Mobile and Sensor Networks*, IET, 2007, pp. 12–15.

[31] Lin H. P., Tseng M. J. and Lin D. B. Modeling fading properties for mobile satellite link channels using Markov model approaches. *Antennas and Propagation Society International Symposium*, 2003, 4(4): 192–195.

[32] Abdi A., Lau W. C., Alouini M. S., *et al.* On the second-order statistics of a new simple model for land mobile satellite channels. *Vehicular Technology Conference. IEEE*, 2001, 1(1): 301–304.

[33] Gutierrez-Dtaz-De-Leon C. A. and Pätzold M. Efficient Sum-of-Sinusoids-based simulation of mobile fading channels with asymmetrical Doppler power spectra. *International Symposium on Wireless Communication System. IEEE*, 2007, 1: 246–251.

[34] Haas E. Aeronautical channel modeling. *IEEE Transactions on Vehicular Technology*, 2002, 51(2): 254-264.

[35] Norbury J. R. Satellite land mobile communication systems. *Electronics & Communication Engineering Journal*, 1989, 1(6): 245–253.

[36] Pätzold M., Killat U. and Laue F. An extended Suzuki model for land Monile satellite channels and its statistical properties. *IEEE Transactions on Vehicular Technology*, 1998, 47(2): 617–630.

[37] Pop M. F. and Beaulieu N. C. Limitations of sum-of-sinusoids fading channel simulators. *IEEE Transactions on Communications*, 2001, 49(4): 699–708.

[38] Lutz E., Cygan D., Dippold M., *et al.* The land mobile satellite communication channel-recording, statistics, and channel model. *IEEE Transactions on Vehicular Technology*, 1991, 40(2): 375–386.

[39] Rec. ITU-R P.676-9. Attenuation by atmospheric gases. 2012.

[40] Palmer L. Computer modeling and simulation of communications satellite channels. *IEEE Journal on Selected Areas in Communications*, 1984, 2(1): 89–102.

[41] Milojevic M., Haardt M. and Heuberger A. Measurement based satellite to outdoor channel modeling for multiple satellite systems. *2008 14th European Wireless Conference*. IEEE, 2008, pp. 1–7.

[42] Loo C. and Butterworth J. S. Land mobile satellite channel measurements and modeling. *Proceedings of the IEEE*, 1998, 86(7): 1442–1463.

[43] Lau A. H. L. A five-state financial distress prediction model. *Journal of Accounting Research*, 1987, 25(1): 127–138.

[44] Zheng Y. R. and Xiao C. Simulation models with correct statistical properties for Rayleigh fading channels. *IEEE Transactions on Communications*, 2003, 51(6): 920–928.

[45] Rec. ITU-R P. 676-9. Attenuation by atmospheric gases. 2012.

[46] Zheng Y. R. and Xiao C. Simulation models with correct statistical properties for Rayleigh fading channels. *IEEE Transactions on Communications*, 2003, 51(6): 920–928.

[47] Li X. and Wu S. A generalized model for mobile satellite channels and performance analysis. *International Conference on Communication Technology Proceedings*, 2003, 2(2): 1070–1073.

Chapter 3

Multi-State Markov Chain Model for Satellite Channels

Summary

This chapter firstly analyzes the satellite channel two-state Markov chain model and then analyzes the satellite channel state and model in the atmospheric environment and the ground environment. Meanwhile, the satellite channel three-state Markov chain model, simulation method, and Simulink system module are established. The satellite channel five-state Markov chain model and the six-state Markov chain model outage probability of the satellite channel are analyzed by using the light shadowing, moderate shadowing, and heavy shadowing of the ground environments. The satellite channel Markov chain model and its simulation based on principal component analysis and fuzzy clustering method are further studied in transmitting.

It can be seen from Chapter 2 that the establishment of the satellite channel model is usually based on the envelope or power of the received signal, and the statistical model is used to describe the channel transmission characteristics [1, 2]. In traditional channel modeling, researchers often only consider the influence of one or two factors on the channel transmission characteristics and establish a channel model that obeys certain statistical characteristics under certain conditions. However, in actual satellite communication, the transmitting signals are often affected by many external factors during the transmission process, which may result in different degrees of signal attenuation. In addition, the relative motion of the satellite and the receiving terminal will cause continuous changes in the signal transmission environment and the channel transmission characteristics. Therefore, it is necessary to divide the channel state according to the dynamics of the signal transmission environment, so that the channel state corresponds to the transmission environment, and different channel states are characterized by different statistical characteristics, which can establish the corresponding model and set the corresponding model parameters. At the same time, considering the transition of the channel state with the actual environmental changes of the receiving terminal, it is necessary to establish a satellite channel model in a wide-area environment [3–6].

The C. Loo channel model, Suzuki channel model, Corazza channel model, and Lutz channel model are all based on multipath fading and shadowing fading in the ground mobile environment without considering the influence of atmospheric propagation fading factors on signal transmission. However, as the operating frequency of satellite communication systems continues to increase, the signals are increasingly affected by various weather conditions in the atmosphere [7–10]. Therefore, in order to describe the satellite channel transmission characteristics more comprehensively, it is necessary to comprehensively consider the satellite channel modeling problems in the atmospheric propagation environment and the ground mobile environment.

3.1 Satellite Channel Two-state Markov Chain Model

The channel characteristics are first analyzed and then the channel model is established [11] when studying satellite mobile communication systems. The satellite channel model is divided into two types: single-state model and multi-state model [12–14]. The single-state model assumes that the envelope or power of the received signal is subject to a statistical probability distribution with certain fixed parameters [15, 16]. A multi-state channel model can be established if different probability densities or different probability distributions with different parameter settings are used to describe multiple different environmental states [17–19]. The most effective method to verify the accuracy of a channel model is to compare the statistical properties of the proposed model with those of the measured data in the actual environment. Under the premise of ensuring the accuracy of the model, the complexity of the model implementation should be as low as possible [20]. At present, the main technical gap between domestic and foreign research in this field is the lack of analysis and modeling of the test data, and it is unable to provide models and parameters for the actual channel environment. Therefore, based on the experimental data of the Iridium system measured by the Communication Research Center of Harbin Institute of Technology, a two-state Markov chain model was introduced based on the difference between the single-state model and the multi-state model and the model parameters were fitted using the actual channel environment.

Through the analysis of the experimental data of the Iridium system, it was concluded that the main fading mechanism of satellite mobile communication is the ground shadowing effect [21]. In the light shadowing environment (for example, there are some telegraph poles on the roadside, occasionally trees, or low-density forest areas), the fading obeys the Rice distribution with a Rice factor of 8 dB. However, when the mobile terminal moves in a relatively dense forest area, small villages, or narrow streets, the signal attenuation is obvious, up to 20–40 dB. Especially when mobile terminals inevitably pass through some diverse physical environments (e.g., suburbs, cities, and open areas), the signal attenuation is more pronounced. Therefore, there will be shadows and no shadows in the propagation path, but their ratios are different.

3.1.1 *Satellite channel two-state Markov chain model in ground environment*

Considering that the single-state model can only describe the single-channel environment, the multi-state model has many parameters and high complexity of simulation. This section presents a two-state channel model with "ideal state" and "non-ideal state" in a wide-area environment to overcome the shortcomings of the multi-state model. The "ideal state" is defined as the state in which the communication link is unaffected by shadows or only slightly shaded, in which case communication is not interrupted. A "non-ideal state" is defined as a state in which a communication link is severely affected by a shadow, in which case communication is interrupted. The probability density function (PDF) of the total received signal envelope is written as:

$$f(r) = p_g f_g(r) + p_b f_b(r) \tag{3.1.1}$$

where p_g is the probability with the "ideal" channel, p_b is the probability with the "non-ideal" channel, $f_g(r)$ is the probability density function of the signal envelope under "ideal state", and $f_b(r)$ is the probability density function of the signal envelope under "non-ideal state".

3.1.1.1 *"Ideal state" channel statistical characteristics*

When the channel is "ideal", the received signal is not affected by the shadow or only the direct view component is affected by the slight shadow, meanwhile, the communication is not interrupted. At this point, the Loo model can be used to describe the channel characteristics. The probability density function of the received signal envelope r is:

$$\begin{aligned} f_g(r) = &\frac{r}{b_0\sqrt{2\pi\sigma_0^2}} \int_0^\infty \frac{1}{z} \exp\left[-\frac{(\ln z - m_{LN})^2}{2\sigma_0^2} - \frac{(r^2+z^2)}{2b_0}\right] \\ &\cdot I_0\left(\frac{rz}{b_0}\right) dz \end{aligned} \tag{3.1.2}$$

where b_0 is the average power of multipath scattering and m_{LN} and σ_0^2 are the mean and variance of the direct view component, respectively, $I_0(\cdot)$ is the first type of modified zero-order Bessel function.

In the "ideal state", the cumulative distribution function (CDF) of the received signal envelope r is given by:

$$\begin{aligned} F_g(R) &= \Pr(r < R) = \int_0^R f(r)dr \\ &= \frac{r}{b_0\sqrt{2\pi\sigma_{LN}}} \cdot \int_0^R \int_0^\infty \frac{r}{y} \cdot \exp\left(-\frac{(r^2+y^2)}{2b_0} - \frac{(\ln y - m_{LN})^2}{2\sigma_{LN}}\right) \\ &\quad \cdot I_0\left(\frac{ry}{b_0}\right) dy\, dr \end{aligned}$$

$$= 1 - \int_0^\infty Q_1\left(\frac{y}{\sqrt{b_0}}, \frac{R}{\sqrt{b_0}}\right) \frac{1}{\sqrt{2\pi\sigma_{LN}y}} \times \exp\left(-\frac{(\ln y - m_{LN})^2}{2\sigma_{LN}}\right) dy \tag{3.1.3}$$

where $Q_1(a,b) = \int_b^\infty z \cdot \exp\left(-\frac{z^2+a^2}{2}\right) \cdot I_0(az)dz$ represents the first step of the Marqum function.

3.1.1.2 *"Non-ideal state" channel statistical characteristics*

When the channel is "non-ideal", the communication link is affected by heavy shadows or full shadow fading, and the communication is interrupted. At this time, the Corazza model is used to describe the channel characteristics [16]. The multipath components and the direct view component of the received signal in the Corazza model are affected by shadowing fading, so the probability density function of the received signal envelope r is written as:

$$f_g(r) = \frac{r}{\sqrt{2\pi\sigma_0^2\sigma_L}} \cdot \int_0^\infty \frac{1}{y^3} \cdot \exp\left[-\frac{(r/y)^2}{2\sigma_0^2} - \frac{(\ln y - m_L)^2}{2\sigma_{LN}\sigma_L^2}\right] \cdot I_0\left(\frac{r\rho}{y\sigma_0^2}\right) dy \tag{3.1.4}$$

where σ_0^2 is the average power of multipath scattering, ρ is the magnitude of the direct view component, m_L is the average value of the direct view component, and σ_L is the standard deviation of the direct view component.

In the "non-ideal state", the cumulative distribution function (CDF) of the received signal envelope r is written as:

$$F_g(R) = \int_0^R f(r)dr = \frac{1}{\sqrt{2\pi\sigma_0^2\sigma_3}} \cdot \int_0^R \frac{r}{y^3} \cdot \int_0^\infty \frac{1}{y^3} \cdot \exp\left(-\frac{(r/y)^2+\rho^2}{2\sigma_0^2} - \frac{(\ln y - m_L)^2}{2\sigma_L^2}\right) \cdot I_0\left(\frac{r\rho}{y\sigma_0^2}\right) dy\, dr \tag{3.1.5}$$

where $K = \rho^2/(2\sigma_0^2)$ is the Rician factor, and the received signal power is normalized, that is, let $\rho^2 + 2\sigma_0^2 = 1$, then the cumulative distribution function (CDF) under "non-ideal state" becomes

$$F_b(r) = \frac{2(K+1)}{\sigma_3\sqrt{2\pi}}\exp(-K) \cdot \int_0^R r\int_0^\infty \frac{1}{y^3}\cdot\exp\left(-\frac{(K+1)r^2}{y^2} - \frac{(\ln y - m_L)^2}{2\sigma_L^2}\right) \cdot I_0\left(\frac{2r\sqrt{K(K+1)}}{y}\right) dy\, dr \tag{3.1.6}$$

The above analyses have shown that the PDF and the CDF of the two-state channel model are a function of b_0, m_{LN}, σ_{LN}, K, m_{L}, and σ_{L}. Therefore, by adjusting these parameters, the model can characterize the fading characteristics of the satellite mobile channels in different channel environments.

3.1.1.3 *Two-state switching*

Since the two-state Markov chain model needs to dynamically switch between "rational state" and "non-ideal state", it is necessary to have a switching method that can better simulate the actual situation. Generally, the state of a random signal at a certain moment is mainly affected by the state of the signal at the nearest neighboring moment and little affected or even not affected by the state in which the signal is far away from the moment. Therefore, the mathematical abstracted Markov chain model is used to implement this random switching process. The Markov chain has two important matrixes, one is the state probability matrix $\boldsymbol{A}$, which is a set of probabilities that each state may occur under certain circumstances. Another is the state transition matrix $\boldsymbol{P}$, which consists of the state transition probabilities. The condition probability $p\{X_{n+1} = j|X_n = i\}$ indicates that when the system is in the state i at time n, the probability of the state j at the time $n+1$ is recorded as $p_{ij}(n)$. This condition probability, also called the transition probability, is a one-step transition probability at time n in the Markov chain $\{X_n, n \in T\}$. Therefore, as long as the state probability matrix $\boldsymbol{A}$ and the one-step transition

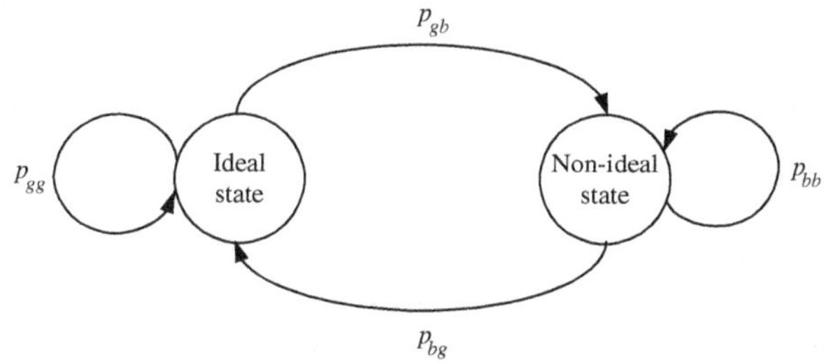

Figure 3.1 Markov chain switching between two states.

probability matrix $\boldsymbol{P}$ are determined, the channel state switching can be described [18].

Figure 3.1 is a state transition diagram between "ideal state" and "non-ideal state". If the current time is in a particular state, the next state of the channel will stay in its original state or switch to the other state. In Figure 3.1, p_{gg} indicates the probability that the channel remains in the "ideal state", p_{bb} represents the probability that the channel stay in a "non-ideal state", p_{gb} represents the probability that the channel switches from "ideal state" to "non-ideal state", and p_{bg} represents the probability that the channel will switch from "non-ideal state" to "ideal state".

$$p_{gb} = 1 - p_{gg} \qquad p_{bg} = 1 - p_{bb} \tag{3.1.7}$$

3.1.2 *Simulation verification*

Another important aspect in the study of satellite mobile communication system is the simulation of the channel model and studying the differences between the simulation model and theoretical model of the channel. The simulation model is a model that the computer can simulate, which is transformed from the theoretical model using software or hardware methods, to do system performance analysis, optimization and evaluation test, and so on. Based on the proposed channel two-state Markov chain model, the characteristics of the channel in "ideal state" can be the superposition of a Rayleigh random process and a lognormal random process. While the characteristics of the channel in "non-ideal state" can be the product of a Rician random process and a lognormal random process. Therefore, the simulation of Rayleigh, Rician, and lognormal random process must be done first to simulate the two-state channel model. In the study of the wireless fading channel characteristics, these three random processes can be achieved by colored Gaussian process.

Figure 3.2 is a dynamic simulation block diagram of the probability model of the mobile satellite channel two-state Markov chain. Figure 3.2 includes the Loo stochastic process under "ideal state" and the Corazza stochastic process under "non-ideal state", and the Markov chain simulation module is in the lower left corner. Two uniformly distributed random number generators can generate the decimal between 0 and 1, switch 1 is the switch which controls the "ideal state" to hold or switch, its threshold is p_{gb}, while switch 2 is the switch which controls the "non-ideal state" to hold or convert, its threshold is p_{bg}. The two switches are independent of each other, and the actual Markov chain conversion is achieved by switching the switch. It can choose to switch output signal above or below according to the previous state. The final output signal can be used to control the switch between "ideal state" and "non-ideal state". When the output is "1", "ideal state" is selected, and when the output is "0", "non-ideal state" is selected. Under the control of the Markov chain, the channel characteristics can be dynamically switched between "ideal state" and "non-ideal state" to achieve dynamic simulation of the fading characteristics.

3.1.3 *Channel model parameter fitting*

The most efficient method to verify the channel model is to compare the statistical properties of the channel model with the measured data. At present, most research on the satellite channel model in China analyzes and improves the foreign measured data and channel model. The main technology gap in this area is the lack of analysis and modeling of the test data, resulting in being unable to provide models and parameters for the actual channel environment. From April 1999 to January 2000, the Harbin Institute of Technology Communication Research Center used the measured Iridium system experimental data to obtain three typical channel model parameters. Figure 3.3(a) is a part of the open space between the Harbin and Mudanjiang Expressway (speed of 100 km/h). Figure 3.3(b) shows the road in the suburbs of Mudanjiang (speed of 60 km/h) with a

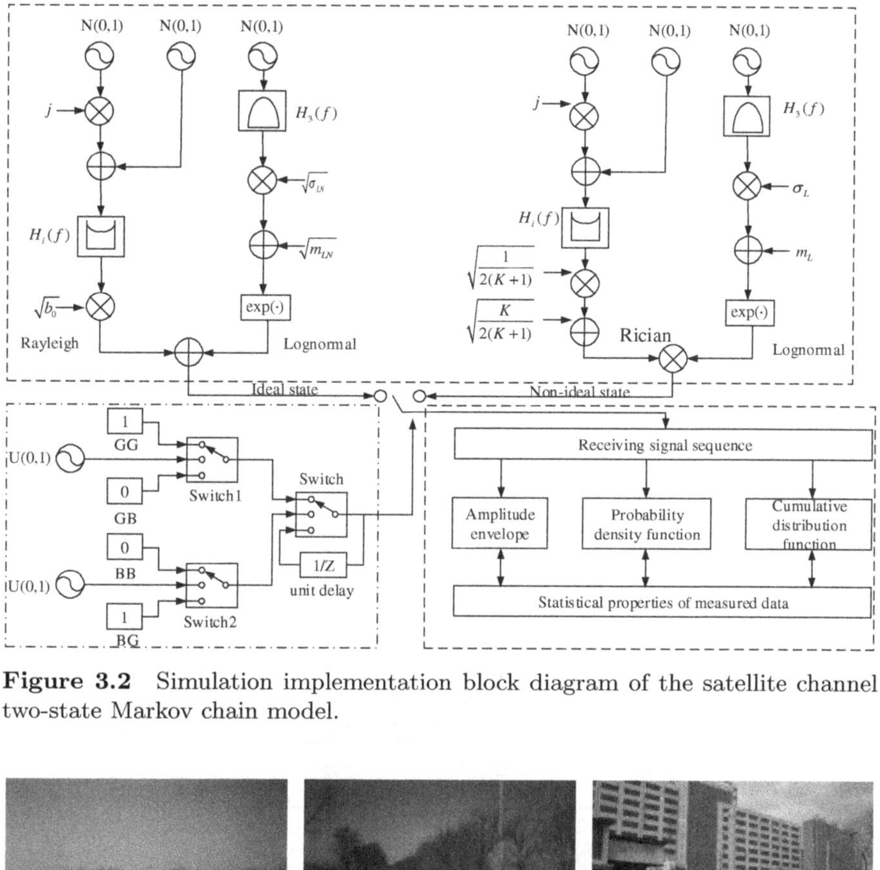

Figure 3.2 Simulation implementation block diagram of the satellite channel two-state Markov chain model.

(a) (b) (c)

Figure 3.3 Measurement environment for channel characteristics: (a) Open space; (b) rural area and (c) urban area.

small number of trees with a height of 3–4 m. Figure 3.3(c) shows the campus of Harbin Institute of Technology (speed of 20 km/h).

Figure 3.4 shows the Iridium system channel characteristic measurement system, which includes receiving antennas, high gain and low noise amplifiers, spectrum analyzers, high-speed GPIB interface card, acquisition control computers, handheld GPS receivers, and

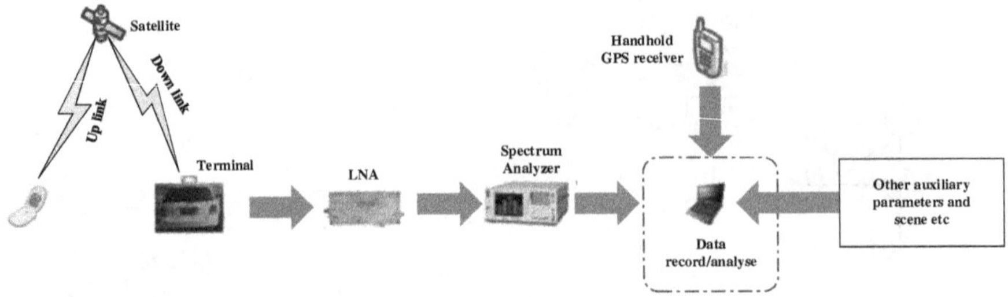

Figure 3.4 Measurement system of Iridium system channel characteristics.

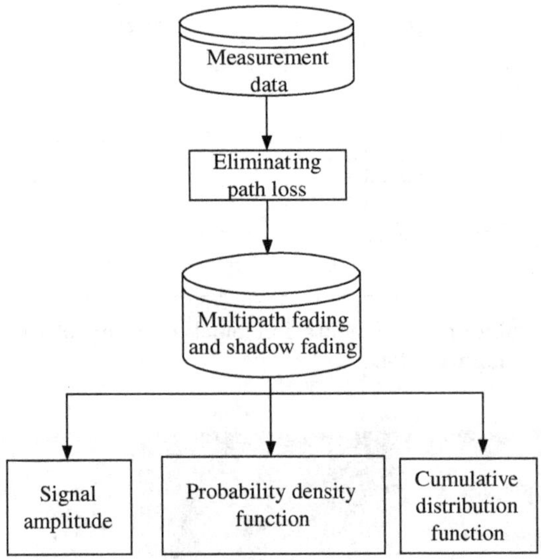

Figure 3.5 The process of measurement data.

some other recording modules for auxiliary parameters and scenarios. During the measurement, the signal is received by a specific antenna. The computer controls the instrument via a high-speed GPIB card and records the measurement data in a file. To ensure accurate recording of the corresponding test data, the GPS receiver is used to determine the specific location of the measurement sites.

Here, the channel shadowing fading and multipath fading characteristics are mainly studied. Therefore, the direct view component of the received signal is first normalized to eliminate the path loss of the measured data. The data processing is shown in Figure 3.5.

Table 3.1 "Ideal state" channel model fitting parameters.

Environment	Height (°)	Ideal State Signal Model Parameter		
		b_0	m_{LN}	σ_{LN}
Open area	20	0.0050	0.0115	0.0116
	40	0.0020	0.0102	0.0002
	60	0.0035	−0.0115	0.0010
	80	0.0026	0.0121	0.0008
Rural area	20	0.0190	−0.0456	0.0061
	40	0.0151	−0.0312	0.0075
	60	0.0090	−0.0839	0.0083
	80	0.0067	−0.0576	0.0415
Urban area	20	0.0079	−0.0403	0.0105
	40	0.0056	−0.0525	0.0058
	60	0.0039	−0.0258	0.0194
	80	0.0038	−0.0312	0.0112

In addition to the state probability matrix $\boldsymbol{A}$ and the state transition matrix $\boldsymbol{P}$, for the channel two-state Markov chain model, it is necessary to determine the six parameters, which are b_0, m_{LN}, σ_{LN} in "ideal state" and K, m_L, σ_L in "non-ideal state". These six parameters are not the same in different physical environments. However, they are always related to the satellite elevation regardless of the geographical environment. Therefore, the first-order and second-order origin moment should be derived through numerical integration first and then least squares error criterion and the method of linear least squares must be used to achieve the fitting parameters of the channel simulation model.

As is shown in Tables 3.1 and 3.2, the model parameters of different elevation in three typical channel environments, that is, light shadowing environment (open space), moderate shadowing environment (rural area), and heavy shadowing environment (urban area) were derived from quantities of repeated calculations, tests, and corrections by using the least squares fitting method. Table 3.3 shows the state probability matrix $\boldsymbol{A}$ and the state transition matrix $\boldsymbol{P}$ corresponding to the three channel environments.

Table 3.2 "Non-ideal state" channel model fitting parameters.

		Ideal State Signal Model Parameter		
Environment	**Height (°)**	K	m_L	σ_L
Open area	20	0.6495	−0.3011	0.0173
	40	1.3089	−0.1532	0.0368
	60	3.1623	−0.0652	0.0518
	80	7.9433	−0.0827	0.0656
Rural area	20	0.4995	−0.3572	0.2590
	40	0.8943	−0.4326	0.2072
	60	2.8733	−0.8456	0.2878
	80	6.9051	−0.6396	0.2303
Urban area	20	0.3451	−1.8962	0.4260
	40	0.8638	−1.7960	0.4835
	60	2.1276	−1.3585	0.5411
	80	4.8794	−1.1743	0.9026

Table 3.3 State probability matrix and state transfer matrix.

	State Probability Matrix		State Transition Matrix	
Environment	p_g	p_b	p_{gg} p_{bb}	p_{gb} p_{gb}
Open area	0.892	0.108	0.956	0.044
			0.109	0.891
Rural area	0.624	0.376	0.832	0.168
			0.253	0.747
Urban area	0.297	0.703	0.382	0.618
			0.821	0.179

3.1.4 *Channel model simulation*

Figures 3.6, 3.8, and 3.10 represent the signal amplitude measured of the Iridium system under three channel environments, that is, light shadowing, medium shadowing and heavy shadowing by Communication Research Center, Harbin Institute of Technology, China, respectively. In contrast, the signal envelope changes with the extent

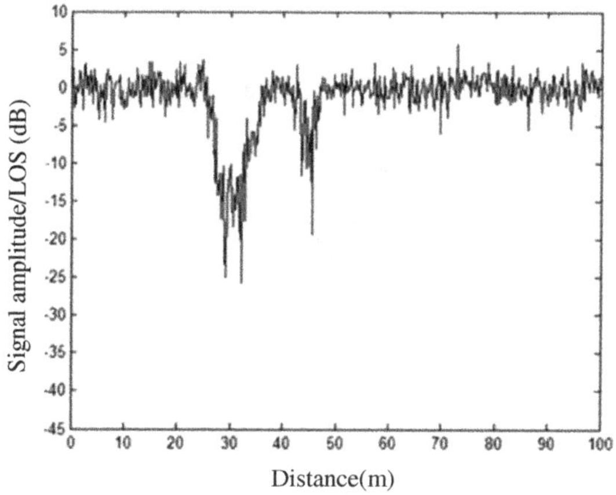

Figure 3.6 Signal amplitude measured in open area.

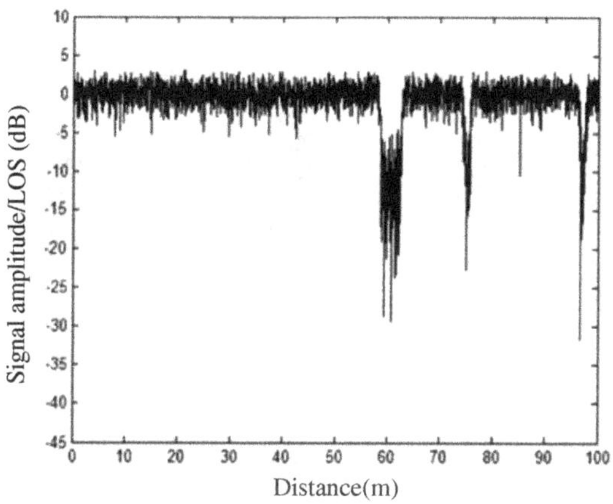

Figure 3.7 Signal amplitude of open area simulation model.

of shadowing. The larger the shadowing is, the higher the fading is. When the shadowing increases, the average value of the received signal envelope decreases. The received signal in the open space is mainly the direct view component, and the several occasional deep fading is mainly caused by the block of telegraph poles on both sides of the road. Figures 3.7, 3.9, and 3.11 show the signal amplitudes obtained by the simulation model under the three environments correspondingly. It can be seen through comparison that the signal

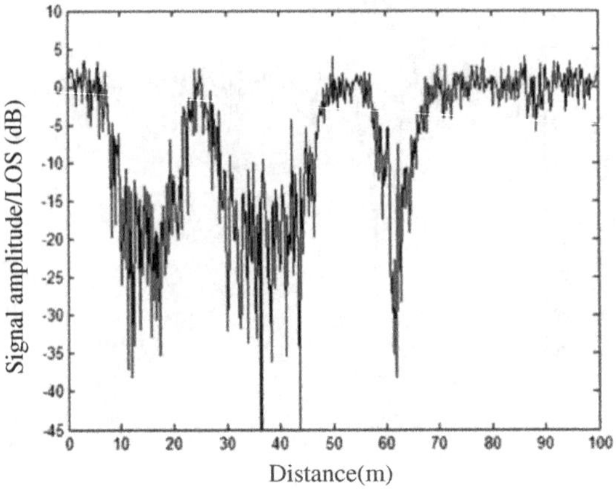

Figure 3.8 Signal amplitude of suburban measurements models.

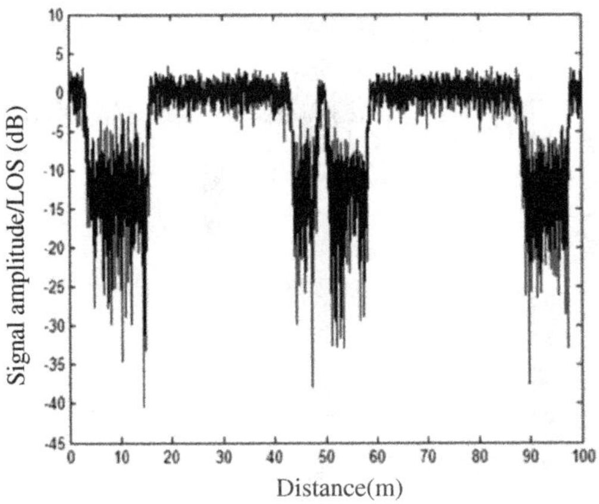

Figure 3.9 Signal amplitude of suburban simulation.

amplitude derived from the simulation model is consistent with the range of the measured data.

Figure 3.12 is a comparison of the probability density function of the received signal amplitude between the simulation model and the measured data in three typical channel model environments. Figure 3.13 is a comparison of the cumulative distribution function of the received signal amplitude between the measured data and simulation model in three typical channel model environments. It can be seen that the signal amplitude between amplitude statistical

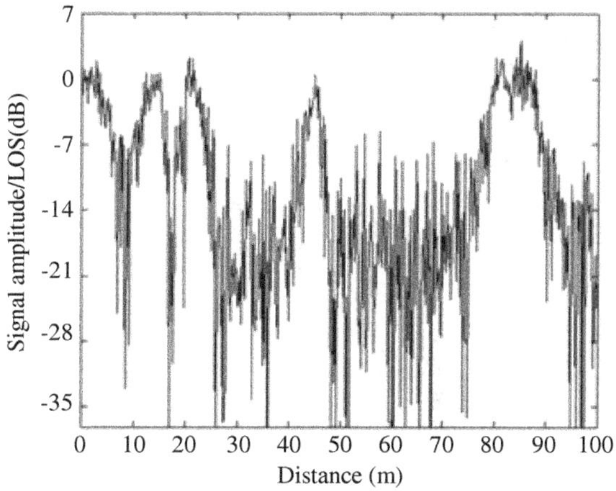

Figure 3.10 Signal amplitude measured in urban area.

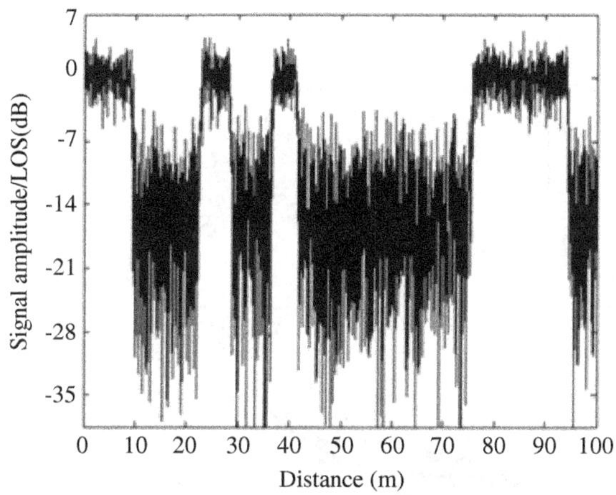

Figure 3.11 Signal amplitude of simulation model in urban area.

characteristics derived from the simulation model is consistent with the range of measured data. In addition, the proposed two-state Markov chain model compensates for the inadequacy of the direct view component of the received signal that the Lutz model cannot simulate shadowing shading [7]. Thus, it is verified that the channel two-state Markov chain model and the corresponding model parameters can accurately depict the characteristics of the mobile satellite channels in typical environments and can be used in the research

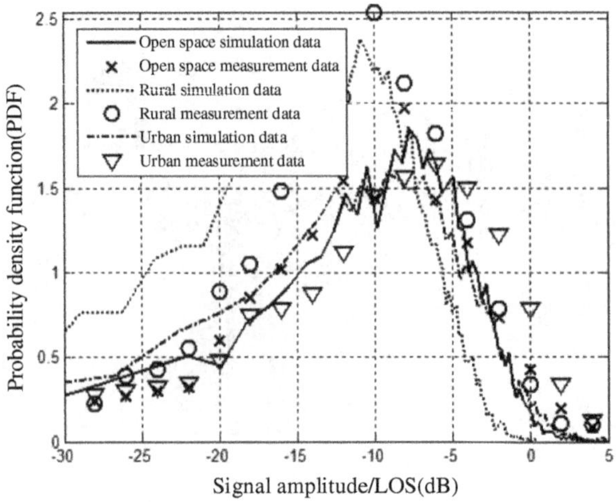

Figure 3.12 PDF comparison between simulation model and measured data.

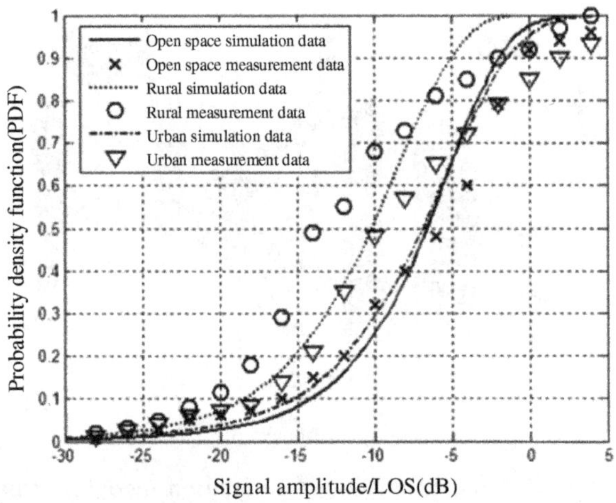

Figure 3.13 Comparison of CDF between simulation model and measured data.

of mobile satellite communication system and its ground simulation verification system in China.

3.2 Satellite Channel Three-state Markov Chain Model

This section will consider the influence of atmospheric environments and ground environments on channel modeling, analyze the characteristics of atmospheric propagation environments and ground mobile environments, and establish a channel state model corresponding to the different environments. According to the statistical characteristics of each state of the channel, the channel state transition condition is studied, and the three-state Markov chain model is established. In the modeling process, only the influence of signal amplitude variation is considered, and the influence of signal phase change is ignored.

3.2.1 *Channel model in atmospheric environment*

Atmospheric factors, such as atmospheric gases, clouds, and rainfall, and so on, can cause signal attenuation in satellite communication links. According to data analysis obtained by C. Loo *et al.* in Canada for extensive testing of the Olympus satellites [22], the envelope of signals received by the satellite is affected by the atmospheric environment (such as clouds and rainfall), which obeys the Gaussian distribution, so the probability density function of the received signal envelope r is written as:

$$f_w(r) = \frac{1}{\sqrt{2\pi}\sigma_w} \exp\left(\frac{-(r - m_w)^2}{2\sigma_w^2}\right) \tag{3.2.1}$$

where $f_w(r_w)$ is the probability density function of the signal envelope under the influence of atmospheric environments. m_w is the mean of the signal envelope, and σ_w^2 is the variance of the signal envelope. The probability distribution parameters of the signal envelope in four typical atmospheric environments are shown in Table 3.4.

Thus, the proposed satellite channel model in the atmospheric environments is shown in Figure 3.14, where $C(t)$ represents the

Table 3.4 Probability distribution parameters of signal envelope in atmospheric environment.

Weather	m_w	σ_w^2
Cloudy day	0.44	0.00041
Cumulus	0.346	0.00272
Thunderstorm	0.436	0.01386
Light rain	0.483	0.00003

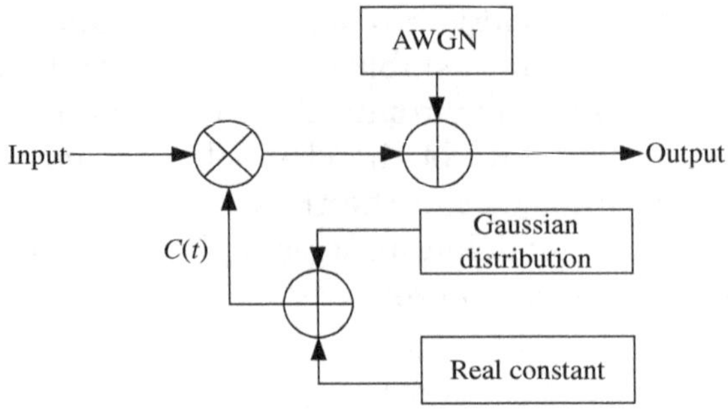

Figure 3.14 Satellite channel model in atmospheric propagation environment.

multiplicative interference of the channel in the atmospheric environments and there is a Gaussian distribution with a mean of zero and a variance of σ_w^2. When the Gaussian distribution is added to the real constant m_w, a Gaussian distribution with a mean of m_w and a variance of σ_w^2 is obtained. As shown in Table 3.4, AWGN is the additive white Gaussian noise of the channel.

3.2.2 *Channel model in ground environment*

In the terrestrial environment, the received signal is mainly blocked by various obstacles on the surface of the earth to cause shadowing fading, multipath fading caused by signal reflection, and scattering or diffraction caused by obstacle obstructions. In order to reflect the actual mobile satellite communication environments, in this section, the ground mobile transmission environments are divided into three typical physical environments: rural, suburban, and urban. In addition, channel modeling is performed in these three environments from the perspective of the received signal envelope [23]. At the same time, the Markov chain is used to simulate the dynamic characteristics of the channel in the terrestrial wide-area environments.

According to the characteristics of the three typical terrestrial environmental channels and the degree to which the direct view component is blocked and the degree of shadowing fading, the channel states are divided into three states: no shadowing, light shadowing, and heavy shadowing.

The unshaded state is a rural environment with a simple environment and less obstructions. Signal propagation is unaffected by any shadowing fading, and the terminal received signal is a composite of multipath components and direct view components. At this time, the envelope of the received signal obeys the Rice distribution, and its probability density function is given by:

$$f_1(r) = \frac{r}{\sigma_1^2} \exp\left(-\frac{r^2 + z_1^2}{2\sigma_1^2}\right) I_0\left(\frac{r z_1}{\sigma_1^2}\right) \tag{3.2.2}$$

where z_1 indicates the magnitude of the direct view component, and $2\sigma_1^2$ is the average power of the multipath components.

A light shadowing state is a relatively complicated suburban environment with obstacles. The propagation process of the signal will be blocked by obstacles such as trees and buildings, and multipath fading and shadowing fading will occur. The received signal of the terminal includes both multipath components and direct view component affected by shadowing fading. Its probability density function is given by:

$$f_2(r) = \frac{r}{\sigma_2^2 \sigma_{\mathrm{LN}} \sqrt{2\pi}} \int_0^{\infty} \frac{1}{s^3} \exp\left(-\frac{(r/s)^2 + z_2^2}{2\sigma_2^2} - \frac{(\ln s - m_{\mathrm{LN}})^2}{2\sigma_{\mathrm{LN}}^2}\right) \times I_0\left(\frac{r z_2}{s\sigma_2^2}\right) ds \tag{3.2.3}$$

where σ_2 is the Gaussian component standard deviation of the Rice distribution, σ_{LN} is the standard deviation of the lognormal distribution, and m_{LN} is the mean of the lognormal distribution.

The heavy shadowing state is very complicated urban environments with many obstructions. During the propagation of the signal, the direct view component is completely occluded, and only the multipath components due to obstacle obstructions exist. Meanwhile, the multipath components are also affected by shadowing fading. At this time, the received signal envelope can be regarded as the product of the Rayleigh distribution and the lognormal distribution, and its probability density function is given by:

$$f_3(r) = \frac{r}{\sigma_3^2 \sigma_{\mathrm{m3}} \sqrt{2\pi}} \int_0^{\infty} \frac{1}{s^3} \exp\left(-\frac{(r/s)^2}{2\sigma_3^2} - \frac{(\ln s - m_{\mathrm{LN}})^2}{2\sigma_{\mathrm{LN}}^3}\right) ds \tag{3.2.4}$$

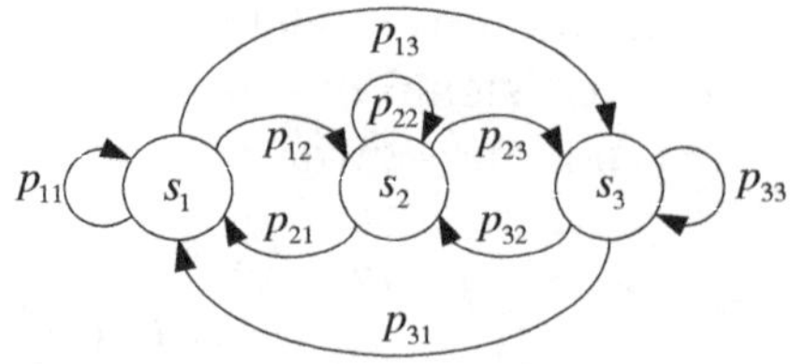

Figure 3.15 Three-state Markov chain conversion process.

where σ_3 is the Gaussian component standard deviation of the Rayleigh distribution, σ_{LN} is the standard deviation of the lognormal distribution, and m_{LN} is the mean of the lognormal distribution.

In the ground mobile environment, as the location of the mobile terminal changes continuously, the signal transmission environment changes among three typical physical environments, and the corresponding channel state also changes. The state conversion between two states can be realized by the Markov chain. The conversion is only related to the previous state. The specific conversion process between two states is shown in Figure 3.15.

The Markov chain can be described by a state transition matrix and a state probability matrix [24], whose state transition matrix is defined as:

$$\mathbf{P} = \begin{bmatrix} p_{11} & p_{12} & p_{13} \\ p_{21} & p_{22} & p_{23} \\ p_{31} & p_{32} & p_{33} \end{bmatrix} \tag{3.2.5}$$

where p_{ij} indicates the transition probability from the state i to the state j, and

$$0 \leq p_{ij} \leq 1, \sum_{j=1}^{3} p_{ij} = 1 \tag{3.2.6}$$

where i =1,2,3.

As time goes by, the Markov chain will gradually evolve to a stable state. If the state transition matrix $\boldsymbol{P}$ is known, the state probability matrix $\mathbf{A} = [A_1 \quad A_2 \quad A_3]$ can be obtained by calculating under a given initial probability matrix. A_1, A_2, and A_3 represent the ratio of the durations of the three states in the channel simulation, respectively, and the durations of the states in the three environments are different. As long as the state transition matrix of the actual communication environment is given, the channel statistical characteristics in the ground environment can be simulated according to the channel model.

According to the above analyses, the probability density function of the received signal envelope of the satellite channel three-state Markov chain model under the ground environment is obtained by the Equations (3.2.2), (3.2.2), and (3.2.4), i.e.,

$$f_g(r) = A_1 f_1(r) + A_2 f_2(r) + A_3 f_3(r) \tag{3.2.7}$$

where A_1, A_2, and A_3 are the proportions of the three states, respectively, and $A_1 + A_2 + A_3 = 1$.

Figure 3.16 shows a block diagram of the satellite channel three-state Markov chain model in a terrestrial mobile environment, where $C(t)$ represents the multiplicative interference of the channel under the influence of the ground mobile environment. The Rice distribution module describes the multipath fading characteristics of the channel in the unshaded state, the Rice–lognormal distribution module describes the channel fading characteristics in the relatively complicated environment with light shadowing fading, and the Rayleigh–lognormal distribution module describes the fading characteristics of the channel in heavy shadowing states. The Markov chain conversion process enables switching between the two states of the channel caused by constant changes in the ground environment. AWGN represents additive white Gaussian noise of the channel.

3.2.3 *Satellite channel three-state Markov chain model*

In the satellite communications, atmospheric and ground environments can cause signal attenuation or distortion. This section considers the signal attenuation caused by the atmospheric environment and the ground environment as independent. The w indicates the influence of the atmospheric environment on the signal transmission

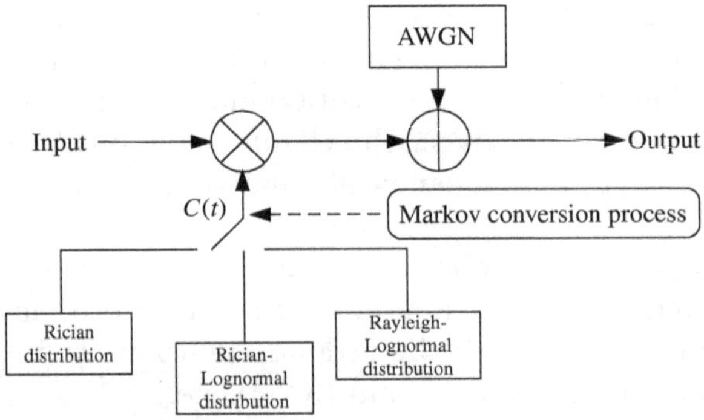

Figure 3.16 Satellite–ground link channel three-state Markov chain model.

and g denotes the influence of the ground environment on the signal transmission and the terminal receiving signal can be expressed as $r = wg$. According to probability theory, the probability density function of the total received signal envelope r is:

$$f_{w-g}(r) = \int_0^{\infty} f_g(r|w) f_w(w) dw \tag{3.2.8}$$

where $f_w(w)$ is the probability density function under the influence of atmospheric environment, and $f_g(r|w)$ is the probability density function affected by ground environment under certain atmospheric conditions. g and w are independent of each other, so

$$f_g(r|w) = \frac{1}{w} f_g\left(\frac{r}{w}\right) \tag{3.2.9}$$

In an actual ground mobile environment, as the signal transmission environment changes, the channel state is constantly changing among no shadows, light shadows, and heavy shadow states. Therefore, the probability density function of the received signal envelope affected by the ground environments in the atmospheric environment is defined as:

$$f_g(r|w) = \frac{1}{w}\left[A_1 f_1\left(\frac{r}{w}\right) + A_2 f_2\left(\frac{r}{w}\right) + A_3 f_3\left(\frac{r}{w}\right)\right] \tag{3.2.10}$$

Substituting Equations (3.2.1), (3.2.2), and (3.2.9) into Equation (3.2.8), the probability density function of the received signal envelope affected by the atmospheric environments in the unshaded ground environments is obtained by:

$$
\begin{aligned}
f_{w-n}(r) &= \int_0^{\infty} \frac{1}{w} f_1\left(\frac{r}{w}\right) f_w(w) dw \\
&= \frac{r}{\sqrt{2\pi}\sigma_1^2 \sigma_w} \int_0^{\infty} \frac{1}{w^2} \exp\left(-\frac{(r/w)^2 + z_1^2}{2\sigma_1^2} - \frac{(w - m_w)^2}{2\sigma_w^2}\right) \\
&\times I_0\left(\frac{r z_1}{w\sigma_1^2}\right) dw
\end{aligned} \tag{3.2.11}
$$

where m_w and σ_w^2 are the mean and variance of the envelopes of the Gaussian distribution affected by the atmospheric environments, respectively. σ_1^2 is the variance of the signal envelope of the unshaded ground environment obeying the Rice distribution, and z_1 is the strength of the direct view component.

Substituting Equations (3.2.1), (3.2.3), and (3.2.9) into Equation (3.2.8), the probability density function of the received signal envelope affected by the atmospheric environment in a light shadow ground environment is:

$$
\begin{aligned}
f_{w-l}(r) &= \int_0^{\infty} \frac{1}{w} f_2\left(\frac{r}{w}\right) f_w(w) dw \\
&= \frac{r}{2\pi\sigma_2^2 \sigma_{\mu 2} \sigma_w} \int_0^{\infty} \int_0^{\infty} \frac{1}{w^2 s^3} \exp\left(-\frac{(r/(ws))^2 + z_2^2}{2\sigma_2^2}\right) \\
&\times I_0\left(\frac{r z_2}{s w \sigma_2^2}\right) \cdot \exp\left(-\frac{(\ln s - m_{\mu 2})^2}{2\sigma_{\mu 2}^2} - \frac{(w - m_w)^2}{2\sigma_w^2}\right) ds\, dw
\end{aligned} \tag{3.2.12}
$$

where σ_2^2 is the variance of the signal envelope under the light shadowing ground environment, obeying the Rice distribution, z_2 represents the size of the direct view component, and $m_{\mu 2}$ and $\sigma_{\mu 2}^2$ are the mean and variance of the signal envelope affected by the shadowing fading, respectively.

Substituting Equations (3.2.1), (3.2.4), and (3.2.9) into Equation (3.2.8), the probability density function of the received signal envelope affected by the atmospheric environment in a heavy shadowing ground environment is:

$$\begin{aligned} f_{w-h}(r) &= \int_0^\infty \frac{1}{w} f_3\left(\frac{r}{w}\right) f_w(w)dw \\ &= \frac{r}{2\pi\sigma_3^2\sigma_{\mu3}\sigma_w} \int_0^\infty \int_0^\infty \frac{1}{w^2 s^3} \\ &\quad \times \exp\left(-\frac{(r/w)^2}{2\sigma_3^2 s^2} - \frac{(\ln s - m_{\mu3})^2}{2\sigma_{\mu3}^2} - \frac{(w-m_w)^2}{2\sigma_w^2}\right) ds\, dw \end{aligned} \tag{3.2.13}$$

where σ_3^2 is the variance of the multipath fading effect under the influence of heavy shadowing ground environments, where $m_{\mu3}$ and $\sigma_{\mu3}^2$ are the mean and variance, respectively, affected by shadowing fading.

According to the above analysis, the probability density function of the received signal envelope of the satellite channel Markov chain model, considering the atmospheric environment and the ground environment influence, is defined as:

$$\begin{aligned} f_{w-g}(r) &= \int_0^\infty \frac{1}{w}\left[A_1 f_1\left(\frac{r}{w}\right) + A_2 f_2\left(\frac{r}{w}\right) + A_3 f_3\left(\frac{r}{w}\right)\right] f_w(w)dw \\ &= A_1 f_{w-n}(r) + A_2 f_{w-l}(r) + A_3 f_{w-h}(r) \end{aligned} \tag{3.2.14}$$

Figure 3.17 shows a block diagram of the Markov chain model system for the satellite channels. Among them, the Gaussian distribution and the real constant are added up to simulate the influence of weather conditions on signal transmission in the atmospheric environments, the Rice distribution describes the fading characteristics of the channel in the ideal environment with fewer obstacles, and the Rice–lognormal distribution module describes the fading characteristics of the channel when the ground environments are relatively complicated. The Rayleigh–lognormal distribution describes the fading characteristics of the channel when the ground environments are very complicated. AWGN is an additive white Gaussian noise of the channel.

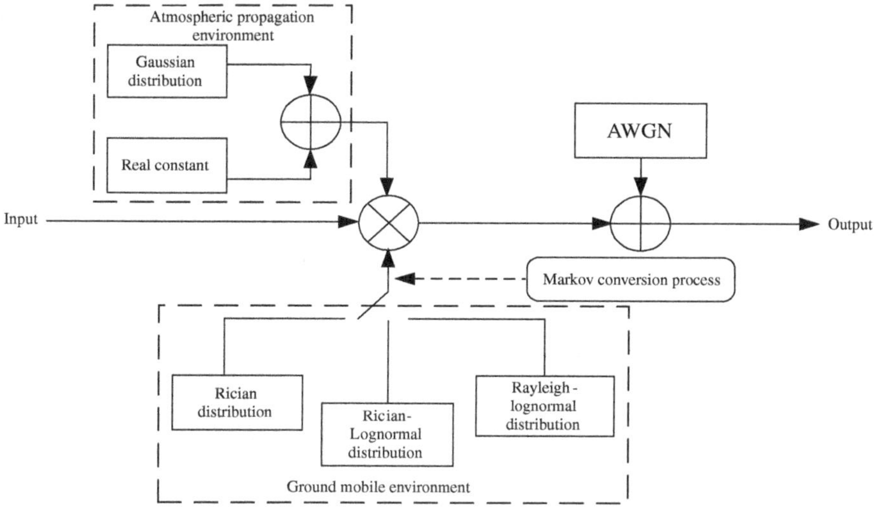

Figure 3.17 Satellite channel three-state Markov chain model system block diagram.

3.2.4 *Satellite channel three-state Markov chain model statistical characteristics*

The statistical characteristic parameters of the channel model are the main indicators to measure channel performance [25]. The statistical characteristic parameters of the built-in satellite channel Markov chain model are analyzed as follows:

The cumulative distribution function (CDF) of the satellite channel Markov chain model is determined by its probability density function (PDF) as the upper limit integration, i.e.,

$$\begin{aligned} F_{-}(r \leq r_0) &= \int_0^{r_0} f_{w-g}(r)dr \\ &= \int_0^{r_0} [A_1 f_{w-n}(r) + A_2 f_{w-l}(r) + A_3 f_{w-h}(r)]dr \end{aligned} \tag{3.2.15}$$

where $f_{w-n}(r)$, $f_{w-l}(r)$ and $f_{w-h}(r)$ are determined by Equations (3.2.11), (3.2.12), and (3.2.13), respectively, and A_1, A_2, and A_3 represent the proportion of each of the three shaded states in the channel.

According to the definition of the level crossover rate (LCR) $N_\lambda(r)$ in Equation (2.5.10), the horizontal crossover rate (HCR) of the satellite channel Markov chain model is defined as:

$$N(r) = \int_0^\infty (\dot{x} f_{\lambda_1 \dot{\lambda}_1}(r, \dot{x}) + \dot{x} f_{\lambda_2 \dot{\lambda}_2}(r, \dot{x}) + \dot{x} f_{\lambda_3 \dot{\lambda}_3}(r, \dot{x}))d\dot{x} \tag{3.2.16}$$

$$N(r) = A_1 N_1(r) + A_2 N_2(r) + A_3 N_3(r) \tag{3.2.17}$$

where

$$N_1(r) = \sqrt{\frac{\beta_1}{2\pi}} \int_0^\infty K(r, w) f_w(w) f_1(r/w) dw \tag{3.2.18}$$

$$N_2(r) = \sqrt{\frac{\beta_2}{2\pi}} \int_0^\infty K(r, w) f_w(w) f_2(r/w) dw \tag{3.2.19}$$

$$N_3(r) = \sqrt{\frac{\beta_3}{2\pi}} \int_0^\infty K(r, w) f_w(w) f_3(r/w) dw \tag{3.2.20}$$

where f_c is the 3 dB cutoff frequency, $f_w(\cdot)$ is the probability density function under the influence of the atmospheric environments, and $f_1(\cdot)$, $f_2(\cdot)$, and $f_3(\cdot)$ are the probability density functions for the effects of three different shaded ground environments, respectively.

$$K(r, w) = \sqrt{1 + \frac{\gamma}{\beta}\left(\frac{r\sigma_w}{w}\right)^2}$$

$$\beta_i = 2\sigma_i^2(\pi f_{\max})^2, \quad i = 1, 2, 3$$

$$\gamma = (2\pi f_c/\sqrt{2\ln 2})^2$$

The average fading duration (AED) of the satellite channel Markov chain model is calculated by Equation (2.5.17).

3.2.5 *Satellite channel three-state Markov chain model simulation method*

The establishment of the theoretical analysis model of the satellite channel is the mathematical reflection of the transmission characteristics of the satellite channels, which must be implemented by software or hardware in practical applications. Therefore, the simulation implementation of the theoretical model is also an indispensable key

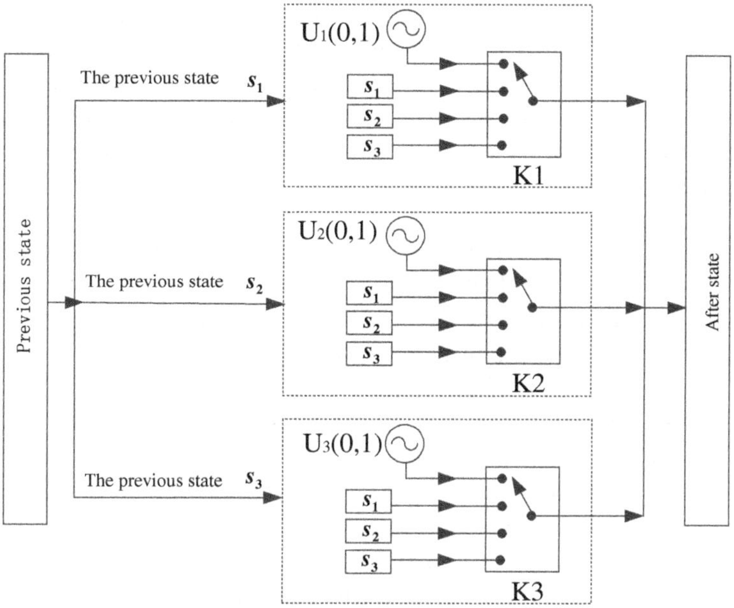

Figure 3.18 Implementation method of Markov chain state transition.

link in channel research, and the designed channel simulation model must be infinitely close to the theoretical model, so as to obtain an accurate and effective simulation of the satellite channel [26].

3.2.5.1 *Markov chain state transition implementation*

When the weather conditions have a certain impact on signal transmission in the atmospheric propagation environments, as the mobile receiving terminal switches among three physical environments with different topographical features, the transmission characteristics of the channel are continuously changing. In the modeling and simulation of the channel multi-state, the changing process of the channel states is realized by Markov chain [27].

The state transition process of the channel three-state Markov chain model of the satellite channel in the ground environments, shown in Figure 3.16, can be divided into three sub-modules, as shown in Figure 3.18.

In Figure 3.18, $U_1(0,1)$, $U_2(0,1)$, and $U_3(0,1)$ are random number generators, which are uniformly distributed and independent of each other in the interval [0 1] and whose generation frequency of the random number is determined by the change speed of the mobile terminal environment. s_1, s_2, and s_3 are the channel states in the three ground environments, respectively, and the selection thresholds of switches K1, K2, and K3 are the transition probabilities of the different channel states, respectively. The transition implementation of the three shadowing states is as follows:

(1) When the current state is s_1, the selection switch K1 is operated, and the switches K2 and K3 are suspended. If the random number generated by $\mathrm{U}_1(0,1)$ is less than or equal to p_{11}, the s_1 state remains unchanged. If the random number generated by $\mathrm{U}_1(0,1)$ is greater than p_{11} and less than $p_{11}+p_{12}$, the switch K1 is turned to the s_2 state, otherwise the switch K1 is turned to the s_3 state.
(2) When the current state is s_2, the selection switch K2 is operated, and the switches K1 and K3 are suspended. If the random number generated by $\mathrm{U}_2(0,1)$ is greater than p_{23}, then switch K2 turns to the s_3 state. If the random number generated by $\mathrm{U}_2(0,1)$ is less than p_{21}, then switch K2 turns to the s_1 state, otherwise it continues to maintain the s_2 state.
(3) When the current state is s_3, the selection switch K3 is operated, and the switches K2 and K1 are suspended. If the random number generated by $\mathrm{U}_3(0,1)$ is less than or equal to p_{31}, the switch K3 is turned to the s_1 state. If the random number generated by $\mathrm{U}_3(0,1)$ is greater than p_{31} and less than $p_{31}+p_{32}$, the switch K1 is turned to the s_2 state, otherwise the switch K3 is turned to the s_3 state.
(4) Referring to above from step (1) to step (3), the mutual conversion among three different channel states can be achieved.

Thus, according to the given initial channel state and the corresponding state transition matrix, the simulation of the channel dynamics can be achieved in the manner shown in Figure 3.18.

3.2.5.2 *Implementation method of satellite channel Markov chain model*

The simulation of the satellite channel Markov chain model is obtained by the combined calculation of the real Gaussian process, as shown in Figure 3.19.

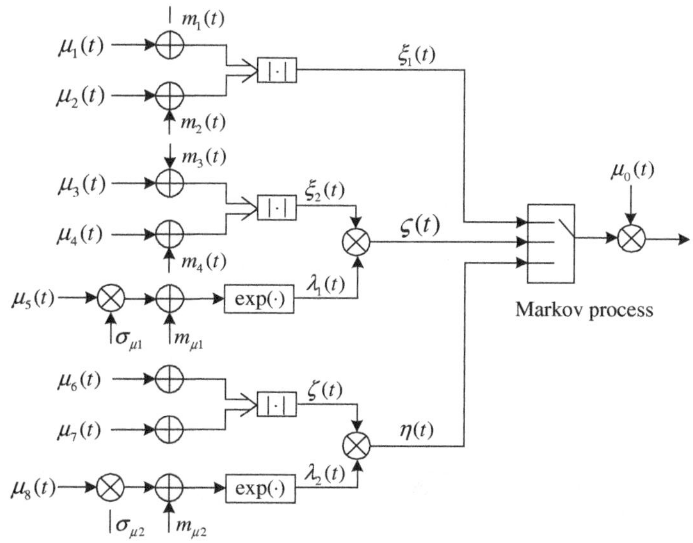

Figure 3.19 Implementation of Markov chain model for satellite channels.

In Figure 3.19, $\mu_i(t)(i = 0, 1, 2, \ldots, 8)$ is a mutually uncorrelated real Gaussian process, $\mu_0(t)$ denotes the process of the signal affected by the atmospheric propagation environment, $m_i(t)(i = 1, 2, 3, 4)$ is the direct view component, $\xi_i(t)(i = 1, 2)$ represents the Rice process, and $\xi_1(t)$ is the process of the signal affected by the ground-free shadow environment. $\lambda_i(t)(i = 1, 2)$ represents a lognormal process, and the product of $\xi_2(t)$ and $\lambda_1(t)$ yields a Rice–lognormal process to describe the effect of a light shadowing ground environment on the signal, $\zeta(t)$ is the Rayleigh process, and the product of $\zeta(t)$ and $\lambda_2(t)$ is a Rayleigh–lognormal process to describe the effect of a heavy shadowing ground environment on the signal, $\sigma_{\mu 2}$ and $\sigma_{\mu 3}$ are the standard deviations of the lognormal distribution, $m_{\mu 2}$ and $m_{\mu 3}$ are the mean of the lognormal distribution. $|\cdot|$ represents the modulus of the complex stochastic signal and $\exp(\cdot)$ represents the exponential operation.

3.2.5.3 *Simulation verification*

In order to verify the validity of the satellite channel three-state Markov chain model, the simulation of the channel three-state Markov chain model in the ground mobile environment and the

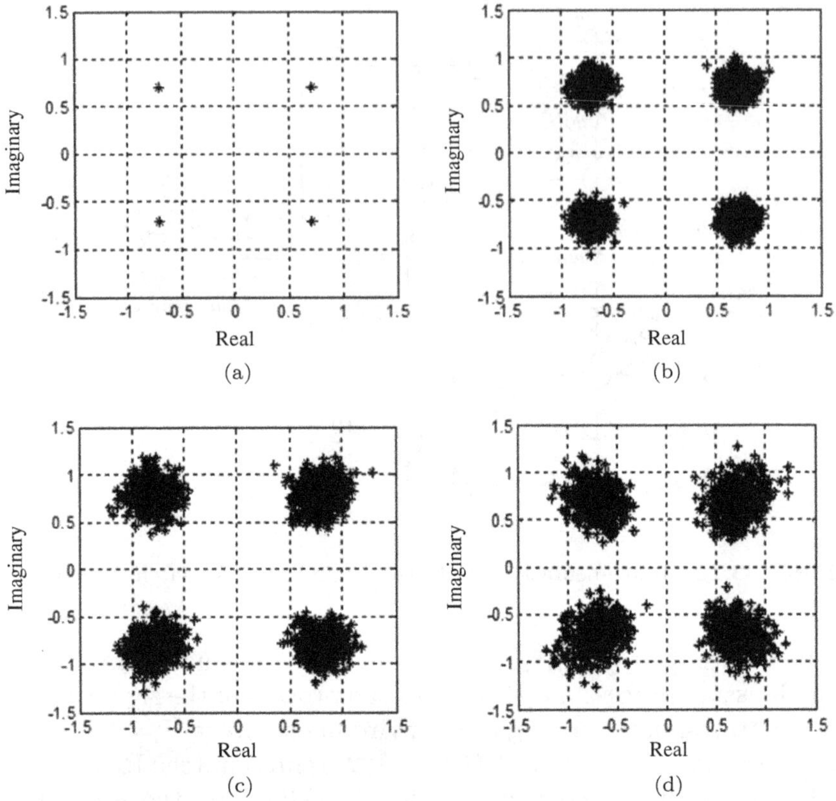

Figure 3.20 Signal constellations: (a) Input signal QPSK; (b) Output under shadowless state; (c) Output under light shadowing state and (d) Output under heavy shadowing state.

satellite channel Markov chain model considering the various weather conditions in the atmospheric propagation environment was carried out and analyzed.

Experiment 3.1: For the different channel models established by the influence of three typical ground environments on the channel characteristics, the parameters of the corresponding channel simulation model were selected, and the QPSK modulated signals were used as the input to the channels of three different shadowing levels to obtain the output signals, as shown in Figure 3.20.

It can be seen from Figure 3.20 that there is the attenuation or distortion of signal in different physical environments. In the shadowless

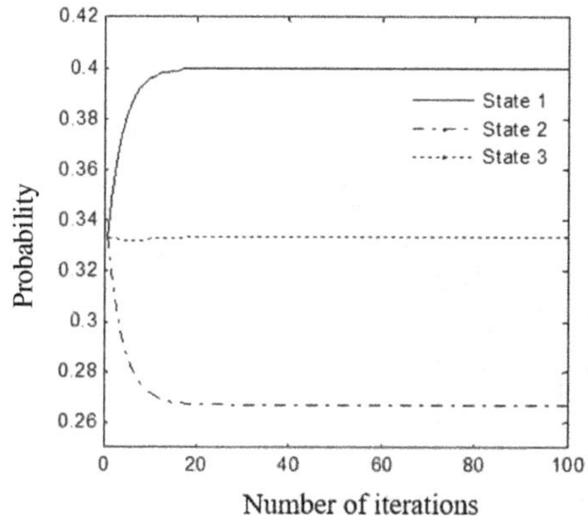

Figure 3.21 Convergence of state probability distribution to steady-state distribution.

rural environment, the constellation points of the received signals in the simulation model are relatively compact and the signal fading is small, in the suburban environment with light shadows, the constellation points of the received signals in the simulation model are more divergent, and the degree of signal fading increases to some extent, and under heavy shadowing urban environment, the constellation points of the received signals in the simulation model are heavily divergent and the signal attenuation is greater.

In designing simulation of the actual system, the transition between channel states should be determined by the specific actual environment, and the changing physical environment will have a great impact on state transition. Therefore, the parameters in [28] are selected here as simulation data. The state transition matrix is given by:

$$\boldsymbol{P} = \begin{bmatrix} 0.8628 & 0.0737 & 0.0635 \\ 0.1247 & 0.8214 & 0.0539 \\ 0.0648 & 0.0546 & 0.8806 \end{bmatrix}$$

Under the initial distribution of the equal probability, the Markov chain will gradually evolve into a steady state over time, and the distribution curves of the state probability converges to the steady-state distribution process, as shown in Figure 3.21. It is known from Figure 3.21 that the state probability matrix consisting of the proportion of the channel state durations of the three different degrees of the

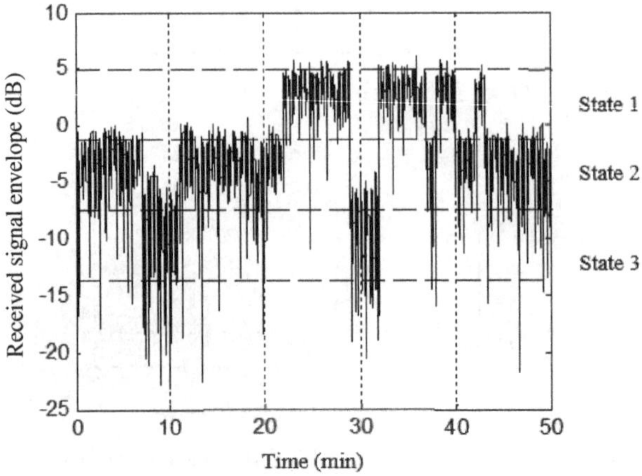

Figure 3.22 Received signal envelope.

shadowing is given by:

$$\boldsymbol{A} = [0.4 \quad 0.2667 \quad 0.3333]$$

As mobile receiving terminals continue to change, the channel transmission environment will change. Figure 3.22 shows the simulation results of the received signal envelope of the three-state Markov chain model in the ground mobile environment. Figure 3.22 shows that there are three distinct degrees of the received signal fading, which simulates the transmission characteristics of the different channels in a wide-area environment.

Figure 3.23 shows the probability density function (PDF) and the complementary cumulative distribution function (CCDF) for the three-state Markov chain model. It can be seen from Figure 3.23 that the built-in three-state Markov chain simulation model has a good agreement with the statistical characteristics of the theoretical model and the rationality of the channel three-state Markov chain model in the ground mobile environment is effectively verified.

Experiment 3.2: The built-in satellite channel Markov chain model is simulated according to the simulation method shown in Figure 3.19. In the simulation, the relevant parameters in the atmospheric environments were given in Table 3.4, along with the simulation parameters in the ground environment without shadowing, light shadowing state, and heavy shadowing state, as shown in Table 3.5.

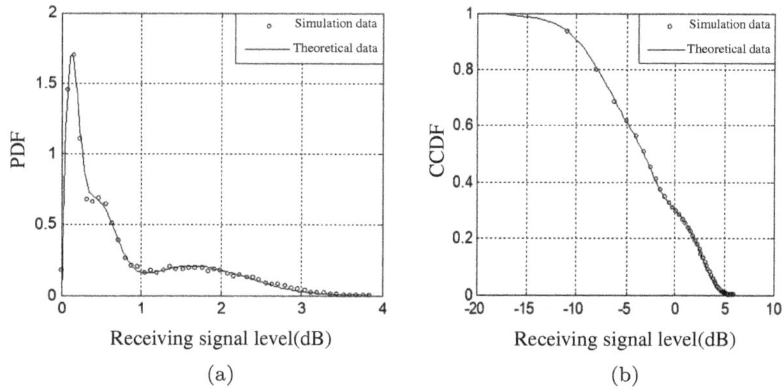

Figure 3.23 PDF and CCDF for ground mobile environment model (a) Probability density function (PDF); (b) Complementary cumulative distribution function (CCDF).

Table 3.5 Simulation parameters of ground mobile environment model.

Environment Conditions	z	σ	σ_μ	m_μ
No shadowing	1	0.1	–	–
Light shadowing	0.9383	0.3273	0.0205	0.1882
Heavy shadowing	—	0.1847	0.186	0.3515

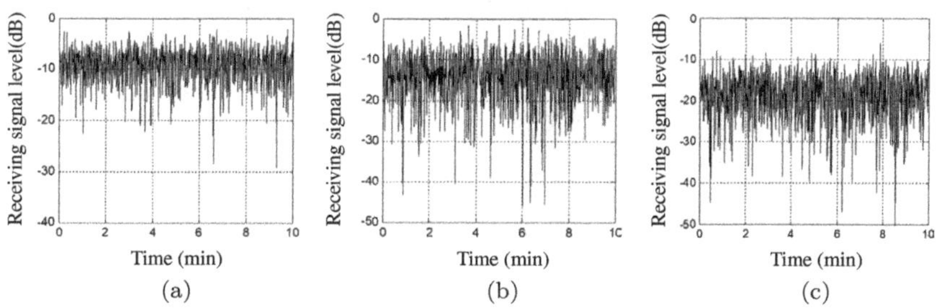

Figure 3.24 Received signal envelope under weather conditions: (a) Shadowless environment under cumulus weather; (b) Light shadowing environment under cumulus weather and (c) Heavy shadowing environment under cumulus weather.

Figure 3.24 shows the envelope of the received signal affected by different shadowing fading in the cumulus weather. Figure 3.24 shows that the signal attenuation in the same atmosphere condition is determined by the ground environments. As the influence of shadowing

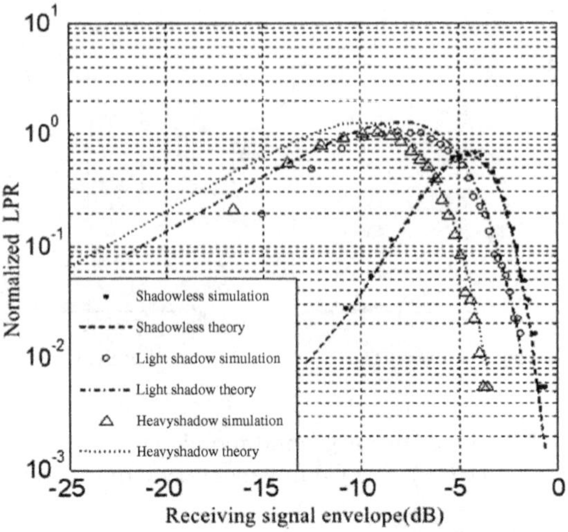

Figure 3.25 Normalized level pass rate.

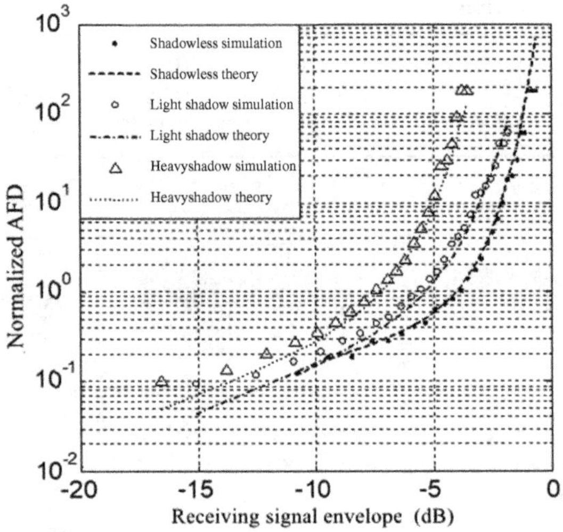

Figure 3.26 Normalized average fade duration.

fading increases, the signal attenuation becomes more and more serious.

Figures 3.25 and 3.26 show the statistical characteristics of the normalized level pass rate (LPR) and average fading duration (AFD) of the satellite channel Markov chain model affected by different

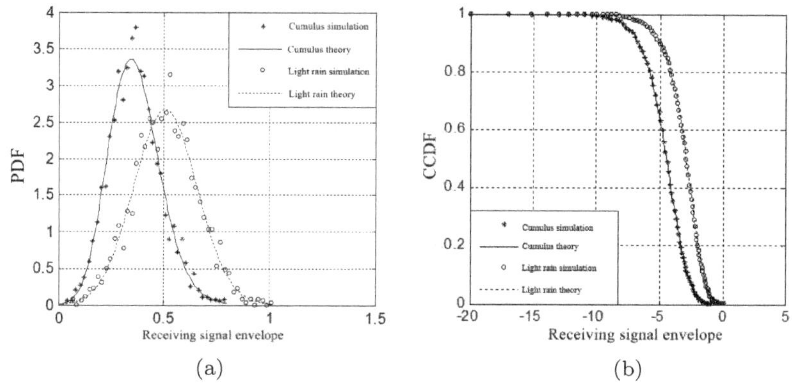

Figure 3.27 PDF and CCDF curves when affected by atmospheric environment without shadowing: (a) Probability density function (PDF); (b) Complementary cumulative distribution function (CCDF).

shadowing fading in the cumulus weather, which reflect the communication quality of the satellite channel. It can be seen from the simulation results that the larger the shadowing fading is, the more serious the signal attenuation is and the worse the satellite communication quality is.

The statistical characteristics of the channel affected by different atmospheric environments in three different shaded ground environments, that is, probability density function (PDF) and complementary cumulative distribution function (CCDF) are shown from Figure 3.27 to Figure 3.29. It has been seen from Figure 3.27 to Figure 3.29 that the simulation results are very similar to the theoretical analysis results, and in the same ground environment, the signal fading under cumulus weather is more serious than in the light rain weather.

Figures 3.30 and 3.31 show the statistical characteristics of the satellite channel Markov chain model and the satellite channel model without being affected by the atmospheric environment. It can be seen from Figures 3.30 and 3.31 that the simulation results of the two channel models are basically consistent with the theoretical model, and the signal attenuation ratio of the satellite channel Markov chain model is serious when the atmospheric environment and the ground environment are considered, that is, in case the atmospheric environment and the ground environment are also considered, the satellite channel Markov chain model is reasonable.

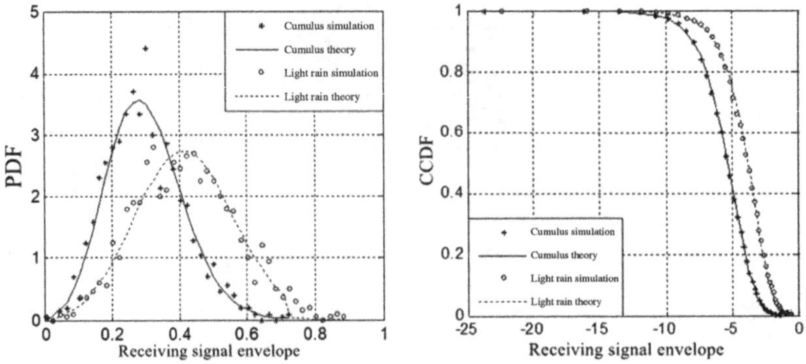

(a) (b)

Figure 3.28 PDF and CCDF curves under light shadowing affected by atmospheric environment: (a) Probability density function (PDF); (b) Complementary cumulative distribution function (CCDF).

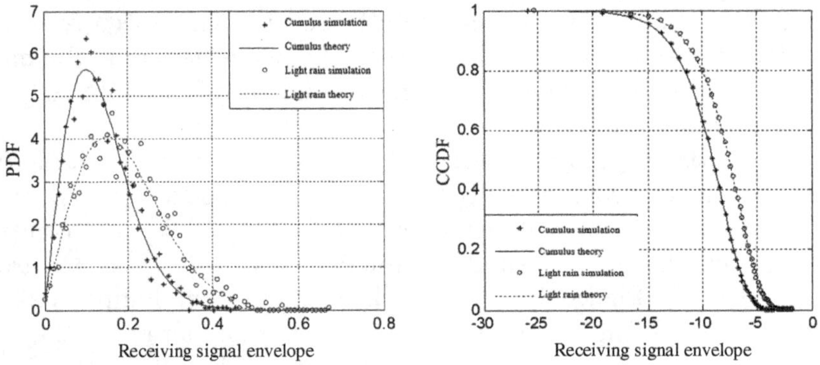

(a) (b)

Figure 3.29 PDF and CCDF curves under severe shadowing affected by atmospheric environment: (a) Probability density function (PDF); (b) Complementary cumulative distribution function (CCDF).

3.2.5.4 *Simulink implementation of satellite channel three-state Markov chain model*

In this section, the Simulink simulation of the satellite channel Markov chain model is carried out, and the satellite communication system simulation module is established to provide a better software platform for satellite communication system research.

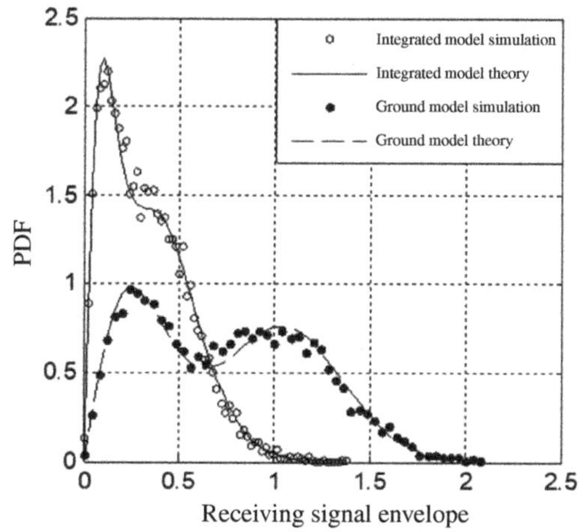

Figure 3.30 PDF of the received signal.

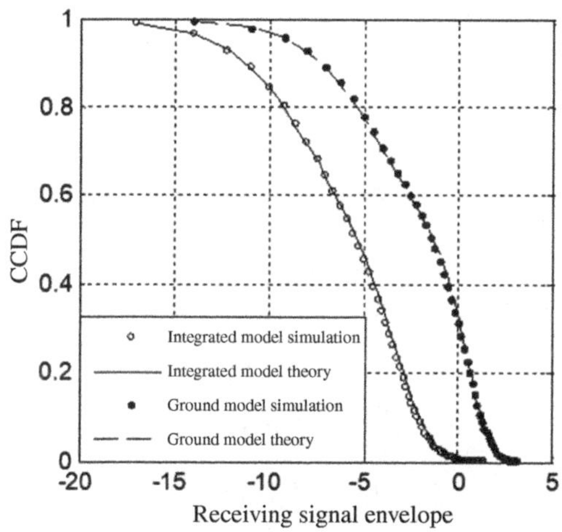

Figure 3.31 CCDF of the received signal.

3.2.5.4.1 Simulation module of probability distribution function

According to the previous analyses, the satellite channel Markov chain model can be established by appropriately combining different probability distribution functions based on specific influencing factors to simulate the channel transmission characteristics. The satellite channel Markov chain model mainly includes a Gaussian distribution describing the influence of the atmospheric environment, a

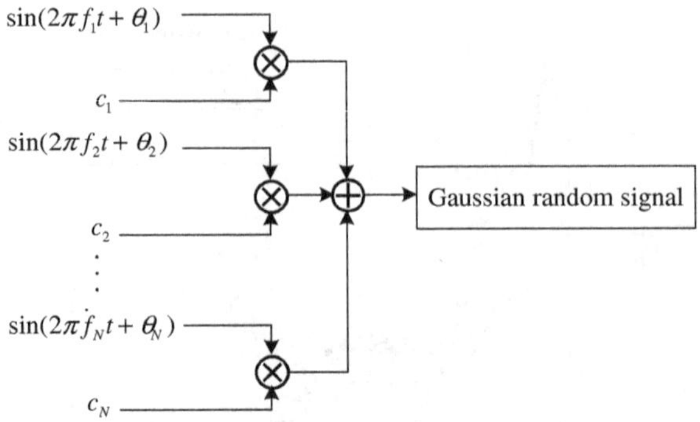

Figure 3.32 Gaussian distribution signal generated by sine wave superposition.

Rice (Rayleigh) distribution describing the multipath fading, and a lognormal distribution describing the shadowing fading. This section will implement the construction of the above various probability distribution function simulation modules on the Simulink platform [29], which are described as follows:

(1) *Simulation module of Gaussian distribution*: The simulation of the Gaussian distribution signals is realized by the idea of sine wave superposition. The Gaussian distribution signal can be generated by the sinusoidal function superposition, as shown in Figure 3.32. The model parameters are, respectively, determined by the following formulas.

$$c_n = \sigma\sqrt{\frac{2}{N}} \tag{3.2.21}$$

$$f_n = f_m \sin\left(\frac{n\pi}{2N}\right) \tag{3.2.22}$$

$$\theta_n = \frac{2\pi n}{N+1} \tag{3.2.23}$$

where N is the number of the used sinusoidal functions, $1 \leq n \leq N$. f_m is the maximum Doppler shift of the system, and σ represents the standard deviation of the Gaussian distribution signals.

Through multiple simulation comparisons, when $N = 8$, the signal envelope of the simulation model is very close to the signal envelope of the theoretical model, so the random signal that

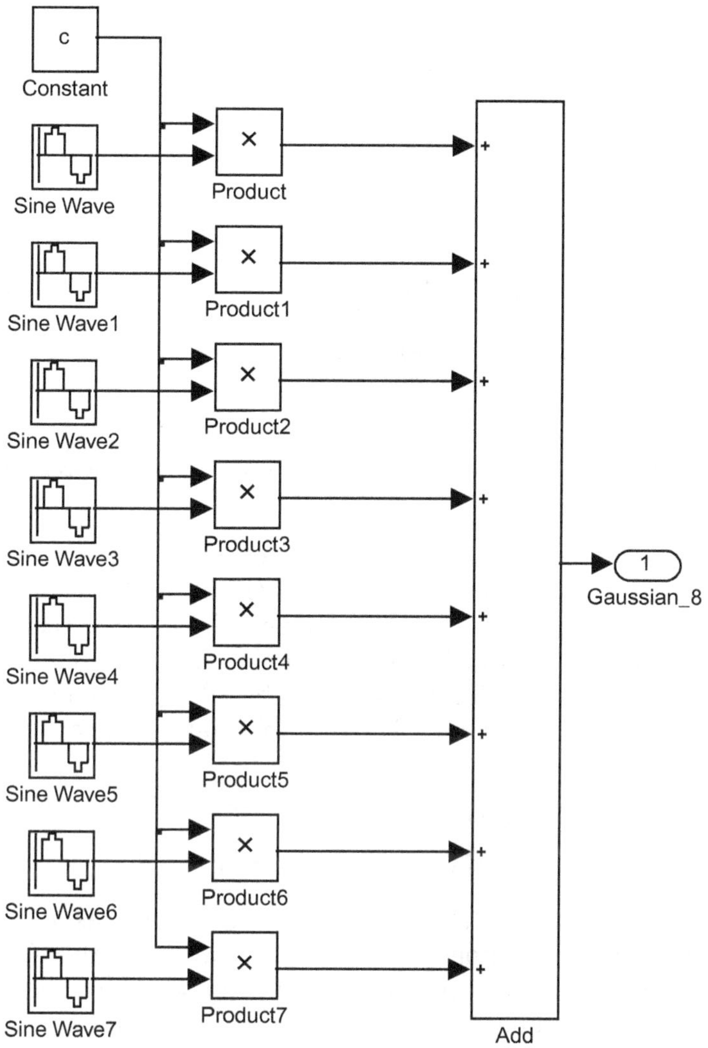

Figure 3.33 Gaussian distribution simulation module in Simulink platform.

obeys Gaussian distribution can be simulated through the superposition of eight sine waves [22]. In the Simulink platform, the random signal simulation module based on Gaussian distribution is implemented, as shown in Figure 3.33.

(2) *Simulation module of Rice distribution*: When the terminal receives the signal as a combination of the multipath components

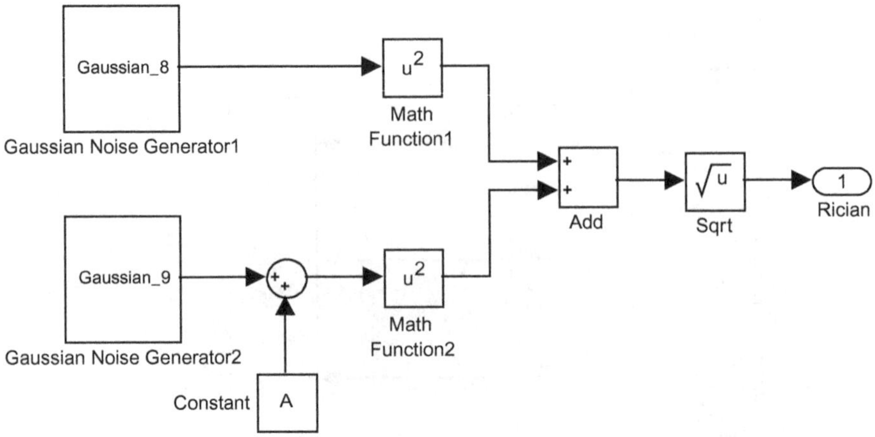

Figure 3.34 Rice (Rayleigh) distribution simulation module in Simulink environment.

and the direct view component, the envelope of the synthesized signal at this time obeys the Rice distribution. If the direct view component does not exist, then the envelope of the received signal is subject to the Rayleigh distribution. Both the Rayleigh process and the Rice process are based on two uncorrelated real Gaussian processes. The specific mathematical expression is written as:

$$\zeta(t) = |\mu_1(t) + j\mu_2(t)| \tag{3.2.24}$$

$$\xi(t) = |\mu_1(t) + j\mu_2(t) + m(t)| \tag{3.2.25}$$

where $\mu_1(t)$ and $\mu_2(t)$ are real Gaussian processes and $m(t)$ is a direct view component. In order to facilitate the simulation implementation, in the establishment of the Rice distribution simulation module, the direct view component $m(t)$ is regarded as a unchanged constant. Therefore, the implementation method of the Rice distribution simulation module in the Simulink platform is shown in Figure 3.34. When the constant module $\boldsymbol{A}$ is taken to zero, a simulation module of the Rayleigh distribution can be obtained.

The implementation of the Gaussian signal generation module in Figure 3.34 is shown in Figure 3.33. Among them, the Doppler parameters c_n, f_n, and θ_n are calculated by Equations (3.2.21) and (3.2.22).

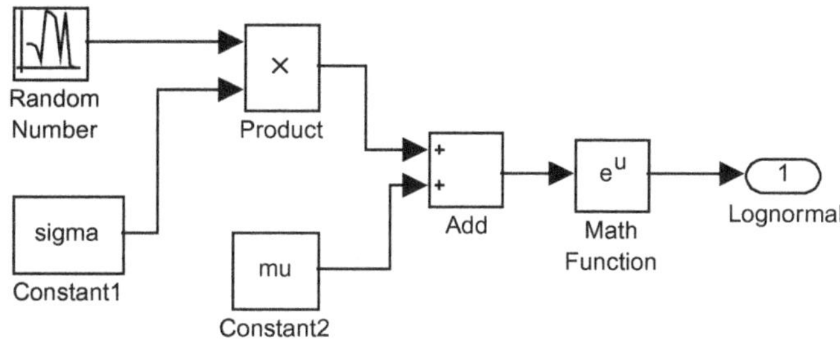

Figure 3.35 Lognormal distribution simulation module in Simulink environment.

(3) *Simulation module of lognormal distribution*: It is known from the relationship between the lognormal distribution and the Gaussian distribution that if the stochastic process x obeys the Gaussian distribution, $y = e^x$ obeys a lognormal distribution. Therefore, the specific method, generating the lognormal distribution process, is to first generate a Gaussian distribution process and then perform an exponential transformation to obtain the lognormal distribution process.

Therefore, in the Simulink environment, the simulation module of random signal obeying lognormal distribution is obtained by using Gaussian distribution, as shown in Figure 3.35. The Gaussian random signal generator x with the zero mean and unity variance in the Simulink module library is transferred to get the lognormal distribution simulation module with the specified mean value and variance through $\boldsymbol{\sigma}\nu(t) + m_u$ and exponential operation.

3.2.5.4.2 Satellite channel three-state Markov chain model

In this section, the Simulink implementation of the channel module in the atmospheric environments and the ground environments is carried out, and the simulation model of the satellite channel Markov chain model is also established based on the consideration of the atmospheric environments and the ground environments.

(1) *Simulation module of the channel model in the atmospheric environments*: The channel statistical properties affected by the atmospheric environments can be described by a Gaussian distribution. Therefore, the simulation module of the channel model under the atmospheric environments in the Simulink platform

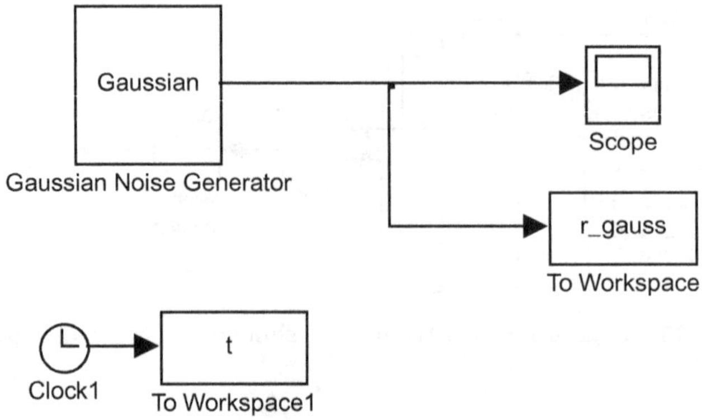

Figure 3.36 Simulation module of propagation model under atmospheric environment.

depicted in Figure 3.33 is shown in Figure 3.36. The channel characteristics simulation under different atmospheric conditions can be achieved by setting the parameters of the Gaussian noise generator module.

In Figure 3.36, set the simulation time to 10 s, double-click the "Gaussian Noise Generator module", set the sampling time $t_s = 0.01$, and the initial seed number is 61. The channel model parameters under the thunderstorm weather are selected as $m = 0.436$ and $\sigma^2 = 0.01386$ from Table 3.4. The Simulink simulation results can be observed from the oscilloscope module scope, as shown in Figure 3.37.

(2) *Simulation module of the channel model under the ground environment*: The channel statistical properties affected by the ground environment can be described by a combination of Rayleigh distribution, Rice distribution, and lognormal distribution. Therefore, the simulation module of the channel model under the ground propagation environment in Simulink platform can be established by using the simulation module of both the Rice or Rayleigh distribution and the lognormal distribution, constructed in Figures 3.34 and 3.35, as shown in Figure 3.38.

In Figure 3.38, the Rice distribution module is used to describe the statistical characteristics of the ground-free shadowing channel, the product of the Rice distribution and the lognormal distribution is used to describe the channel statistical characteristics

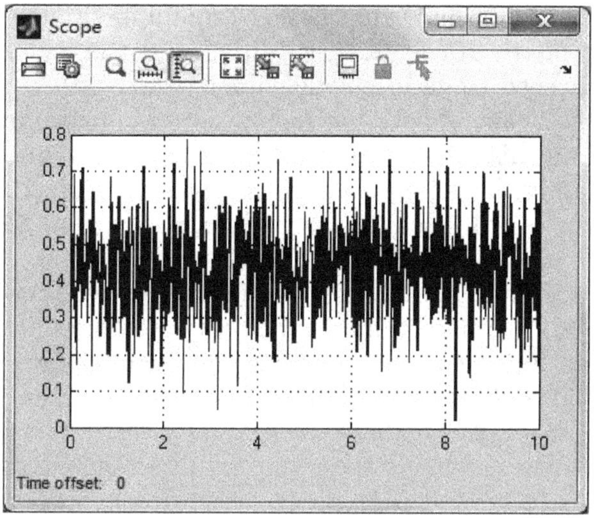

Figure 3.37 Received signal amplitude of Simulink simulation.

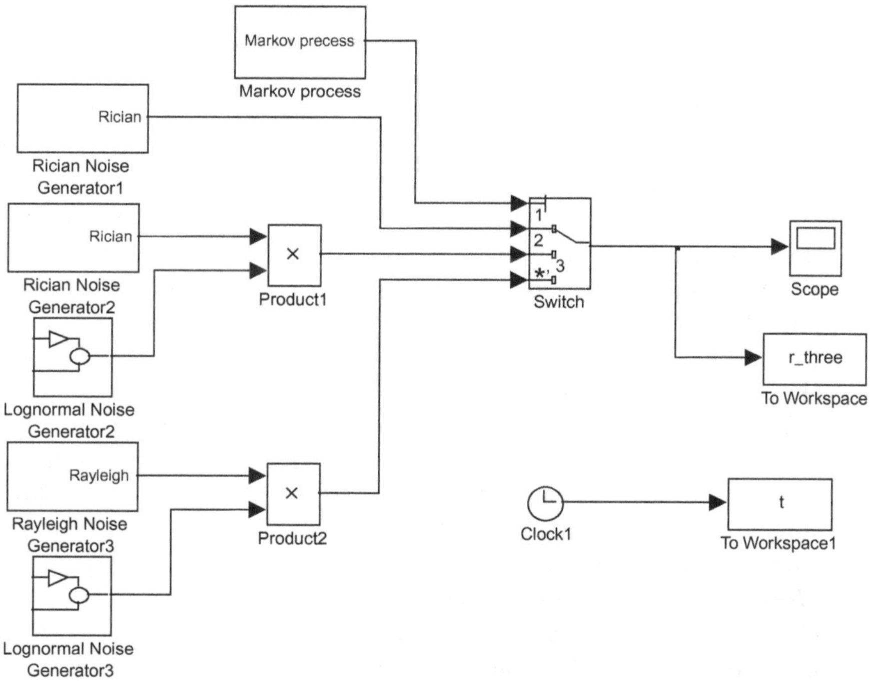

Figure 3.38 Channel model simulation module in ground environment.

under the light shadowing fading ground environment, and the product of the Rayleigh distribution and the lognormal distribution is used to describe the channel statistical characteristics under the heavy shadowing fading ground environment. As the signal transmission environment changes, the channel state transitions among unshadowing, light shadowing, and heavy shadowing states will occur. This transformation is a Markov chain process, as shown in Figure 3.39.

The random number generator in the Simulink module library is selected to generate three numbers that obeys uniform distribution in [0,1]. Switch1, switch2, and switch3 are the channel state hold and transfer switches in the three shadowing states, the threshold of the three switches is related to the transition matrix of the channel states, and the three switches switch independently. The real Markov transformation process is realized by switch0. According to the previous state, switch0 selects the output signal of the switch1, the switch2, or the switch3, and its output is used to control the channel state transition under no shadowing, light shadowing, and heavy shadowing states. If the output of switch0 is 1, the channel model without shadowing is selected. If the output of switch0 is 2, the channel model with light shadowing is selected. Otherwise, the channel model with heavy shadowing is selected.

In Figure 3.39, "1–1" indicates that the channel remains unchanged in the unshadowing state, "1–2" represents the transition from the unshadowing state to the light shadowing state, and "1–3" denotes the transition from the unshadowing state to the heavy shadowing state. The meaning of other modules is similar.

In Figure 3.39, set the simulation time to 10 s, sampling time $t_s = 0.01$, and maximum Doppler frequency $f_m = 91\,\text{Hz}$, and the initial seed number is 61. Double-click *Rician Noise Generator1 module*, set $\sigma^2 = 0.1, A = 1$. Double-click *Rician Noise Generator2 module*, set $\sigma = 0.3273$, $A = 0.9383$. Double-click *Lognormal Noise Generator2 module*, set $\sigma = 0.0205$, $m = 0.1882$. Double-click the *Rayleigh Noise Generator3 module,* set $\sigma = 0.1847$, and double-click *the Lognormal Noise Generator3 module*, set $\sigma = 0.186$, $m = 0.3515$. Double-click *the Markov process module* to set the switch threshold. The Simulink simulation results

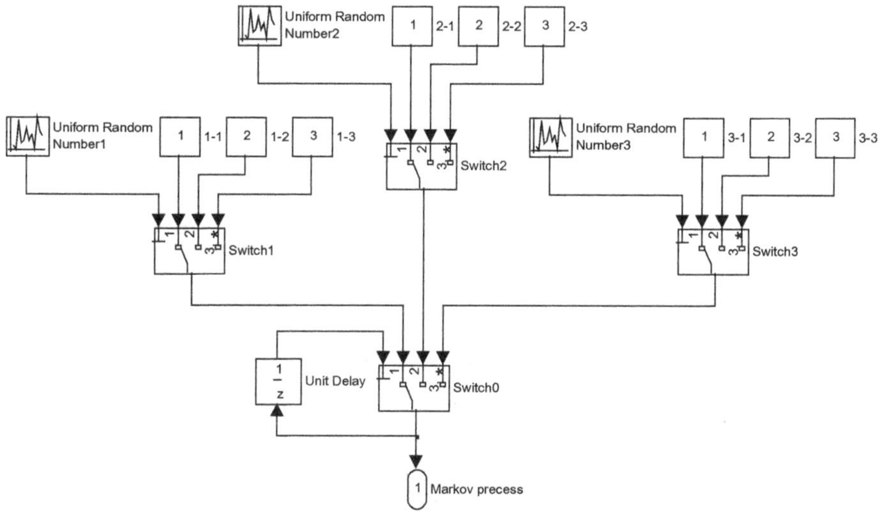

Figure 3.39 Simulation module of Markov chain process.

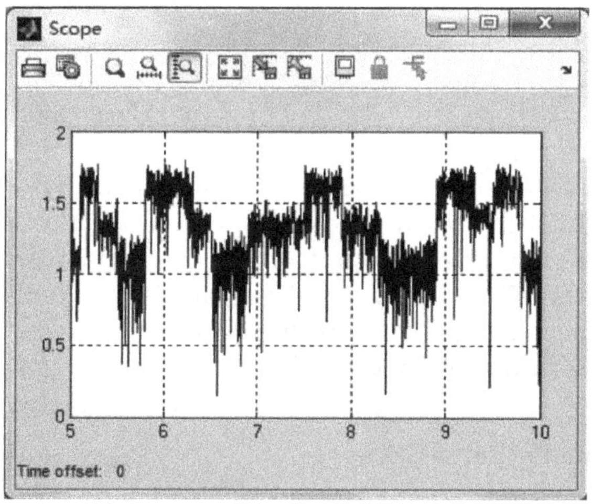

Figure 3.40 Received signal amplitude of Simulink simulation.

can be observed from the oscilloscope module scope, as shown in Figure 3.40. It has been seen from Figure 3.40 that there are three different degrees of fading in the received signal, corresponding to three different shadowing states in the ground environments.

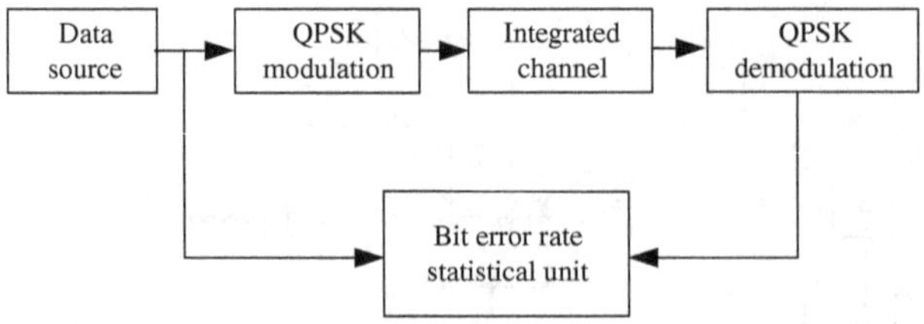

Figure 3.41 Block diagram of the satellite communication system.

(3) *Satellite channel Markov chain model simulation system*: In order to verify the correctness and reasonableness of the satellite channel Markov model considering both the atmospheric environments and the ground environments, a simulation model of the satellite communication system based on the channel model is established, and the bit error rate (BER) characteristics of the output signal are analyzed. The overall block diagram of the satellite communication system is shown in Figure 3.41.

As shown in Figure 3.41, the data sources of the transmitting end are generated by the *Random Integer Generator module*, and the signals are modulated by the QPSK modulator and entered into the channel model for comprehensive description. Here, it is assumed that the channel fading is slow and there is no other fading effect. Finally, the fading signal output of the channel is demodulated by QPSK and then output to the BER statistical unit, which analyzes the generated bit error results. The specific implementation of Simulink is shown in Figure 3.42.

3.2.5.4.3 Simulation verification

In the simulation of the satellite communication system, the QPSK signal is affected by both channel multiplicative interference and additive white Gaussian noise. The multiplicative interference is generated by the established channel model, and the additive white Gaussian noise is generated by the AWGN module. After the output of the channel is demodulated by the QPSK demodulator, the output signal is obtained and input to the BER calculation module to analyze the BER characteristics.

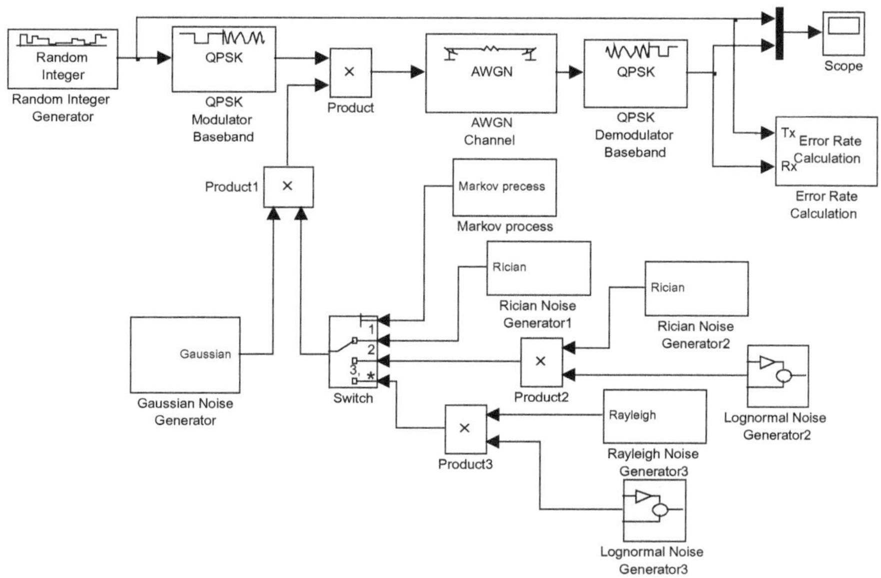

Figure 3.42 Simulink implementation of satellite communication system.

In Figure 3.42, the data source module, QPSK modulation or demodulation module, AWGN module, and BER calculation module can all be called from the Simulink model library.

The simulation time is set to 10 s, the sample value of the signal generator and AWGN module is set to 0.001, the relevant parameter settings of the integrated channel are shown in Tables 3.4 and 3.5, and the Bit Error Rate Analysis Toolbox (BERTool) in Simulink is called to analyze satellite communication system simulation result. The specific simulation results are shown in Figure 3.43.

Figure 3.43(a) shows the BER curves of the channel model simulation system under four typical weather conditions: cloudy, cumulus, thunderstorm, and light rain when the signal is transmitted without considering the ground environment. Compared with the AWGN channel, the signal affected by four typical weather conditions will be fading, and the signal attenuation is minimal in light rain conditions, and the signal attenuation is much heavier in thunderstorms and cumulus weathers, when the signal-to-noise ratio (SNR) of the simulation system is greater than 15 dB, and also the signal attenuations in the thunderstorm weather are greater than the cumulus

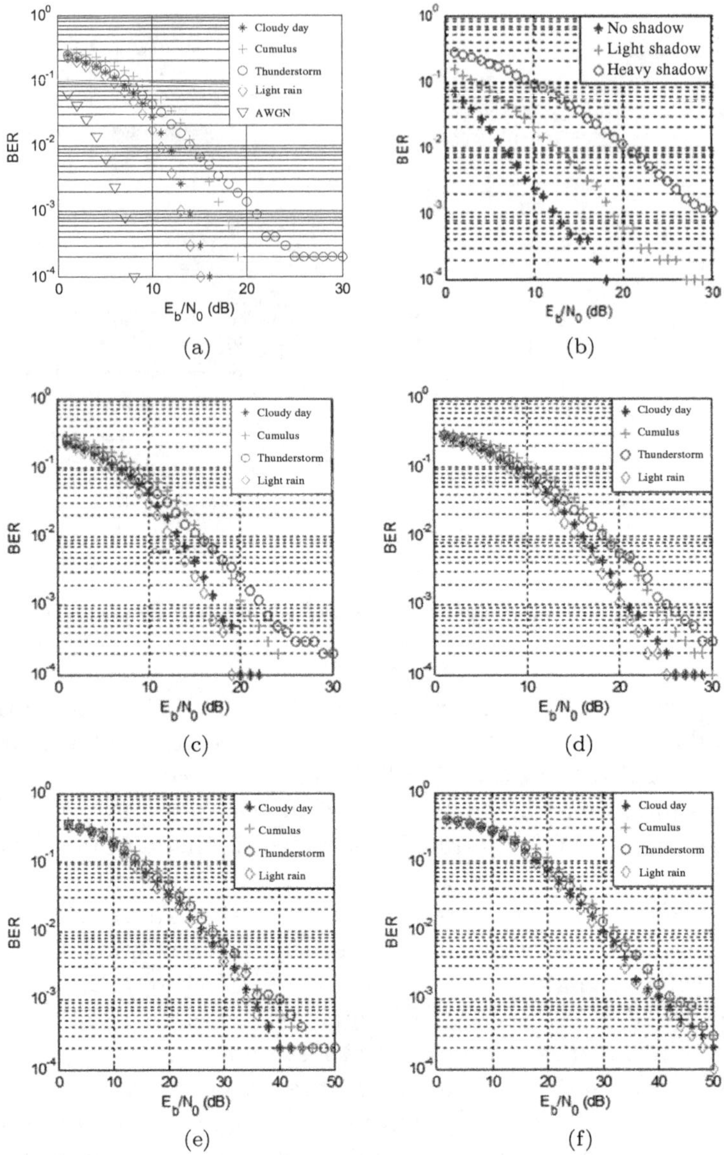

Figure 3.43 Simulation of BER performance of satellite communication system (a) BER only affected by the atmospheric environment; (b) BER only affected by the ground environment; (c) BER affected by atmospheric environment without shadowing; (d) BER affected by atmospheric environment in light shadows; (e) BER affected by atmospheric environment in severe shadows and (f) BER affected by the integrated environmental impact.

weather. Therefore, rainfall attenuation should be taken into account when studying the transmission characteristics of satellite channel under high-frequency band.

Figure 3.43(b) shows the BER curves of the channel model simulation system under the ground environments when the signal is not affected by weather conditions. From the BER curves, as the signal is affected by the shadowing fading, the BER is gradually increased, and the signal attenuation is more and more serious.

Figure 3.43(c) is the BER curves of the channel model simulation system under four typical weather conditions: cloudy, cumulus, thunderstorm, and light rain when the signal is transmitted under without shadowing ground environment, the signal attenuation is still minimal in light rain, but more serious under thunderstorms and cumulus weather. The larger the SNR is, the more serious the thunderstorm weather signal attenuation is, at this time, the BER curves tend to be horizontal and straight, which indicates that the system's BER performance can no longer be significantly improved.

Figures 3.43(d) and 3.43(e) are the BER curves of the channel model simulation system under four typical weather conditions: cloudy, cumulus, thunderstorm, and light rain when the signal is transmitted in the light shadowing and heavy shadowing ground environment. From the BER curves, the signal fading law is consistent with the unshadowing environment under four typical weather conditions, but the signal attenuation is significantly increased compared with the unshadowing environment.

Figure 3.43(f) is the BER curves of the channel model simulation system under four typical weather conditions: cloudy, cumulus, thunderstorm, and light rain when the signal is transmitted in the terrestrial wide-area environment. From the approximation trend of the BER curves under four typical weather conditions, the influence of the atmospheric environments on signal attenuation is relatively small but cannot be neglected compared with the signal attenuation effect of the ground environments.

Based on the analyses of the BER characteristics of the satellite communication simulation system, when considering both the atmospheric environments and the ground environments, the established Markov chain model of the satellite channel is reasonable and feasible and can be used as the theoretical basis for the design and improvement of the satellite communication system.

3.3 Satellite Channel Five-state Markov Chain Model

In this section, based on the three states of light shadowing, moderate shadowing, and heavy shadowing ground moving environment, the satellite channel five-state Markov chain model is studied. There are "good state", "not bad state", and "bad state", where "good state" and "bad state" have also two sub-states, and the transition probabilities are also derived. This model is easier to be simulated and can be easily applied to complex satellite links [30–35].

3.3.1 *Five-state Markov chain model*

3.3.1.1 *Transfer model*

Figure 3.44 shows a channel five-state Markov chain model. In the Gilbert–Elliot channel Markov model and the three-state Markov chain model, there are three shadowing areas: light shadowing areas (state 1), moderate shadowing areas (state 2), and heavy shadowing areas (state 3). The light shadowing areas (state 1) have two

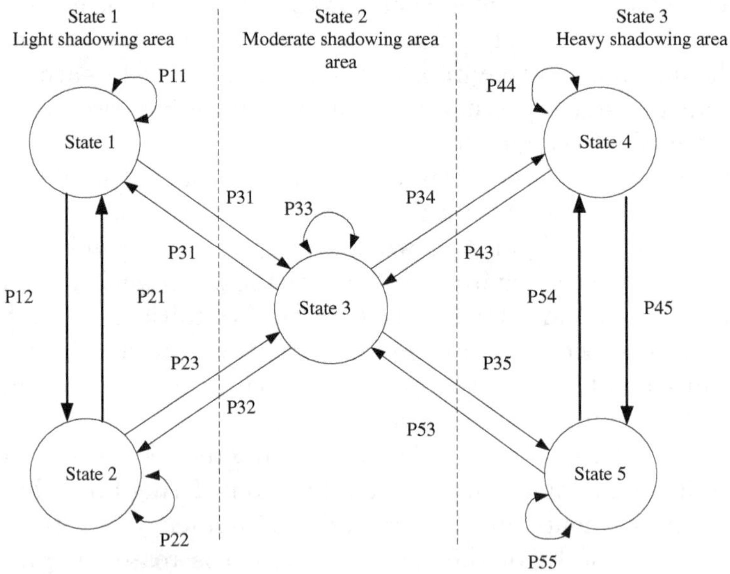

Figure 3.44 Transition of five-state Markov model.

sub-states (sub-states 1, 2), and the heavy shadowing areas (state 3) also have two sub-states (sub-states 4, 5). The light shadowing areas can convert to moderate shadowing area, but not directly to light shadowing area. The heavy shadowing areas can convert to moderate shadowing areas, but not directly to heavy shadowing areas. When the received signal is in the light shadowing area or heavy shadowing area for a period of time, the state can transfer between the two sub-states.

The Markov chain part of the model defines three states for each tape of environment plus elevation angle combination representing the following link conditions:

Light shadowing areas:
State 1: LOS conditions
State 2: Light shadowing conditions

Moderate shadowing areas:
State 3: Moderate shadowing conditions

Heavy shadowing areas:
State 4: Heavy shadowing conditions
State 5: Full shadowing conditions.

At a given time or path location, the Markov chain model can use two matrices to describe the LMS channel:

(1) The state probability matrix is denoted as $\boldsymbol{A}$.
(2) The state transition probability matrix is denoted as $\boldsymbol{P}$.

In the channel five-state Markov chain model, there are five states, and the transition probability matrix is given by:

$$\boldsymbol{P} = \begin{bmatrix} p_{11} & p_{12} & p_{13} & p_{14} & p_{15} \\ p_{21} & p_{22} & p_{23} & p_{24} & p_{25} \\ p_{31} & p_{32} & p_{33} & p_{34} & p_{35} \\ p_{41} & p_{42} & p_{43} & p_{44} & p_{45} \\ p_{51} & p_{52} & p_{53} & p_{54} & p_{55} \end{bmatrix}$$

In Figure 3.44, it is impossible for the state 1 or the state 2 to transfer to the state 4 or the state 5 directly, and the Markov chain has no absorption state, so the transition matrix can be simplified as:

$$\boldsymbol{P} = \begin{bmatrix} p_{11} & p_{12} & p_{13} & 0 & 0 \\ p_{21} & p_{22} & p_{23} & 0 & 0 \\ p_{31} & p_{32} & p_{33} & p_{34} & p_{35} \\ 0 & 0 & p_{43} & p_{44} & p_{45} \\ 0 & 0 & p_{53} & p_{54} & p_{55} \end{bmatrix}$$

where p_{ij} represents the transition probability from the state i to the state j, and $i, j = 1, 2, 3, 4, 5$, and further $\sum_{j=1}^{6} p_{ij} = 1, 0 \leq p_{ij} \leq 1$.

The state probability matrix $\boldsymbol{A}$ indicates the probabilities of all states and is written as $\boldsymbol{A} = \{A_1, A_2, A_3, A_4, A_5\}$. This Markov chain is aperiodic, irreducible, and normal regressive, so its steady-state probability distribution exists and is equal to the limit distribution. Then,

$$\boldsymbol{AP} = \boldsymbol{A}$$

So, if the state transition probability matrix $\boldsymbol{P}$ is known, a state probability matrix $\boldsymbol{A}$ can be obtained. Matrix $\boldsymbol{P}$ and matrix $\boldsymbol{A}$ are related to many factors such as incidence angle, environments, and other factors.

3.3.1.2 *Shadowing fading model*

According to the Lutz model and the satellite channel three-state Markov chain model, the satellite channels are divided into three states: "good state", "not good and not bad state", and "bad state". A "good state" is a clear LOS link without shadowing or light shadowing, which is subject to the Rician distribution. The " not good and not bad state" is affected by moderate shadowing and belongs to the Loo distribution. The "bad state" is affected by heavy shadowing or full shadowing and belongs to the Rayleigh–lognormal distribution.

The probability density function (PDF) of the Rayleigh–lognormal distribution is written as:

$$f_{\text{Rayl-L}}(s) = \frac{10}{\delta \ln 10\sqrt{2\pi}} \int_0^{\infty} \frac{e^{-x^2}}{S_0^2} e^{-\frac{s}{S_0}} dS_0 \tag{3.3.1}$$

where $x = \frac{10\log S_0 - m}{\sqrt{2}\delta}$, m $= E\{10\log(S_0)\}$, $\delta = \sqrt{E\{10\log(S_0)\} - m^2}$.

The PDF of the Loo distribution can be expressed as:

$$f_{\text{Loo}}(S) = \frac{10}{\delta\sigma^2 \ln 10\sqrt{2\pi}} \cdot \int_0^{\infty} e^{\left[\frac{(s+S_0)}{\sigma^2} + \frac{(10\log_2 S_0 - m)^2}{2\delta^2}\right]} \times I_0\left(\frac{2\sqrt{S_0 s}}{\sigma^2}\right) dS_0 \tag{3.3.2}$$

where σ^2 denotes the average received power of multipath.

The PDF of Rice distribution is written as:

$$f_{\text{Rice}}(S) = (1+K)e^{[K+(1+K)s]} I_0(2\sqrt{K(1+K)s}) \tag{3.3.3}$$

where K is a Rice factor and defined as the ratio of the power in the LOS path to the power in the diffuse paths.

The PDF of the received power in satellite channel five-state Markov chain model is the weighted of the PDF $f_{\text{Rice}}(S)$ of the Rice distribution, the PDF $f_{\text{Loo}}(S)$ of the Loo distribution, and the PDF $f_{\text{Rayl-L}}(S)$ of the Rayleigh–lognormal distribution.

$$f(S) = A_1 f_{\text{Rice1}}(S) + A_2 f_{\text{Rice2}}(S) + A_3 f_{\text{Loo}}(S) + A_4 f_{\text{Rayl-L1}}(S) + A_{5\ \text{Rayl-L2}}(S)$$

where $A_1 + A_2 + A_3 + A_4 + A_5 = 1$, A_1, A_2, A_3, A_4, and A_5 are the percentages corresponding to state 1, state 2, state 3, state 4, and state 5, respectively.

3.3.1.3 *State division*

The measured data used in this section was measured in a narrow street in the old town of Munich [36].

In Figure 3.45, the state samples are distinguished by dividing the received signal envelope. The received signals from −10 dB to −20 dB indicate state 3. The received signals from −10 dB to maximum are divided into the state 1 and the state 2 uniformly. The received signals from −20 dB to minimum are divided into the state 4 and the state 5 uniformly.

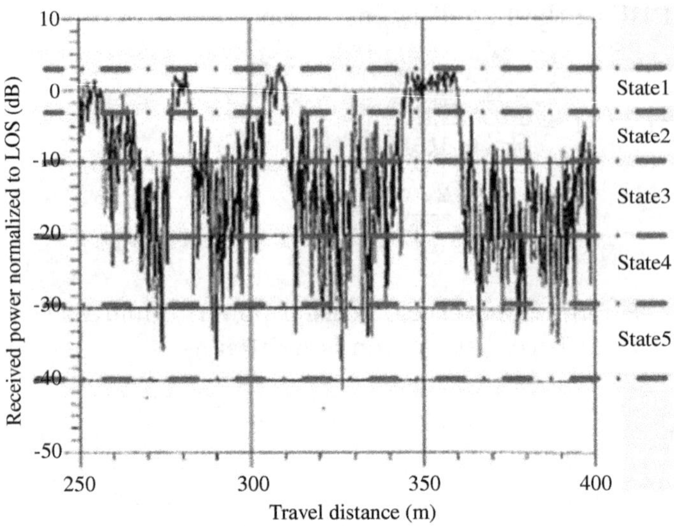

Figure 3.45 State dividing.

The state series are:

$$\mathbf{S} = [2\,1\,1\,1\,2\,3\,3\,2\,2\,3\,4\,4\,5\,3\,1\,1\,3\,2\,4\,5\,3\,4\,3\,3\,2\,1\,1\,3\,4\,3\,2\,3\,3\,4\,4\,5\,4\,3\,2\,3\,4$$
$$5\,4\,3\,2\,3\,3\,4\,3\,2\,1\,1\,1\,1\,1\,1\,1\,1\,1\,1\,3\,3\,4\,5\,4\,3\,3\,4\,3\,3\,4\,3\,4\,3\,3\,4\,5\,4\,3\,3\,2\,3].$$

From state series S, the transition probability matrix can be obtained as:

$$\boldsymbol{P} = \begin{bmatrix} 0.7647 & 0.0588 & 0.1765 & 0 & 0 \\ 0.2727 & 0.0909 & 0.6364 & 0 & 0 \\ 0.0345 & 0.2759 & 0.3103 & 0.3793 & 0 \\ 0 & 0 & 0.5556 & 0.1111 & 0.3333 \\ 0 & 0 & 0.3333 & 0.6667 & 0 \end{bmatrix}$$

The cumulative distribution matrix is given by:

$$\boldsymbol{P}_1 = \begin{bmatrix} 0.7647 & 0.8235 & 1.0000 & 0 & 0 \\ 0.2727 & 0.3636 & 1.0000 & 0 & 0 \\ 0.0345 & 0.3103 & 0.6207 & 1.0000 & 1.0000 \\ 0 & 0 & 0.5556 & 0.6667 & 1.0000 \\ 0 & 0 & 0.3333 & 1.0000 & 1.0000 \end{bmatrix}$$

Table 3.6 Percentage of different substates.

State	A_1	A_2	A_3	A_4	A_5
Percentage	23.55	11.95	36.68	20.87	6.95

Table 3.7 Parameters corresponding to the five-state model.

State								
1	**2**	**3**			**4**		**5**	
K	K	K	m	δ	m	δ	m	δ
13	10	4	7.5	1.5	12	3	14	5

3.3.2 *Simulation tests*

The simulation of the satellite channel five-state Markov chain model is compared with the measured data shown in Figure 3.45 and the simulation of the satellite channel six-state Markov chain model. Matrix $\boldsymbol{A}$ can be obtained from the state transition probability matrix $\boldsymbol{P}$, which is shown in Table 3.6. According to the measured data of each state, the parameters of the satellite channel five-state Markov chain model are obtained by statistical fitting method, as shown in Table 3.7. The five-state Markov chain model and the CDF of the six-state Markov chain model are shown in Figure 3.46. Figure 3.47 shows the normalized level crossover rate. Figure 3.48 showed the normalized average fading duration (AFD). The BER of the new five-state Markov model and the five-state Markov model is simulated by the method in [37]. The average BER is shown in Figure 3.49.

In comparison, the simulated CDF, LCR, and AFD based on the five-state are more consistent with the measurement ones than ones based on the six-state model. Both the five-state Markov chain model and the six-state Markov chain model can accurately represent the actual channel, but the five-state Markov chain model has smaller average BER, which is more suitable for cities and suburbs. So it has practical significance.

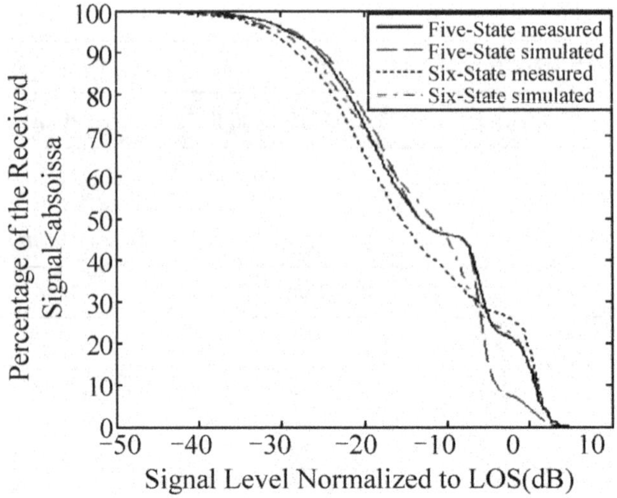

Figure 3.46 Comparison of simulation and measured data CDF.

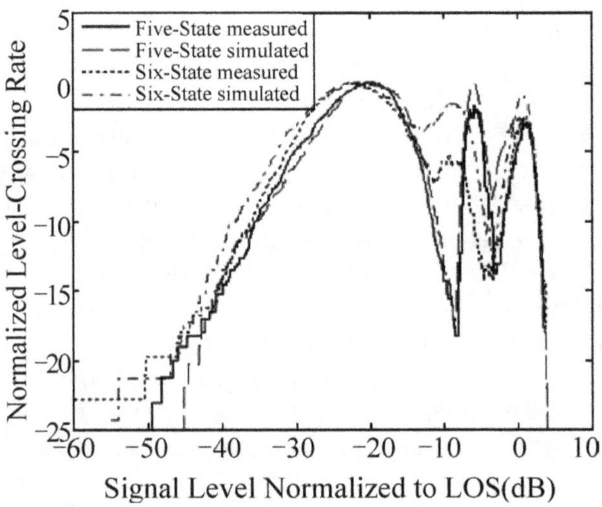

Figure 3.47 Comparison of simulation and measured data LCR.

3.4 Interrupt Probability of Six-state Markov Chain Model for Satellite Channel

In this section, theoretical analysis and simulation of the outage probability for six-state Markov model on Land Mobile Satellite (LMS) channels was given [38].

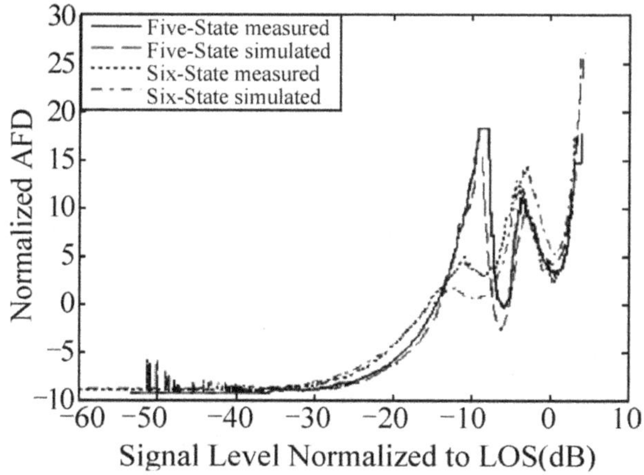

Figure 3.48 Comparison of simulation and measured data AFD.

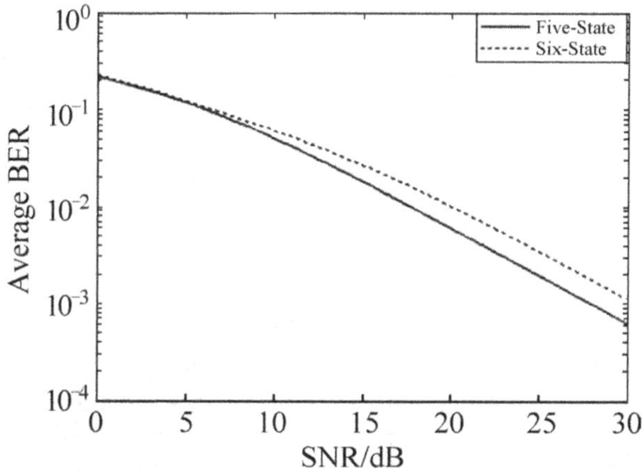

Figure 3.49 Comparison of average BER between five-state and six-state models.

These six states are divided into the "good" or "bad" two-state model, whereas the "good" or "bad" state is also divided into three sub-states. The fading in the "good" state is relatively benign and is represented by the Rician distribution. The fading in the "bad" state is modeled by the Rayleigh–lognormal distribution and called as the Suzuki distribution. However, in these studies, the outage probability is not analyzed, and its effective analysis method is not given. An analysis method of the outage probability and the CDF of the Rician distribution was given in [39]. The described method was used to analyze the outage probability and CDF of the channel six-state Markov chain model of the satellite, and the closed-loop formulas for CDF and outage probability [40] were found finally.

3.4.1 *Analysis of satellite channel six-state Markov chain model*

Assuming slow fading, the received signal of the nth transmitted symbol is written as:

$$y(n) = s(n)x(n) + w(n) \tag{3.4.1}$$

where $x(n)$ is the first received signal and $n = 1,2,\ldots$, $w(n)$ is additive white Gaussian noise, and $s(n)$ is channel gain. The received power in the attenuation envelope is given by $P = |s|^2 = s_I^2 + s_Q^2$. In a non-attenuating environment, the signal-to-noise ratio (SNR) γ is defined as:

$$\gamma = E[|x(n)|^2]/N_0 = E_s/N_0 \tag{3.4.2}$$

It can be seen from Equation (3.4.2) that when attenuation occurs, the signal-to-noise ratio (SNR) is γS. When the channel states are in the substate 1, 2, 3 of the “good” state, the envelope of the attenuation belongs to the Rice distribution, and the fractional times are A_1, A_2, and A_3, respectively. The Rice factors are K_1, K_2, and K_3, respectively. When the channel states are in the substate 1, 2, 3 of the “bad” state (named them as “state 4, 5, 6”), the envelope of the attenuation belongs to the Rayleigh–lognormal distribution, and the mean and standard deviation are m_2, σ_1, m_2, σ_2, m_3, and σ_3, respectively. The probability density functions (PDFs) of the Rice distribution and the Rayleigh–lognormal distribution are denoted as $f_{\text{rice}}(s)$ and $f_{\text{Ray}_\text{L}}(s)$, respectively. So, the joint probability density function of energy (S) is defined as:

$$\begin{aligned} f(s) = {} & A_1 f_{\text{Rice1}}(s) + A_2 f_{\text{Rice2}}(s) + A_3 f_{\text{Rice3}}(s) + A_4 f_{\text{Ray\_L1}}(s) \\ & + A_5 f_{\text{Ray\_L2}}(s) + A_6 f_{\text{Ray\_L3}}(s) \end{aligned} \tag{3.4.3}$$

where $A_1 + A_2 + A_3 + A_4 + A_5 + A_6 = 1$. So, the CDF of S is defined as:

$$F(S) = A_1 F_{\text{Rice1}}(S) + A_2 F_{\text{Rice2}}(S) + A_3 F_{\text{Rice3}}(S) + A_4 F_{\text{Ray\_L1}}(S) + A_5 F_{\text{Ray\_L2}}(S) + A_6 F_{\text{Ray\_L3}}(S) \tag{3.4.4}$$

where F_{Rice1} and $F_{\text{Ray\_L}}$ represent the CDF of the Rice distribution and the Rayleigh–lognormal distribution.

3.4.1.1 *Several distributions*

3.4.1.1.1 Rice distribution

Due to the fading characteristics of the channel, the diffusers and reflectors on the receiver and transmitter can cause LOS path and multipath components. The Rice factor K is the signal power ratio of the LOS component to the multipath component. The amplitude of the received signal on the Rice channel follows the Rice distribution, and its PDF is known [36]. Then, the PDF of the received signal power is calculated by:

$$f_{\text{Rice}}(s) = (1 + K)e^{-[K+(1+K)s]} I_0(2\sqrt{K(1 + K)s}) \tag{3.4.5}$$

Here, it is assumed that the average received power is 1. The CDF of the Rice channel is written as:

$$F_{\text{Rice}}(S) = 1 - Q(\sqrt{2K}, \sqrt{2(1 + K)S}) \tag{3.4.6}$$

where $Q(\sqrt{2K}, \sqrt{2(1 + K)S})$ is the Marcum's Q function in [41]. When K is zero, the standardized PDF and CDF of the Rayleigh distribution are obtained by Equations (3.4.5) and (3.4.6). Equation (3.4.5) can be rewritten as:

$$F_{\text{Rice}}(S) = 1 - e^{-2[K+(1+K)S]} \sum_{k=0}^{\infty} \left[\frac{K}{S + KS}\right]^{\frac{k}{2}} I_k(2\sqrt{K(1 + K)S}) \tag{3.4.7}$$

where I_k is a K-order Bessel function and K is a Rice factor.

3.4.1.1.2 Rayleigh–lognormal distribution

The Rayleigh–lognormal distribution characterizes the combined effects of the Rayleigh fading and lognormal shadowing and establishes a shadowing multipath channel without LOS path. The received power is subject to a Rayleigh distribution with a local average received power of S_0, which obeys the lognormal distribution. Therefore, the PDF of the fading received power in Rayleigh–lognormal is defined as:

$$f_{\text{Ray\_L}}(s) = \int_0^{\infty} f_{\text{Rayl}}(s/S_0) f_{\text{Log}}(S_0) dS_0 \tag{3.4.8}$$

$$f_{\text{Log}}(s_0) = \frac{10}{\ln 10} \frac{e^{-x^2}}{\sqrt{2\pi}\delta} \tag{3.4.9}$$

where $x = \frac{10 \log s_0 - m}{\sqrt{2}\delta}$.

From the literature [36], the PDF of the Rayleigh fading can be obtained. From Equations (3.4.8) and (3.4.9), the Rayleigh–lognormal distribution function can be written by Hermite integration as:

$$F_{\text{Ray\_L}}(S) = \frac{1}{\sqrt{\pi}} \sum_{i=1}^{n} w_i (1 - e^{-\frac{S}{\partial_i}}) + R_n \tag{3.4.10}$$

where $\partial_i = 10^{\frac{\sqrt{2}\delta x_i + \mu}{10}}$. x_i and w_i are the abscissa and weight of the ith term of the nth-order Hermite polynomial, respectively. R_n is a residual term that can be ignored when $\delta < 7\,\text{dB}$ (the proof process is in [41]).

3.4.1.2 *Maximum ratio combined diversity reception*

Of any known linear diversity combiner, the maximum ratio combining (MRC) has the best statistical attenuation of multipath fading. Each signal is weighted proportionally to their SNR and then summed. For L-branch diversity, the total signal power S is the sum of the received signal powers $s_i (i = 1, \ldots, L)$ from each antenna [42].

3.4.1.2.1 Rician channel

When L-channel fading is statistically independent of each other, the PDF of the Rician channel is given by:

$$f_{\text{Rice}}^{\text{MRC}}(s) = \frac{e^{-[LK+(1+K)s]}(1+K)^{\frac{L+1}{2}} s^{\frac{L-1}{2}}}{(LK)^{\frac{L-1}{2}}} \times I_{L-1}(2\sqrt{K(1+K)Ls}) \quad (3.4.11)$$

So, its CDF is represented as:

$$F_{\text{Rice}}^{\text{MRC}}(s) = 1 - e^{-2[LK+(1+K)S]} \sum_{k=0}^{\infty} \left[\frac{LK}{S+KS}\right]^{\frac{k}{2}} I_k(2\sqrt{LK(1+K)S}) \quad (3.4.12)$$

3.4.1.2.2 Rayleigh–lognormal channel

The PDF of the Rayleigh–lognormal channel is defined as:

$$f_{\text{Ray\_L}}^{\text{MRC}}(s) = \int_0^{\infty} f_{\text{Rayl}}(s_1/S_0) \cdots f_{\text{Rayl}}(s_L/S_0) f_{\text{Log}}(S_0) \mathrm{d}S_0 \quad (3.4.13)$$

where $s_1 + s_2 + \cdots + s_L = s$.

From Equation (3.4.13), the CDF of the channel can be written in the desired form as:

$$F_{\text{Ray\_L}}(S) = \frac{1}{\sqrt{\pi}} \sum_{i=1}^{n} w_i \left(1 - e^{-\frac{S}{\partial_i}} \sum_{m=0}^{L-1} \frac{(s/\partial_i)^m}{m!}\right) + R_n \quad (3.4.14)$$

where R_n can be ignored when $\delta < 7\text{dB}$ (the proof process is in [41]).

3.4.1.3 *Outage probability*

In general, a more useful measure of transmission quality is the outage probability P. It represents the percentage of time that the instantaneous BER is above an acceptable threshold ε. Therefore, it can be written as:

$$p_{\text{out}} = \Pr[S < \overline{S}] = F(\overline{S})$$

If using BPSK or DPSK, then

$$\overline{S_B} = \frac{1}{r}[\text{erfc}^{-1}(2\varepsilon)]^2 \tag{3.4.15}$$

$$\overline{S_D} = \frac{\ln(2\varepsilon)}{-\gamma} \tag{3.4.16}$$

When the MRC diversity reception is not used, the outage probability of the six-state Markov chain model is given by:

$$\begin{aligned} p_{\text{out}}(\gamma) = F(\overline{S}) &= A_1 F_{\text{Rice1}}(\overline{S}) + A_2 F_{\text{Rice2}}(\overline{S}) + A_3 F_{\text{Rice3}}(\overline{S}) \\ &\quad + A_4 F_{\text{Ray\_L1}}(\overline{S}) + A_5 F_{\text{Ray\_L2}}(\overline{S}) + A_6 F_{\text{Ray\_L3}}(\overline{S}) \end{aligned} \tag{3.4.17}$$

When using MRC diversity reception, the outage probability is given by:

$$\begin{aligned} p_{\text{out}}^{\text{MRC}}(\gamma) = F(\overline{S}) &= A_1 F_{\text{Rice1}}^{\text{MRC}}(\overline{S}) + A_2 F_{\text{Rice2}}^{\text{MRC}}(\overline{S}) + A_3 F_{\text{Rice3}}^{\text{MRC}}(\overline{S_b}) \\ &\quad + A_4 F_{\text{Ray\_L}_1}^{\text{MRC}}(\overline{S_b}) + A_5 F_{\text{Ray\_L}_2}^{\text{MRC}}(\overline{S}) + A_6 F_{\text{Ray\_L}_2}^{\text{MRC}}(\overline{S}) \end{aligned} \tag{3.4.18}$$

The outage probability can be obtained by combining Equations (3.4.15), (3.4.17), and (3.4.18) when using BPSK modulation. The outage probability can be obtained by combining Equations (3.4.16), (3.4.17), and (3.4.18) when using DPSK modulation.

3.4.2 *Algorithm simulation*

The transition probability matrix $\mathbf{P}$ could be obtained by [40]:

$$\mathbf{P} = \begin{bmatrix} 0.4286 & 0 & 0.5714 & 0 & 0 & 0 \\ 1.0000 & 0 & 0 & 0 & 0 & 0 \\ 0.2222 & 0.2222 & 0.2222 & 0.3334 & 0 & 0 \\ 0 & 0 & 0.5000 & 0 & 0.1667 & 0.3333 \\ 0 & 0 & 0 & 0.2500 & 0.2500 & 0.5000 \\ 0 & 0 & 0 & 0 & 0.4286 & 0.5714 \end{bmatrix}$$

Table 3.8 Six-state stability probability (%).

	A1	**A2**	**A3**	**A4**	**A5**	**A6**
Probability	24	7.0	31.1	20.9	0.40	16.6

Table 3.9 Six-state related parameters.

	K1	**K2**	**K3**	m_1	σ_1	m_2	σ_2	m_3	σ_3
Probability	13	10	6	9	1.5	12	3	14	5

Table 3.10 Two-state related parameters.

	A1	**A1**	**K3**	m_1	σ_1
Probability	0.4	0.6	11	−12.5	6

From the transition probability matrices, the six-state stability probability was obtained, as shown in Table 3.8. The parameters corresponding to the measurement result were shown in Table 3.9. The two-state stability probabilities and the parameters associated with **P** were shown in Table 3.10. Figure 3.50 compared the six-state CDF from theory, simulation, and measurement. Under MRC diversity and no diversity, the outage probability of the BPSK was shown in Figures 3.51 and 3.52. In Figure 3.51, the received SNR was 15 dB. In Figure 3.52, the expected threshold was 10^{-2}. Figures 3.53 and 3.54 used the DPSK model. The other conditions were the same as those in Figures 3.51 and 3.52.

When using BPSK or DPSK modulation, the outage probability of the six-state Markov chain model was theoretically analyzed. The closed-loop equation simplified calculations made simulations easy to be done. When the MRC diversity was used to obtain the closed-loop formula, the average outage probability of the six-state Markov chain model was also theoretically analyzed. This analysis method could be applied to other Markov chain models. When BPSK or DPSK modulation was used, the outage probability of the six-state Markov chain model was compared with that of the two-state Markov chain

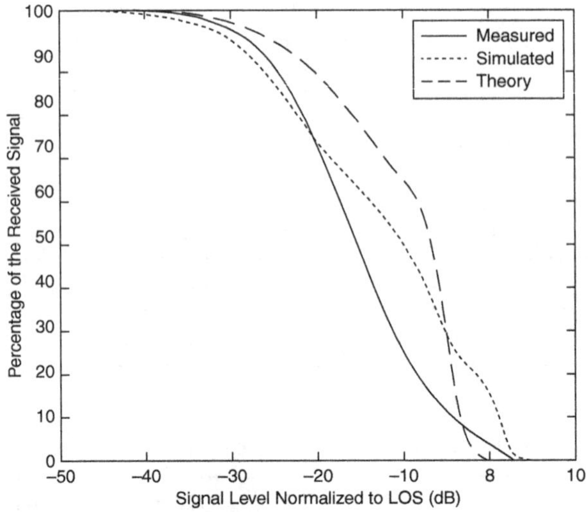

Figure 3.50 Comparison of the CDF of six-state.

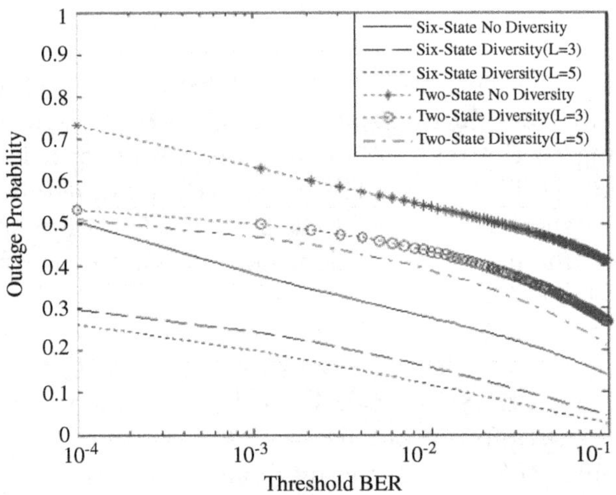

Figure 3.51 Outage probability of BPSK with MRC diversity ($L = 3, 5$) comparison without diversity at SNR=10 dB.

model in the case of MRC and no MRC. The results show that the six-state Markov chain model can be used to model various channel conditions and propagation mechanisms. It can be from Figure 3.51 to Figure 3.54 that the outage probability cannot be reduced when using MRC diversity reception.

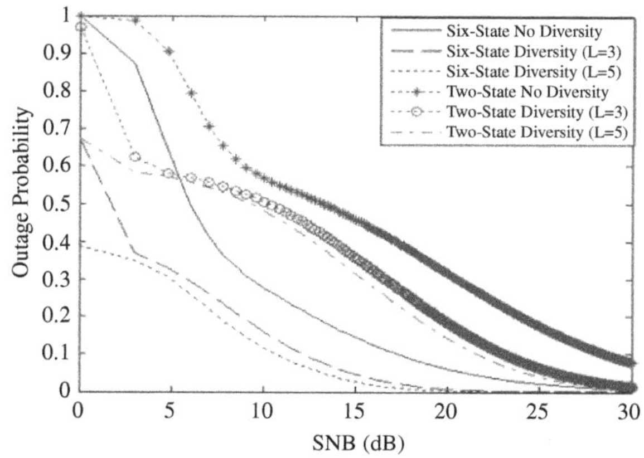

Figure 3.52 Outage probability of BPSK with MRC diversity ($L = 3, 5$) comparison without diversity at $\varepsilon = 10^{-2}$.

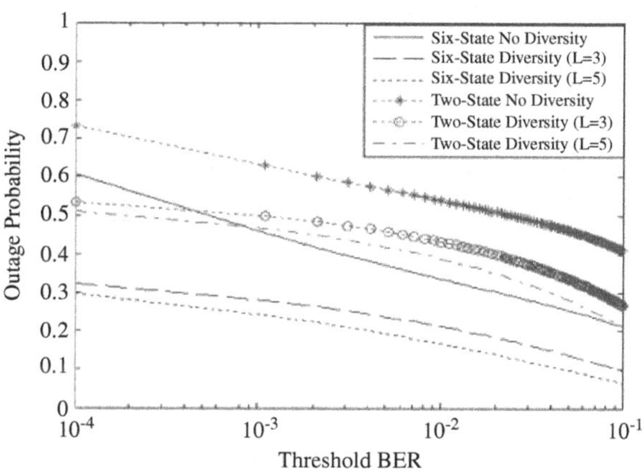

Figure 3.53 Outage probability of BPSK with MRC ($L = 3, 5$) comparison without diversity at SNR = 10 dB.

3.5 Satellite Channel Model Based on Principal Component Analysis and Fuzzy Clustering

Although the traditional satellite channel modeling method can more specifically study the influence of one or two factors on the propagation characteristics of the satellite channel, it is not comprehensive enough. In addition, the traditional methods often rely on subjective experience in determining the main influencing factors and lack

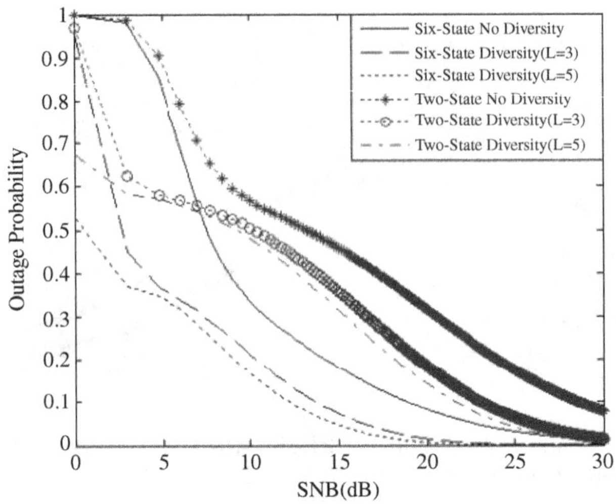

Figure 3.54 Outage probability for DPSK with MRC diversity ($L = 3{,}5$) comparison with diversity at $\varepsilon = 10^{-2}$.

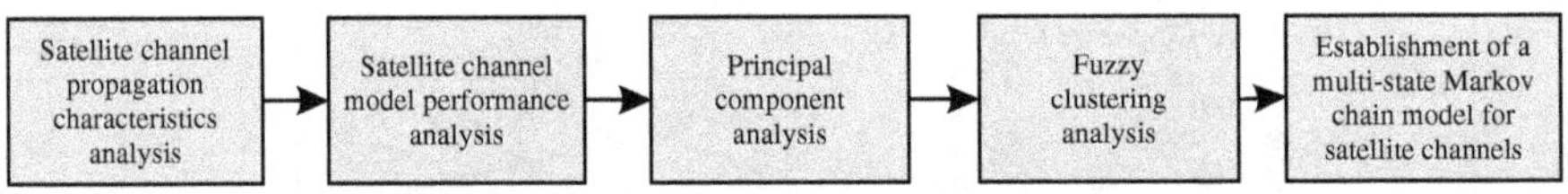

Figure 3.55 Main research ideas.

theoretical support. In view of the above problems, in this section, the principal component analysis (PCA) method is combined with the fuzzy clustering method to study the satellite channel modeling method. The main idea is shown in Figure 3.55.

3.5.1 *Analysis of key influencing factors in satellite channel modeling*

The principal component analysis (PCA) method for the key influencing factors of satellite channel modeling [42, 43] is as follows: firstly, the channel propagation characteristics in a specific case are analyzed, and the relevant factors are quantified; secondly, the PCA method is used to process the quantized data to obtain the main influencing factors on the satellite channel modeling; finally, a satellite channel model is established. The specific steps are as follows:

Step 1: The propagation characteristics of satellite channel under specific conditions are analyzed and the relevant physical quantities are calibrated.

First, the relevant physical quantities are quantitatively labeled as free space loss L_f, Rice factor c_0, amplitude flicker index S_4, atmospheric absorption loss A_{gas}, rain attenuation A_{rain}, cloud and fog attenuation A_c, depolarization effect XPD, Doppler shift f_d, etc.

Step 2: The original parameters and data are collected.

The original parameters and data of the concerned link are collected. Suppose that the data of the original parameters of the link has n samples, and there are p indexes in the samples (that is, the physical quantity needed to study the propagation characteristics of the link). The jth parameter observation value of the ith sample is recorded as x_{ij}, then the original data matrix is written as:

$$\mathbf{X} = (x_{ij})_{n\times p} = \begin{bmatrix} x_{11} & x_{12} & \cdots & x_{1p} \\ x_{21} & x_{22} & \cdots & x_{2p} \\ \vdots & \vdots & & \vdots \\ x_{n1} & x_{n2} & \cdots & x_{np} \end{bmatrix} \tag{3.5.1}$$

where $i = 1, 2, \ldots, n; j = 1, 2, \ldots, p$.

Step 3: Original data standardization

In the PCA, it is necessary for solving the principal component to calculate the covariance matrix. The accuracy of covariance matrix is affected by the dimension and order of magnitude of the parameters. Therefore, when calculating the covariance matrix, the original data is first normalized to make it become a dimensionless quantity. The standardized data is denoted as $(\tilde{x}_{ij})_{n\times p}$ and its element $\tilde{x}_{ij}$ is given by:

$$\tilde{x}_{ij} = \frac{x_{ij} - \bar{x}_j}{\sigma_j} \tag{3.5.2}$$

where

$$\bar{x}_j = \frac{1}{n}\sum_{i=1}^{n} x_{ij} \tag{3.5.3}$$

$$\sigma_j^2 = \frac{1}{n}\sum_{i=1}^{n}(x_{ij} - \bar{x}_j)^2 \tag{3.5.4}$$

Equations (3.5.3) and (3.5.4) are the mean and standard deviation of the jth indicator of the measured samples, respectively, and the standardized indicator variable is recorded as:

$$\tilde{x}_i = \frac{x_i - \bar{x}_i}{\sigma_i} \tag{3.5.5}$$

Step 4: The correlation coefficient matrix of the measured samples is calculated by:

$$\boldsymbol{R} = (r_{ij})_{p\times p} \tag{3.5.6}$$

where

$$r_{ij} = \frac{1}{n}\sum_{k=1}^{n} \tilde{x}_{ki}\tilde{x}_{kj} \tag{3.5.7}$$

where $i, j = 1, 2, \ldots, p, r_{ii} = 1,\ r_{ij} = r_{ji} \cdot r_{ij}$ is the correlation coefficient between the ith indicator and the jth indicator.

Step 5: The eigenvalues and eigenvectors are calculated.

Assume that the p eigenvalues of $\boldsymbol{R}$ are λ_1,$\lambda_2, \ldots, \lambda_p$ and $\lambda_1 \geq \lambda_2 \geq \cdots \geq \lambda_p > 0$, and their respective eigenvectors are $\boldsymbol{a} = [\boldsymbol{a}_1, \boldsymbol{a}_2, \ldots, \boldsymbol{a}_p]^T$, and among them $\boldsymbol{a}_i = [a_{1i}, a_{2i}, \ldots, a_{pi}]^T$. If the cumulative variance contribution rate of the first r eigenvalues reaches 85%, the first r principal components are written as:

$$\begin{cases} F_1 = a_{11}\tilde{x}_1 + a_{21}\tilde{x}_2 + \cdots + a_{p1}\tilde{x}_p \\ F_2 = a_{12}\tilde{x}_1 + a_{22}\tilde{x}_2 + \cdots + a_{p2}\tilde{x}_p \\ \vdots \\ F_r = a_{1r}\tilde{x}_1 + a_{2r}\tilde{x}_2 + \cdots + a_{pr}\tilde{x}_p \end{cases} \tag{3.5.8}$$

where F_r is the rth principal component, and a_{pr} is the correlation coefficient.

Step 6: The scores of the r principal components for n samples are calculated.

The principal component score is the new data of the original data (normalized) in the coordinate system defined by the principal component, i.e.,

$$F_j = \sum_{k=1}^{p} a_{kj}\tilde{x}_k, \quad j = 1,2,\ldots,r \tag{3.5.9}$$

Through the above calculation, the main factors can be determined, which is ready for the next step of channel state clustering.

3.5.2 *Analysis of satellite channel state number*

From the above analyses, the satellite channel single-state model cannot effectively describe the characteristics of the satellite channel in changing environments. Therefore, it is necessary to establish a satellite channel multi-state model. When establishing a multi-state model of the satellite channel, it is very important to determine the number of the channel states. If the number of states is too small, the propagation characteristics of the satellite channel cannot be accurately described. If the number of states is too large, the amount of calculation is large and the real-time performance is poor. In the past, relevant scholars divided satellite channels into specific states based on their experiences, which lacked theoretical basis. In this section, the number of satellite channel states is determined by fuzzy clustering analysis [44] and F-distribution. The specific steps are as follows:

Step 1: Get the original data matrix.

It is known from Equation (3.5.9) that there are the r main factors affecting channel modeling. The coefficient matrix of the simultaneous Equation (3.5.9) can be expressed as $\boldsymbol{W} = [F_1, F_2, \ldots, F_r] = \{w_{ij}\}_{n\times r}$, that is, the original data matrix for the cluster analysis can be expressed as:

$$\boldsymbol{W} = \begin{pmatrix} w_{11} & w_{12} & \cdots & w_{1r} \\ w_{21} & w_{22} & \cdots & w_{2r} \\ \cdots & \cdots & & \cdots \\ w_{n1} & w_{n2} & \cdots & w_{nr} \end{pmatrix} \tag{3.5.10}$$

where w_{ij} represents the jth element of the ith line, n is the number of indicators, and r is the number of principal components.

Step 2: Data standardization.

In establishing the satellite channel model, different factors have different properties and different dimensions, which cannot meet the requirements of fuzzy clustering. Therefore, the original data matrix $\boldsymbol{W}$ needs to be standardized before clustering, that is, it must be transformed into a dimensionless fuzzy matrix. After using a shift and range transformation, the standardization data are obtained as follows:

$$w'_{ij} = \frac{w_{ij} - \min\limits_{1\leq i\leq n}(w_{ij})}{\max\limits_{1\leq i\leq n}(w_{ij}) - \min\limits_{1\leq i\leq n}(w_{ij})} \tag{3.5.11}$$

where min(•)and max(•) denote the minimum and the maximum of w_{ij}, respectively, and w'_{ij} is the transformed data. After transforming, the matrix W' is written as:

$$\boldsymbol{W}' = \begin{bmatrix} w'_{11} & w'_{12} & w'_{1r} \\ w'_{21} & w'_{22} & w'_{2r} \\ & & \\ w'_{n1} & w'_{n2} & w'_{nr} \end{bmatrix} \tag{3.5.12}$$

Step 3: Establish the similar matrix.

There are many methods to establish similar matrices. Here, according to the characteristics of the satellite channel data, the similar matrix is established by the included angle cosine method, and its elements are given by:

$$r_{ij} = \frac{\sum_{k=1}^{m} w'_{ik} \cdot w'_{jk}}{\sqrt{\sum_{k=1}^{m} w'^2_{ik}} \cdot \sqrt{\sum_{k=1}^{m} w'^2_{jk}}} \tag{3.5.13}$$

where i, j has the same meaning as above, and thus the fuzzy similar matrix is obtained and denoted as $\boldsymbol{R}_{ij} = (r_{ij})_{n\times n}$.

Step 4: Dynamic fuzzy clustering.

In the process of fuzzy clustering, the fuzzy equivalent matrix is usually required to have reflexivity, transitivity, and symmetry, while the actual equivalent matrix usually has reflexivity and symmetry. Therefore, it is necessary to transform the existing matrix into an equivalent matrix with reflexivity, transitivity, and symmetry. In this section, the square automorphism method is used to transform the existing similarity matrix R. The specific transformation process is as follows:

$$R \to R^2 \to R^4 \to \cdots \to R^{2(k+1)} = R^{2k} \tag{3.5.14}$$

$$R^2 = R \circ R = (r'_{ij})_{m\times m}, r'_{ij} = \max_{k=1}^{m}(\min(r_{ij}, r_{ji})) \tag{3.5.15}$$

$$t(R) = R^{2k} = (t_{ij})_{m\times m} \tag{3.5.16}$$

The different numerical elements in the fuzzy equivalence matrix $t(R)$ are arranged from large to small, and the threshold λ is selected according to this arrangement, then the dynamic clustering graph of the number of the satellite channel states can be formed.

Step 5: Determine the optimal number of clusters.

In previous related studies, the number of clusters was determined by adjusting the λ value. The clustering results of this method depend on man's subjective will and λ. In this section, the optimal threshold λ is first determined by F-statistics [45], and then the clustering of the samples is carried out according to the λ value in the dynamic clustering graph, and finally the best classification number is obtained. This method can effectively overcome the deficiencies of man's subjective will and make the determined number of clusters be more objective and effective. The specific calculation process is as follows:

First, the center vector $\bar{\boldsymbol{x}}_k$ of the original data matrix sample is calculated by:

$$\bar{\boldsymbol{x}}_k = \frac{1}{n}\sum_{i=1}^{n} \boldsymbol{x}_{ik} \tag{3.5.17}$$

where $k = 1, 2, \ldots, m$.

Let the classification number corresponding to the λ value be r, the number of the samples of the j class be n_j, the sample of the j class be $x_1^{(j)}, x_2^{(j)}, \ldots, x_{n_j}^{(j)}$, and the clustering center of the j class $\bar{\boldsymbol{x}}^{(j)} = (\bar{\boldsymbol{x}}_1^{(j)}, \bar{\boldsymbol{x}}_2^{(j)}, \ldots, \bar{\boldsymbol{x}}_{n_j}^{(j)})$, $\bar{\boldsymbol{x}}_1^{(j)} = \frac{1}{n_j}\sum_{i=1}^{n} x_{ik}^{(j)}$, so F-statistics is defined as:

$$F = \frac{\sum\limits_{j=1}^{r} n_j \left\|\bar{x}^{(j)} - \bar{x}\right\|^2 \cdot (n-r)}{\sum\limits_{j=1}^{r}\sum\limits_{i=1}^{n_j} \left\|x_i^{(j)} - \bar{x}^{(j)}\right\|^2 \cdot (r-1)} \sim F(r-1, n-r) \tag{3.5.18}$$

where the numerator represents the distance between the classes, and the denominator represents the distance between the samples within the class. Therefore, the larger the F value is, the greater the distance between the classes is, and the better the classification effect is. If $F > F_\alpha(r-1, n-r)(\alpha = 0.05)$, the difference between classes is significant based on the theory of variance analysis of mathematical statistics, which indicates that the classification result is reasonable. If there is more than one value of α satisfying the inequality $F > F_\alpha(r-1, n-r)$, the size of $(F - F_\alpha)$ may be further examined to find a satisfactory F value from the larger one.

3.5.3 *Multi-state Markov chain model for satellite channels*

In the actual satellite communication, due to the frequent changes of the surrounding environment (weather changes, clouds, and the movement of the receiving terminal), the channel propagation characteristics are unstable and the channel state is constantly changing. At this time, it is obviously unreasonable to use the channel single-state model to describe the channel characteristics. Many scholars used the channel multi-state model to describe the dynamic changes of the channel, but the traditional channel multi-state model could not accurately predict the channel state at a certain time and skillfully unify multiple states into one model, which has poor practicability [46–57].

In this section, on the basis of fully studying the traditional channel multi-state Markov chain model, the number of satellite channel states is determined by principal component analysis (PCA) and fuzzy clustering method, and the satellite channel multi-state Markov chain model is established, which can not only overcome the sharp increase of computation caused by too many states but also avoid the problem that too few states cannot fully describe the channel characteristics and find the best balance between the two. The process of establishing satellite channel multi-state Markov model is as follows:

Step 1: Determine the channel single-state statistical model.
It is assumed that the optimal cluster number n of the satellite channel states in a certain region is obtained by fuzzy clustering method, that is, the link channel can be divided into n states. Among them, the first satellite channel obeys the Rayleigh distribution, the second obeys the Suzuki distribution, and the third obeys the Rice distribution,..., the nth channel obeys the Suzuki distribution. Their respective parameters can be obtained from the original data matrix (3.5.1).

Step 2: Establish a satellite channel multi-state Markov chain model.
The satellite channel multi-state Markov chain model can be denoted by the transition probability matrix $\boldsymbol{P}$ and the state vector $\boldsymbol{S}$, and the channel n-state Markov transfer

matrix is defined as:

$$P = \begin{bmatrix} p_{11} & p_{12} & \cdots & p_{1n} \\ p_{21} & p_{22} & \cdots & p_{2n} \\ \vdots & \vdots & & \vdots \\ p_{n1} & p_{n2} & \cdots & p_{nn} \end{bmatrix}_{n\times n} \tag{3.5.19}$$

where p_{ij} represents the transition probability from the state i to the state j, and

$$\sum_{j=1}^{n} p_{ij} = 1, (i = 1, 2, \ldots, n) \tag{3.5.20}$$

The state vector is denoted as:

$$\boldsymbol{S} = [s_1, s_2, \ldots, s_n] \tag{3.5.21}$$

Because this Markov chain is aperiodic and irreducible, its steady-state distribution exists and is equal to the limit distribution. Therefore, the state after the k transition in the channel multi-state Markov chain model is written as:

$$\boldsymbol{S}^k = \boldsymbol{S}\boldsymbol{P}^{k-1} \tag{3.5.22}$$

where $\boldsymbol{P}^{k-1}$ represents the k multiplication of the transfer matrix $\boldsymbol{P}$.

3.5.4 *Simulation verification*

In order to verify the proposed satellite channel model, simulation verification was carried out based on the data of the Ka-band satellite communication in Nanjing area [41–44]. The altitude of the satellite was 35786 km, the transmission frequency was 30 GHz, and the moving speed of the vehicle terminal is 36 km/h. According to the ITU-R data and the fitting method from Equation (3.5.23) to Equation (3.5.24), the original data of the satellite channel could be

Table 3.11 Channel-related parameters.

ρ	A_R(dB)	A_ω(dB)	A_c(dB)	ΔTr(K)	XPD(dB)	X(dB)	L_{bf}(dB)
2.28	80.92	2.65	3.73	269.95	20.36	25.8	213.14
2.31	79.61	2.84	3.68	269.95	23.59	23.1	213.01
1.39	58.94	2.43	3.49	269.94	21.37	21.9	213.15
0.96	61.63	2.54	2.98	269.94	19.83	23.6	213.06
1.83	67.19	2.61	3.15	269.94	21.24	24.7	212.93
0.89	71.23	2.64	3.26	269.94	18.67	24.8	212.56
0.91	69.27	2.65	2.87	269.95	19.38	25.3	213.20
2.64	79.58	2.67	3.62	269.94	20.61	24.9	212.89
2.65	64.95	2.73	2.69	269.95	19.49	25.8	212.95
1.58	75.26	2.35	3.54	269.93	17.26	24.3	213.11
0.79	80.16	2.68	3.68	269.94	20.61	25.6	213.08
1.26	73.85	2.64	3.51	269.94	19.34	23.9	212.79
2.98	67.98	2.56	3.26	269.96	20.11	25.7	212.69
1.97	74.31	2.36	2.95	269.93	19.48	25.4	213.17

obtained. Part of the data is shown in Table 3.11.

$$K(\theta) = \mathrm{K}_0 + \mathrm{K}_1\theta + \mathrm{K}_2\theta^2 + \mathrm{K}_3\theta^3 \tag{3.5.23}$$

$$\mu(\theta) = m_0 + m_1\theta + m_2\theta^2 + m_3\theta^3 \tag{3.5.24}$$

$$\sigma(\theta) = \sigma_0 + \sigma_1\theta + \sigma_2\theta^2 + \sigma_3\theta^3 \tag{3.5.25}$$

where θ was an elevation of less than 40°. $\mathrm{K}_0, \mathrm{K}_1, \mathrm{K}_2, \mathrm{K}_3, m_0$, $m_1, m_2, m_3, \sigma_0, \sigma_1, \sigma_2, \sigma_3$ were fitting coefficients. $K(\theta)$, $\mu(\theta)$, and $\sigma(\theta)$ are the fitted values.

Based on PCA of the above channel parameters, the correlation coefficient matrix R is obtained by:

$$\boldsymbol{R} = \begin{bmatrix} 1.000 & 0.433 & 0.162 & 0.033 & 0.449 & 0.278 & 0.086 & -0.160 \\ 0.433 & 1.000 & 0.679 & -0.039 & 0.936 & 0.389 & -0.150 & -0.440 \\ 0.162 & 0.679 & 1.000 & 0.135 & 0.603 & 0.531 & -0.313 & -0.292 \\ 0.033 & -0.039 & 0.135 & 1.000 & -0.070 & 0.318 & -0.062 & -0.023 \\ 0.449 & 0.936 & 0.603 & -0.070 & 1.000 & 0.377 & 0.116 & -0.253 \\ 0.278 & 0.389 & 0.531 & 0.318 & 0.377 & 1.000 & -0.012 & 0.110 \\ 0.086 & -0.150 & -0.313 & -0.062 & 0.116 & -0.012 & 1.000 & 0.220 \\ -0.160 & -0.440 & -0.292 & -0.023 & -0.253 & 0.110 & 0.220 & 1.000 \end{bmatrix}$$

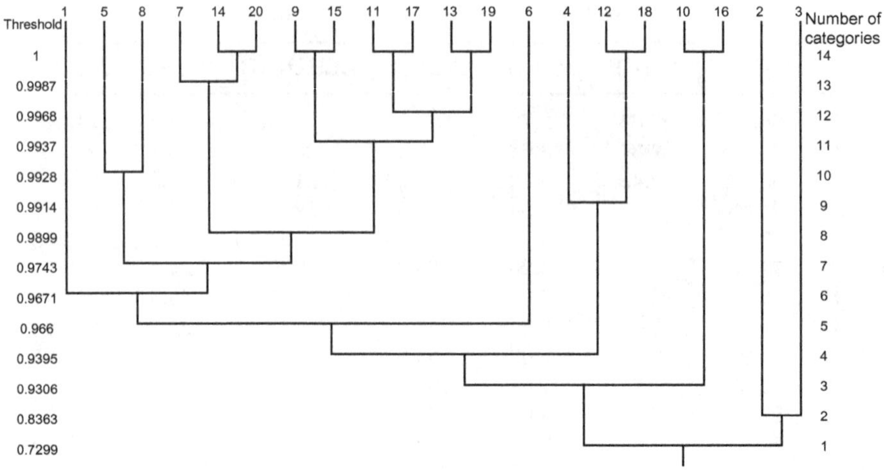

Figure 3.56 Dynamic clustering diagram.

The cumulative contribution rate (CCR) of the first three eigenvalues has reached 86.12%, which indicates that the first three principal components basically contain the information of all the indicators. After taking the first three eigenvalues and calculating the corresponding eigenvectors, the first three principal components are written as:

$$F_1 = -0.036x_1 + 0.264x_2 - 0.112x_3 + 0.392x_4 + 0.379x_5 - 0.586x_6 - 0.206x_7 + 0.481x_8 \quad (3.5.26)$$

$$F_2 = 0.259x_1 - 0.318x_2 + 0.766x_3 + 0.007x_4 - 0.168x_5 - 0.398x_6 + 0.212x_7 + 0.105x_8 \quad (3.5.27)$$

$$F_3 = 0.008x_1 + 0.709x_2 + 0.044x_3 + 0.009x_4 - 0.666x_5 - 0.067x_6 + 0.170x_7 + 0.130x_8 \quad (3.5.28)$$

where F_1, F_2, F_3 are the first three principal components, and x_i is the first column of the matrix. A dynamic clustering diagram for fuzzy clustering for the satellite channel states is shown in Figure 3.56.

From the F distribution, the best classification number and F distribution specific data can be determined, as shown in Table 3.12. According to the above analysis, the optimal cluster number of the satellite channel states was three, that is, the Markov model has

Table 3.12 *F*-distribution-related parameters.

Number of Clusters	**2**	**3**	**4**	**5**	**6**	**7**
λ	0.8363	0.9306	0.9395	0.9660	0.9671	0.9743
F	5.36	4.32	4.11	3.96	3.67	3.55
$F_{0.05}$	3.59	3.24	3.06	2.96	2.92	2.91
Number of Clusters	**8**	**9**	**10**	**11**	**12**	**13**
λ	0.9899	0.9914	0.9928	0.9937	0.9968	0.9987
F	3.12	3.01	2.19	2.08	1.89	1.65
$F_{0.05}$	2.95	3.02	3.14	3.31	3.28	3.97

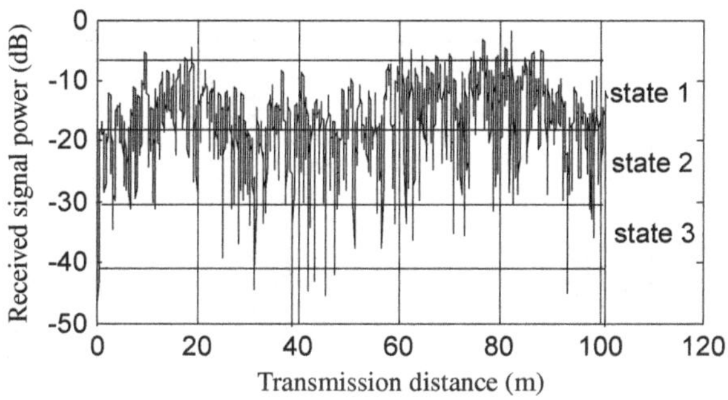

Figure 3.57 Status division.

three states, as shown in Figure 3.57. Based on the measured data of the signal envelope amplitude, the envelope of the received signal is divided into three ranges from −6 dB to −16 dB, from −16 dB to −26 dB, and from −26 dB to −36 dB. The corresponding states are divided into Figure 3.56, $\boldsymbol{S} = \{2, 2, 2, 1, 3, 3, 1, 1, 2, 2, 3, 1, 3, 1, 1, 2, 2, 3, 2, 2, 1, 2, 2, 1, 1, 2, 3, 1, 2, 2, 3, 1, 2, 3, 1, 2, 2, 2, 3, 1, 2, 3, 2, 3, 1, 2, 3, 1, 2, 3, 2, 3, 3, 1, 1, 3, 2, 3, 3, 2, 1, 2, 1, 3, 2, 2\}$.

The number of the divided states is used to calculate the transfer matrix $\boldsymbol{P}$, and the theoretical value of the state $\boldsymbol{S}$ is obtained as:

$$\boldsymbol{P} = \begin{bmatrix} 0.2 & 0.64 & 0.16 \\ 0.273 & 0.333 & 0.394 \\ 0.55 & 0.3 & 0.15 \end{bmatrix}$$

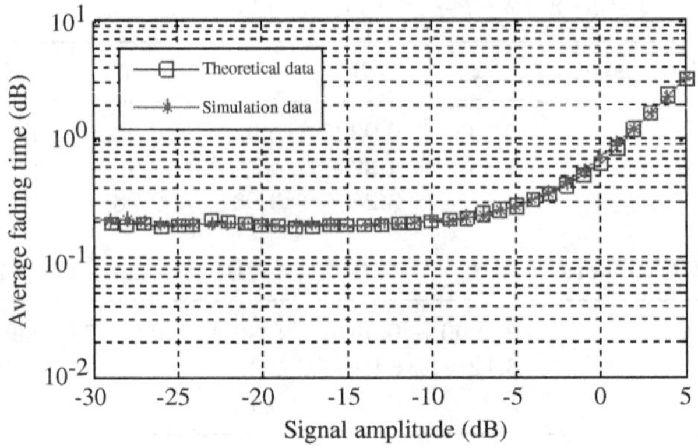

Figure 3.58 Average fading time graph.

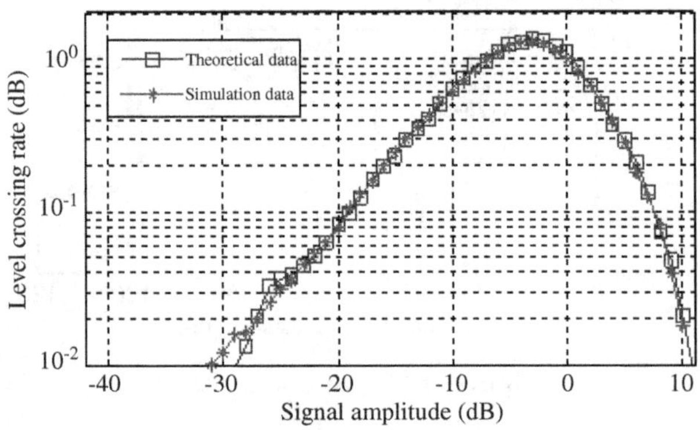

Figure 3.59 Level cross rate.

$$\boldsymbol{S} = \begin{bmatrix} 0.321 \\ 0.423 \\ 0.256 \end{bmatrix}^T$$

Using the above obtained parameters and Matlab simulation of the model, the second-order statistics (average fading time and level crossover rate) of the model are shown in Figures 3.58 and 3.59, respectively.

In the satellite channel modeling, the two most important indicators for evaluating channel performance are the average fading time and level crossing rate, especially when the signal level is below 0 dB. Figures 3.58 and 3.59 show that the average fading time, the level crossover rate, and the measured data produced by the channel model simulation have good consistency and better agreement with the general method at low level, which demonstrate the effectiveness of the simulation model.

References

[1] Mingchuan Y. Research on fading characteristics simulation in mobile satellite channel. Master thesis, Harbin: Harbin Institute of Technology, 2010 (in Chinese).

[2] Jingjun L. Simulation and modeling for two typical kinds of wireless fading channels. Master's thesis, Harbin: Harbin Institute of Engineering, 2011 (in Chinese).

[3] Cui-qin Z., Dong-ya S., Xue-mei Z., *et al.* Simulation for statistical models of mobile satellite channel and Markov two-state models. *Journal of Chongqing University of Posts and Telecommunications (Natural Science Edition)*, 2008, 20(1): 20–25 (in Chinese).

[4] Yang M., Meng F., Shi S., *et al.* Markov chain based two-state satellite mobile channel model. *IEEE 73rd Vehicular Technology Conference (VTC Spring)*, Yokohama, 2011, pp. 1–5.

[5] Po Z. and Zhi-gang C. Markov chain process based satellite mobile channel model and long term prediction method. *Journal of Electronics and Information*, 2011, 33(12): 2948–2953 (in Chinese).

[6] Yecai G., Tao Y., Xiuzai Z., *et al.* The analysis and implementation of geos-ynchronous orbit meteorological satellite. *Application of Electronic Technique*, 2014, 40(8): 98–100 (in Chinese).

[7] Li W., Law C. L., Ong J. T., *et al.* Ka-band land mobile satellite channel model: With rain attenuation and other weather impairments in equatorial zone. *IEEE Vehicular Technology Conference Proceedings. VTC 2000-Spring Tokyo*, 2000, 23(11): 2468–2472.

[8] Zhang X. Z., Guo Y. C. and Yang C. The meteorological satellite composite channel model under cloudy weather conditions. *Journal of Computational Information Systems*, 2013, 9(13): 5337–5344.

[9] Xiuzai Z., Yecai G., Jinli C., *et al.* Statistical characterization for l and X-band meteorological satellite channel. *Journal of Applied Meteorology*, 2012, 23(4): 478–484 (in Chinese).

[10] Da L. An integrated model of satellite mobile channel considering weather effects. Master's degree thesis, Xi'an: Xidian University, 2011 (in Chinese).

[11] Ippolito L. J. *Satellite Communications Systems Engineering: Atmospheric Effects, Satellite Link Design and System Performance.* John Wiley & Sons, 2017.

[12] Fontan F. P., Mayo A., Marote D., *et al.* Review of generative models for the narrowband land mobile satellite propagation channel. *International Journal of Satellite Communications & Networking*, 2008, 26(4): 291–316.

[13] Fontan F. P. Channel modeling for land mobile satellite services. *IEEE Proceedings of the Fourth European Conference on Antennas and Propagation*, 2010, pp. 1–5.

[14] Xu J. E., Jong J. H. and Ravishankar C. Channel modeling for a land mobile satellite system. *IEEE Global Telecommunications Conference*, Washington, DC, 2007, pp. 4596–4600.

[15] Loo C. and Butterworth J. S. Land mobile satellite channel measurements and modeling. *Proceedings of the IEEE*, 1998, 86(7): 1442–1463.

[16] Corazza G. E. and Vatalaro F. A statistical model for land mobile satellite channels and its application to nongeostationary orbit systems. *IEEE Transactions on Vehicular Technology*, 1994, 43(3): 738–742.

[17] Lacoste F., Villacieros B. M., Prieto-Cerdeira R., *et al.* SISO and MIMO enhanced two-statemodelling of the Land Mobile satellite channel for various frequencies, environments and elevation angles. *European Conference on Antennas and Propagation*. IEEE, The Hague, Netherlands, 2014, 2277–2281.

[18] Fontan F. P., Vazquez-Castro M., Cabado C. E., *et al.* Statistical modeling of the LMS channel. *IEEE Transactions on Vehicular Technology*, 2001, 50(6): 1549–1567.

[19] Vucetic B. and Du J. Channel modeling and simulation in satellite mobile communication systems. *IEEE Journal on Selected Areas in Communications*, 1992, 10(8): 1209–1218.

[20] Milojevic M., Haardt M. and Heuberger A. Measurement based satellite to outdoor channel modeling for multiple satellite systems. *IEEE Wireless Conference*, 2008, 2008: 1–7.

[21] Tan H., Zhang N. T. and Zhang Z. Z. The test and study of system propagation characteristic of LEO mobile satellite communication. *Journal of China Institute of Communications*, 2003, 24(7): 86–92.

[22] Loo C. Land mobile satellite channel measurement at K-band using Olympus. *IEEE 44th Vehicular Technology Conference*, 1994, 2: 919–923.

[23] Shang M., Jian-hao H. and Jian W. Design and implementation of satellite channel simulator based on three-state Markov chain. *Journal of System Simulation*, 2007, 19(17): 3961–3965 (in Chinese).

[24] Lin H. P., Tseng M. J. and Lin D. B. Modeling fading properties for mobile satellite link channels using Markov model approaches. *Antennas and Propagation Society International Symposium*, 2003, 4(3): 192–195.

[25] Jianpei C. Research on physical statistics modeling of land mobile fading channel. Master's degree thesis, Kunming: Yunnan University, 2012 (in Chinese).
[26] Tao Y. The research of satellite channel modeling and simulation method. Master's degree thesis, Nanjing: Nanjing University of Information Science and Technology, 2014 (in Chinese).
[27] Yecai G., Weijuan Z. and Xiuzai Z. Design and simulation of satellite channel three-state Markov model. *Computer Applications*, 2016, 36(1): 91–95 (in Chinese).
[28] Fontan F. P., Vazquez-Castro M., Cabado C. E., *et al.* Statistical modeling of the LMS channel. *IEEE Transactions on Vehicular Technology*, 2001, 50(6): 1549–1567.
[29] Song Z., Xiaodong W., Guosheng R., *et al.* Analysis and simulation of composite fading channels based on Simulink. *Journal of Naval Aeronautical Engineering Institute*, 2007, 22(3): 340–344 (in Chinese).
[30] Ming H., Dong S., Yan C., *et al.* A new five-state Markov model for land mobile satellite channels. *2008 8th International Symposium on Antennas, Propagation and EM Theory*, 2008, pp. 1512–1515.
[31] Shen D., Rong J., Yang Y., *et al.* The six-state Markov model for land mobile satellite channels. *IEEE International Symposium on Microwave, Antenna, Propagation and EMC Technologies for Wireless Communications*, 2006, 2: 1619–1622.
[32] Dong-ya S. A novel multi-state channel model on the land mobile satellite channels. *Proceedings of GMC*, 2006, 12(6): 215–223.
[33] Wang H. S. and Moayeri N. Finite-state Markov channel — a useful model for radio communication channels. *IEEE Transactions on Vehicular Technology*, 1995, 44(1): 163–171.
[34] Zhang Q. and Kassam S. A. Finite-state Markov model for Rayleigh fading channels. *IEEE Transactions on Communications*, 1999, 47(11): 1688–1692.
[35] Babich F. and Lombardi G. A Markov model for the mobile propagation channel. *IEEE Transactions on Vehicular Technology*, 2000, 49(1): 63–73.
[36] Tellambura C., Mueller A. J. and Bhargawa V. K. Analysis of M-ary phase-shift keying with diversity reception for land-mobile satellite channels. *IEEE Transactions on Vehicular Technology*, 1997, 46(4): 910–922.
[37] Ming H., Dongya S., Yanni C., *et al.* Theoretical analysis and simulation of the outage probability for six-state Markov model on land-mobile satellite channels. *IEEE Antennas, Propagation and EM Theory*, 2008, pp. 1516–1519.

[38] Tellambura C., Mueller A. J. and Bhargava V. K. BER and outage probability for land mobile satellite channel with maximal ratio combining. *Electronics Letters*, 1995, 31(8): 606–608.

[39] Tellambura C., Mueller A. J. and Bhargawa V. K. Analysis of M-ary phase-shift keying with diversity reception for land-mobile satellite channels. *IEEE Transactions on Vehicular Technology*, 1997, 46(4): 910–922.

[40] Abramowitz M., Stegun I. and Mcquarrie D. A. *Handbook of Mathematical Functions*. Dover Publications, New York, NY, 1970.

[41] Jakes W. C. and Cox D. C. *Microwave Mobile Communications*. IEEE Press, 1993.

[42] Li-hong Z., Yu-ge S., Yu C., Xin-he X., *et al.* Face recognition based on kernel PCA. *Journal of Northeastern University (Natural Science)*, 2006, 27(8): 847–850 (in Chinese).

[43] Xiangyang M., Taiyi Z., and Yatong Z. A robust probability principle component analysis method. *Journal of Xi'an Jiaotong University*, 2008, 42(10): 1217–1220 (in Chinese).

[44] Yecai G. *Biomedical Information and Image Processing*. Science Press, 2009 (in Chinese).

[45] Hong-wei C., Jun-yong T., Xun C., *et al.* Research on environmental factor of exponential product based on F distribution. *Acta Armamentarii*, 2012, 33(10): 1217–1221 (in Chinese).

[46] Li W., Law C. L., Dubey V. K., *et al.* Ka-band land mobile satellite channel model incorporating weather effects. *IEEE Communications Letters*, 2001, 5(5): 194–196.

[47] Babich F., Lombardi G., and Valentinuzzi E. Variable order Markov modelling for LEO mobile satellite channels. *Electronics Letters*, 1999, 35(8): 621–623.

[48] Lin H. P. and Tseng M. C. Modelling fading properties for mobile satellite link channels using non-stationary hidden Markov model. *Microwaves Antennas & Propagation*, 2009, 3(1): 171–180.

[49] Guo Y. C., Yuan T. and Zhang X. Satellite channel Markov model of ka-band based on principal component. *JNW*, 2014, 9(7): 1701–1706.

[50] Guo Y. C., Yuan T., Zhang X. Z. Ka-band satellite channel multi-state Markov model based on dust storm weather and shadow multi-path effects. *Advanced Materials Research Trans Tech Publications*, 2013, 1786–1789.

[51] Guo Y., Yuan T. and Zhang T. Simulator design and its simulation of meteorological satellite channel. *Journal of Computer Applications*, 2013, 33(9): 2650–2652 (in Chinese).

[52] Guo Y., Yuan T., Runzhi Z., *et al.* The analysis and implementation of geosynchronous orbit meteorological satellite. *Electronic Technology Application*, 2014, 40(8): 98–100 (in Chinese).

[53] Hu L., Guo Y. C., Yuan T., *et al.* Orthogonal wavelet transform dynamic weighted multi-modulus blind equalization algorithm based on dynamic particle swarm. *JCP*, 2013, 8(8): 2011–2017.
[54] Guo Y. C., Zhang T. and Yuan T. The design of digital down conversion system in satellite channel simulator. *Science, Technology and Engineering*, 2014, (1): 54–58 (in Chinese).
[55] Zhang X., Guo Y., Yang C., *et al.* Markov time-varying channel modeling and simulation of meteorological satellite. *Information Technology*, 2017, (9): 49–53 (in Chinese).
[56] Guo Y., Cai L., Zhao W., *et al.* Research and simulation of satellite channel three-state Markov model under fog-and-haze weather. *Journal of System Simulation*, 2017, 29(5): 941–949 (in Chinese).
[57] Guo Y., Zhao W., Runzhi Z., *et al.* Research and implementation of the satellite channel two-state Markov model under the cloudy and foggy weather. *Science Technology and Engineering*, 2015, 35(31): 112–116 (in Chinese).

Chapter 4

Nonlinear Satellite Channel Model Based on Different Backgrounds

Summary

In this chapter, the structural characteristics of the traveling wave tube amplifier (TWTA) model, the group delay model, and the Wiener? — Hammerstein model are analyzed, and the nonlinear satellite channel model and equalization system with Gaussian noise background are established. The characteristics of alpha-stable distributed noise are analyzed, and the adaptive neuro-fuzzy inference system model of a nonlinear satellite channel is proposed by using the ability of the adaptive neuro-fuzzy inference system to approach nonlinear function arbitrarily. By combining the TWT amplifier model with the group delay effect and using the channel prior information, a nonlinear satellite channel model based on channel prior information is presented.

There are all kinds of fading and distortions in satellite channels, among which nonlinearity is a typical distortion effect. The nonlinearity of a satellite channel mainly comes from the nonlinearity of the traveling wave tube amplifier (TWTA) in the satellite and the nonlinear effect caused by group delay fading of satellite transmission signals.

4.1 Nonlinear Satellite Channel Model

4.1.1 *TWTA model*

The theoretical research and practical measurement results show that the downlink noises of satellite communication systems are much larger than those of the uplink. In order to combat strong noises and other downlink channel distortions, the satellite internal transponder must amplify the received signal [1]. The TWTA, which is generally used as the internal high-power amplifier (HPA) of the satellite, has the typical characteristics of a general power amplifier, that is, its power characteristic curve includes an amplification region and a saturation region [2]. When the TWTA is working in the amplification region, its input and output signals have an approximately linear relationship, that is, the TWTA linearly amplifies the power of its input signal and its output signal does not generate nonlinear distortion. However, satellites work in outer space, and the internal energy is difficult to be replenished, so the internal power is very limited. In order to make full use of the limited power available in the satellite, the TWTA generally works at or near the saturation point, which inevitably introduces nonlinear distortions into the channel. This nonlinear distortion is characterized by amplitude–amplitude conversion effect (AM/AM) and amplitude–phase conversion effect (AM/PM). Salah proposed two models to describe these two effects: the amplitude–phase conversion model and the orthogonal model.

In the amplitude–phase conversion model of Salah [3], the input signal of the TWTA is written as

$$x(t) = r(t)\cos[\omega_0 t + \varphi(t)], \tag{4.1.1}$$

where $r(t)$ is the modulated signal envelope, $\varphi(t)$ is the phase, ω_0 is the carrier frequency, and $x(t)$ is the modulated signal. The output of the corresponding amplifier is defined as

$$y(t) = A(r(t))r(t)\cos\{\omega_0 t + \varphi(t) + \phi[r(t)]\}, \tag{4.1.2}$$

where $A(r(t))$ is the AM/AM distortion and $\phi[r(t)]$ is the AM/PM distortion. Salah proposed a two-parameter model to describe

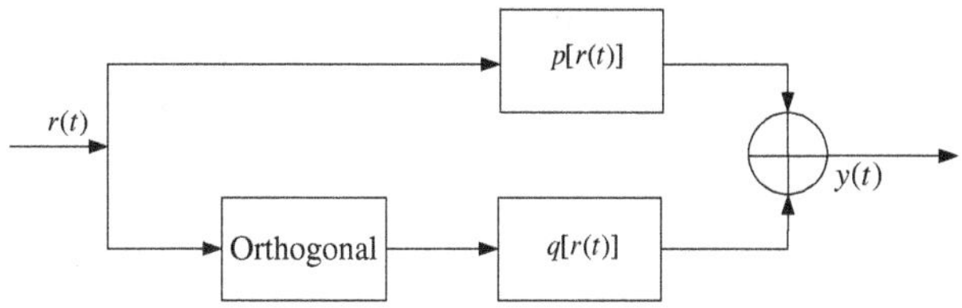

Figure 4.1 Orthogonal model proposed by Salah.

AM/AM and AM/PM distortion [3], which is given as follows:

$$A[r(t)] = \frac{\alpha_a r(t)}{1 + \beta_a r^2(t)}, \tag{4.1.3}$$

$$\phi[A(t)] = \frac{\alpha_\phi r^2(t)}{1 + \beta_\phi r^2(t)}, \tag{4.1.4}$$

where α_a,β_a, α_ϕ, and β_ϕ are parameters of the model. In the orthogonal model of Salah, the output signal of the amplifier can be expressed as the sum of the in-phase and quadrature components, as shown in Figure 4.1.

In Figure 4.1, $p[r(t)]$ is the in-phase gain component of the orthogonal mode and $q[r(t)]$ is the orthogonal gain component:

$$y(t) = p(t) + q(t), \tag{4.1.5}$$

$$p(t) = p[r(t)]\cos[\omega_0 t + \varphi(t)], \tag{4.1.6}$$

$$q(t) = \quad q[r(t)]\cos[\omega_0 t + \varphi(t)], \tag{4.1.7}$$

where $p(t)$ and $q(t)$ are the in-phase component and orthogonal component of the output signal, respectively. According to Equations (4.1.2) and (4.1.5), $p(t)$ in Equation (4.1.6) and $q(t)$ in Equation (4.1.7) are written as

$$p[r(t)] = A[r(t)]\cos\{\phi[r(t)]\}, \tag{4.1.8}$$

$$q[a(t)] = A[a(t)]\sin\{\phi[a(t)]\}. \tag{4.1.9}$$

If Salah's two-parameter model is used to represent the in-phase gain and quadrature gain components, then

$$p[r(t)] = \frac{\alpha_p r(t)}{\beta_p r^2(t)}, \tag{4.1.10}$$

$$q[r(t)] = \frac{\alpha_q r^3(t)}{[1 + \beta_q r^2(t)]^2}. \tag{4.1.11}$$

Table 4.1 Parameters of Salah model.

Model	Parameter	Function	α	β
Amplitude–phase conversion model	First group	$A[r(t)]$	2.159	1.152
		$\phi[r(t)]$	4.003	9.104
	Second group	$A[r(t)]$	1.964	0.994
		$\phi[r(t)]$	2.529	2.817
	Third group	$A[r(t)]$	1.662	0.552
		$\phi[r(t)]$	0.153	0.346
Orthogonal model	First group	$p[r(t)]$	2.092	1.247
		$q[r(t)]$	5.529	2.709
	Second group	$p[r(t)]$	1.640	0.062
		$q[r(t)]$	0.204	0.133

Equations (4.1.10) and (4.1.11) are orthogonal representations of the Salah model. Table 4.1 shows the parameters of the two Salah models, each of which is obtained from different measured data through several statistical experiments. In this section, the Salah parameters are the first set of parameters for the amplitude–phase conversion model.

4.1.2 *Group delay model*

The group delay is the phase rate (phase shift) of the system with respect to frequency, which is a parameter to describe the phase response of the system. The group delay is defined as

$$\tau(\omega) = -\frac{d\theta(\omega)}{d\omega}, \tag{4.1.12}$$

where $\theta(\omega)$ is the phase frequency response of the system. When this parameter is constant, the delay generated by each frequency component of the signals which pass through the system is a fixed constant. At this time, the group delay acts as a linear filter on the signal. If $\tau(\omega)$ is a variable, the different delays corresponding to each frequency component of the signals will be produced when it passes through the system. Thus, the waveform of the output signals is inevitably distorted with respect to the input signals, and the original signals cannot be recovered by the synchronization of the receiver. It is pointed out in [4] that the phase frequency characteristic of the channel was expressed as

$$\theta(\omega) = b_1(\omega - \omega_0) + b_2(\omega - \omega_0)^2 + b_3(\omega - \omega_0)^3 + b_4(\omega - \omega_0)^4 + L. \tag{4.1.13}$$

After deriving the frequency from Equation (4.1.13), the following equation is obtained:

$$\tau(\omega) = -c_1 - 2c_2(\omega - \omega_0) - 3c_3(\omega - \omega_0)^2 + L, \tag{4.1.14}$$

where ω_0 is the carrier frequency, $-c_1$ is the constant term, $-2c_2(\omega - \omega_0)$ is the linear group delay term, and the other items "behind" are the high-order group delay terms. The constant term represents the linear effect of the channel, whereas the remaining terms represent the nonlinear effect. In practical systems, the linear group delay term and quadratic nonlinear group delay term are generally studied. The object of study depends on the problem to be studied [5].

It should be noted that in the nonlinear modeling of TWTAs, it is usually assumed that the uplink noises are zero or additive white Gaussian noise with a specified signal-to-noise ratio (SNR). However, the effects of the different SNRs on modeling have not been studied; channel modeling under α-stable distributed noise has also not been studied, and the group delay effects have not been considered, as well as the effects of other known channel information on modeling have not been utilized fully.

4.2 Nonlinear Satellite Channel Model and Equalization System under Gaussian Noise Background

The nonlinear effect of HPA in a satellite causes serious distortion of the signal, which affects the quality of communication. The key to the research of channel distortion compensation is to use the appropriate nonlinear model to approximate the nonlinearity of the channel.

The traditional method uses the Volterra series to represent the nonlinearity of the satellite channel, and on this basis, the corresponding equalization algorithm is studied.

If the Volterra series model of the nonlinear satellite channel is adopted, the equalization model based on the Volterra series structure must be adopted in order to compensate for this effect. However, the Volterra equalizer has numerous coefficients. If the order of the tap coefficients is higher than three, the computational complexity is very large and the convergence speed is slow, which is not conducive to real-time processing of data. The algorithm proposed on this basis, even if it can improve the operation speed to a certain extent, is still subject to the slow operation speed of the Volterra equalizer itself, and it is also limited to the study of equalization algorithm based on the Volterra model. Researchers proposed to use the nonlinear models of other structures to approximate the nonlinearity of the satellite channel, but only the modeling method is given and the model parameters are obtained by system identification, and the influence of channel noise on channel modeling is not further considered.

In satellite communication, the channel noises are mainly atmospheric noises, which are a α-stable distributed noise with very thick trailing [6–12]. Its properties are completely different from those of white Gaussian noise (AWGN). If the actual channel environment is considered, the influence of uplink noises on the modeling cannot be ignored. At present, there are no reports on the modeling methods of the nonlinear satellite channels under the background of α-stable distributed noise.

In this section, the nonlinear satellite channel modeling method is discussed while considering α-stable distributed noise. First, the Wiener model is proposed to model the nonlinear satellite channel and the Hammerstein equalizer is used to equalize the channel. On this basis, the performance of the Wiener model in the Gaussian white noise environment with different SNRs is studied, and then this modeling method is compared with predistortion and adaptive equalization compensation method. After that, the characteristics of α-stable distributed noise are introduced. Under the α-stable distributed noise, the adaptive neuro-fuzzy interference system (ANFIS) [13] is used to model the channel and it is compared with the Wiener model.

4.2.1 *Wiener and Hammerstein models for nonlinear satellite channel*

4.2.1.1 *Wiener and Hammerstein models*

The Wiener model is a commonly used nonlinear model for system identification and system modeling. As shown in Figure 4.2, h_{wi} is the linear coefficient of the Wiener model and g_{wj} is the nonlinear coefficient of the Wiener model. In addition, the signal $x(n)$ cannot be measured.

The input and output of the Wiener model are denoted as

$$\hat{x}(n) = \sum_{i=0}^{N} h_{wi} a(n-i) \tag{4.2.1}$$

and

$$\hat{y}(n) = \sum_{j=1}^{M} g_{wj} (\hat{x}(n))^j, \tag{4.2.2}$$

respectively.

The Hammerstein model is shown in Figure 4.3, where g_{Hk} denotes the nonlinear coefficient, h_{Hl} denotes the linear coefficient, and the input z(n) and output s(n) of Hammerstein model are written as in Equations (4.2.3) and (4.2.4), respectively. Among them, the linear and nonlinear orders of the Hammerstein model must be greater than the linear and nonlinear orders of the corresponding Wiener model:

$$z(n) = \sum_{k=1}^{K} g_{Hk}(y(n))^k \quad (K > M) \tag{4.2.3}$$

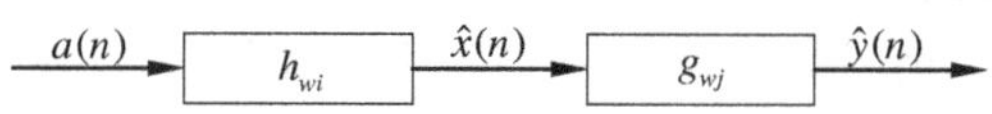

Figure 4.2 Wiener model.

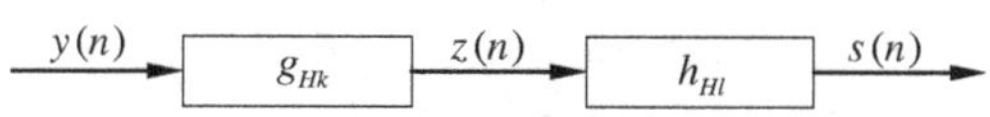

Figure 4.3 Hammerstein model.

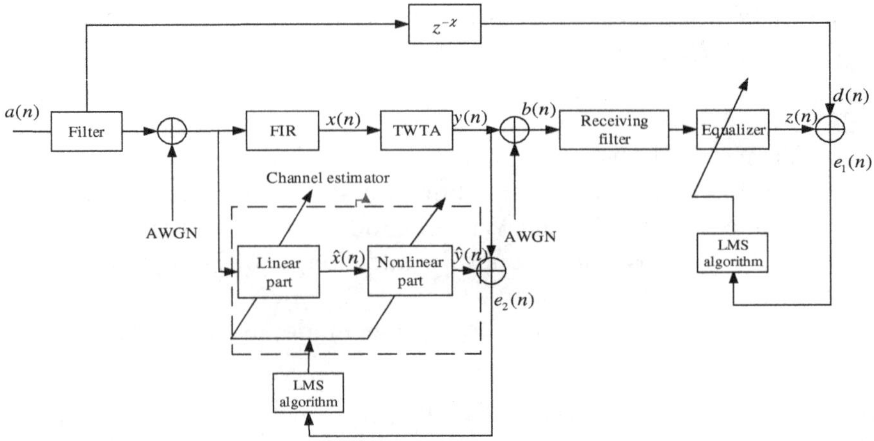

Figure 4.4 The equivalent model of Wiener–Hammerstein communication system.

and

$$s(n) = \sum_{l=0}^{L} h_{Hl} z(n-l) \quad (L > N), \tag{4.2.4}$$

where $y(n)$ represents the input signal of the Hammerstein model.

4.2.1.2 *Wiener–Hammerstein equalizer for nonlinear satellite channels*

The Wiener model is used to describe a nonlinear satellite channel. The compensation device can be a Hammerstein model equalizer or predistorter (Hammerstein–Wiener predistortion). In this section, the two compensation devices are discussed separately and the weight vector iteration formula (Hammerstein–Wiener predistorter) of the Wiener–Hammerstein equalizer is derived. An equivalent baseband model for a satellite communication system, including a channel estimator and an adaptive equalizer, is shown in Figure 4.4. $a(n)$ is the transmitted signal from the ground, which passes through the transmitting filter and the uplink, and reaches the satellite receiving end after superimposing the uplink noise. $x(n)$ is the output signal of the linear filter part of the high-power amplifier, which is generally not measurable $y(n)$ is the output signal of the nonlinear part of HPA, and $\hat{x}(n)$ and $\hat{y}(n)$ are the output signals of the linear and nonlinear

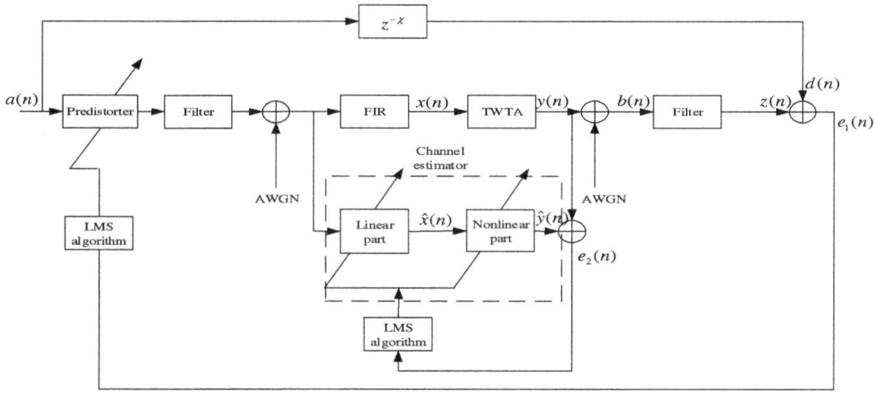

Figure 4.5 Equivalent baseband model of satellite communication system including channel estimator and predistorter.

parts of the Wiener model, respectively. $b(n)$ is the received signal. $z(n)$ is the output signal of the adaptive equalizer based on the Hammerstein model. The channel estimator based on the Wiener model is located inside the satellite and is connected parallel with the HPA. The adaptive equalizer based on the Hammerstein model is located after the receiving filter and before the decision device. Figure 4.4 shows that, since the HPA can be represented as a cascaded form with a memory linear filter (FIR) and a memoryless nonlinear filter (TWTA), the nonlinearity of the satellite channel is somewhat similar to the input–output characteristics described by the Wiener model structure.

The equivalent baseband model for a satellite communication system, including a channel estimator and a predistorter, is shown in Figure 4.5. Its basic structure is the same as in Figure 4.4, except that the compensation device is placed before the transmitting filter. The error signal used to update the tap coefficients must be transmitted to the transmitter through the receiver to update the predistorter coefficients. During the transmission process, a large amount of channel noises is inevitably superimposed and the performance of the predistorter is seriously degraded. This noise superimposed on the error signal is not considered in [14], but the error signal without noises is directly used to update the tap coefficient of the predistorter. In this section, the mean square errors (MSEs) of the

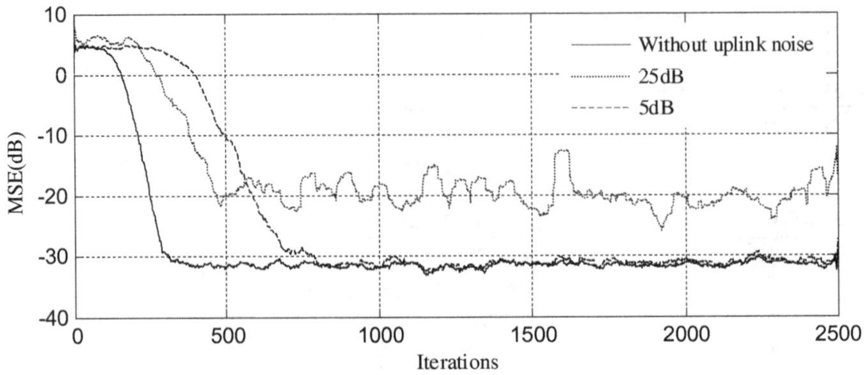

Figure 4.6 MSE curves of Wiener model estimator under different uplink noises.

predistorter and adaptive equalizer are compared without considering the superimposed noise of the error signal of the predistorter to illustrate the superiority of the modeling-equalization method over the predistortion method.

Here, the parameters of the Wiener model need to be estimated for a specific actual channel. When estimating the Wiener model parameters, the uplink channel noises are superimposed on the satellite received signal, which causes the Wiener model parameter estimation quality to be affected by the uplink noises, and the model convergence speed varies with the different uplink noises. Figure 4.6 shows the relationship between MSE curves of the Wiener channel estimator with noises under different conditions, such as 5 and 25 dB uplink noises and without considering uplink noises. As seen in Figure 4.6, if the uplink noises are considered, the Wiener model performance decreases. When the uplink noises are small, the Wiener model can still converge with a small MSE, but the convergence speed is slowed down. When the uplink noises are large, the channel estimator has a large jitter after the convergence of the MSE curves and the steady MSE increases. This shows that the performance of the Wiener model is affected by the uplink noises: the greater the uplink noises, the worse the performance of the Wiener model is. Experiment results have shown that the uplink noises are almost negligible compared to the downlink noises. Since the signal passes through the downlink, the uplink SNR is assumed to be large.

In this section, a deformation model of the TWTA model proposed by Salah is used to describe the nonlinearity of the actual satellite channel. The input–output relationship of the deformation model is

$$S_{\text{out}}(t) = S_{\text{in}}(t)A(r)e^{j\phi(r)}, \tag{4.2.5}$$

with

$$A(r) = \frac{\alpha_A r_0}{1 + \beta_A (r_0 r)^2} \tag{4.2.6}$$

and

$$\phi(r) = \frac{\alpha_\phi (r_0 r)^2}{1 + \beta_\phi (r_0 r)^2}, \tag{4.2.7}$$

where $S_{\text{out}}(t)$ and $S_{\text{in}}(t)$ are the input signal and output signal, respectively. $A(r)$ and $\phi(r)$ denote the amplitude–amplitude conversion effect and amplitude–phase conversion effect of the satellite channel, respectively. $r = \sqrt{P(t)/\overline{P}}$, where $P(t)$ is the instantaneous power of the input signals and $\overline{P}$ is the average power of the input signals. r_0 represents the amplitude fallback of the input signals. $\alpha_A = 2.1587$, $\beta_A = 1.1517$, $\alpha_\phi = 4.0033$, $\beta_\phi = 9.1040$. The execution steps of the entire system are as follows:

Step 1: After the transmitted signal passes through the uplink, it reaches the satellite receiver, and after the satellite receives the signal, it starts the channel estimator.

Step 2: Selecting the linear and nonlinear orders of the Wiener model, and initializing the model parameters. The receiving signal of the satellite is output by the model as $\hat{y}(n)$ and by the HPA as $y(n)$ at the same time.

Step 3: Calculate the error denoted by $e(n) = y(n) - \hat{y}(n)$, estimating model parameters by stochastic gradient descent method based on instantaneous $|e(n)|^2$, where $\boldsymbol{f}_{\text{lin}} = \{f_{\text{lin1}}, f_{\text{lin2}}, \text{L}, f_{\text{linM}}\}$ and $\boldsymbol{f}_{\text{non}} = \{f_{\text{non1}}, f_{\text{non2}}, \text{L}, f_{\text{nonN}}\}$ are the linear filter and the nonlinear filter weight vectors of the Wiener model.

The iterative formulas for linear partial and nonlinear partial weight vectors are written as

$$\boldsymbol{f}_{\text{lin}}(n+1) = \boldsymbol{f}_{\text{lin}}(n) + \mu e_1(n)\nabla_1^*(n), \tag{4.2.8}$$

$$\boldsymbol{f}_{\text{non}}(n+1) = \boldsymbol{f}_{\text{non}}(n) + \mu e_2(n)\nabla_2^*(n), \tag{4.2.9}$$

where

$$\nabla_1^*(n) = \frac{d(\hat{y}(n))}{d\boldsymbol{f}_{\text{lin}}} = \left\{\frac{\partial(\hat{y}(n))}{\partial f_{\text{lin1}}}, \frac{\partial(\hat{y}(n))}{\partial f_{\text{lin2}}}, \text{L}, \frac{\partial(\hat{y}(n))}{\partial f_{\text{linN}}}\right\}^*, \tag{4.2.10}$$

$$\frac{\partial(\hat{y}(n))}{\partial f_{\text{lin,i}}} = f_{\text{lin,1}}\hat{x}(n-i) + 2f_{\text{lin,2}}\hat{x}(n)\hat{x}(n-i) \tag{4.2.11}$$

$$+ \cdots + Mf_{\text{lin,M}}\hat{x}^{M-1}(n)\hat{x}(n-i),$$

$$\nabla_2^*(n) = \frac{d(\hat{y}(n))}{d\boldsymbol{f}_{\text{non}}} = \{\hat{x}(n),\ \hat{x}^2(n),\ \hat{x}^3(n), \ldots, \hat{x}^M(n)\}^*. \tag{4.2.12}$$

Step 4: $y(n)$ passes the downlink to the ground receiver as the input signal of the equalizer after passing the receiving filter again, the output signal of the equalizer is $z(n)$, $d(n)$ is the desired signal of the equalizer, the χ-order delay signal of the transmitted signal (generally 1/2 of the nonlinear order of the equalizer) is taken. The Hammerstein model parameters are estimated by using the stochastic gradient descent method; among them, $\boldsymbol{g}_H = \{g_1, g_2, L, g_K\}$ and $\boldsymbol{h}_H = \{h_0, h_1, L, h_L\}$ are the nonlinear filter and linear filter weight vectors of the Hammerstein model equalizer, respectively. The error function is denoted as $e_1(n) = d(n) - z(n)$.

The tap coefficients of the nonlinear and linear filters are updated by

$$\boldsymbol{g}(n+1) = \boldsymbol{g}(n) + \mu e_1(n)\nabla_3^*(n), \tag{4.2.13}$$

$$\boldsymbol{h}(n+1) = \boldsymbol{h}(n) + \mu e_1(n)\nabla_4^*(n), \tag{4.2.14}$$

where μ denotes a step size.

The gradient of the nonlinear part is obtained by

$$\nabla_3^*(n) = \left\{ \frac{\partial(z(n))}{\partial g_1}, \frac{\partial(z(n))}{\partial g_2}, \ldots, \frac{\partial(z(n))}{\partial g_K} \right\}^*, \tag{4.2.15}$$

where

$$\frac{\partial(z(n))}{\partial g_k} = h_0 y^k(n) + h_1 y^k(n-1) + \mathrm{L} + h_L y^k(n-L), \tag{4.2.16}$$

where $k = 1, 2, \ldots, K$.

If the Wiener model is fully converged, $y(n)$ can be replaced with $\hat{y}(n)$ approximately:

$$y(n) \approx \hat{y}(n). \tag{4.2.17}$$

So, Equation (4.2.16) can be converted into

$$\frac{\partial(z(n))}{\partial g_k} = h_0 \hat{y}^k(n) + h_1 \hat{y}^k(n-1) + \mathrm{L} + h_L \hat{y}^k(n-L). \tag{4.2.18}$$

The gradient of the linear part is estimated as

$$\nabla_4^*(n) = \frac{d(z(n))}{d\boldsymbol{h}} = \{z_h(n), z_h(n-1), \mathrm{L}, z_h(n-L)\}^*. \tag{4.2.19}$$

4.2.2 *Simulation tests*

The transmitted signal was 8PSK and its length was set to 2500. A memoryless nonlinear model of HPA using FIR filters in series was used. IBO (input backoff) was set to 4, the linear filter parameters were given by [0.7962 0.1538 0.0769]. Both the linear and the nonlinear partial orders of the Wiener model were set to 3. The linear and nonlinear partial orders of the Hammerstein model were set to 4. The delay was given by $\chi = 2$, the step size of the equalizer was set to $\mu = 0.0008$.

The Wiener model was initialized to a center tap, and the weight vectors of the linear and nonlinear parts of the equalizer were initialized to $[1, 0, 0, 0]$. Figure 4.7 shows the verification curves of the Wiener model channel estimator and the actual channel output obtained by the method. It can be seen from Figure 4.7 that the Wiener model channel estimator fits well with the actual nonlinear

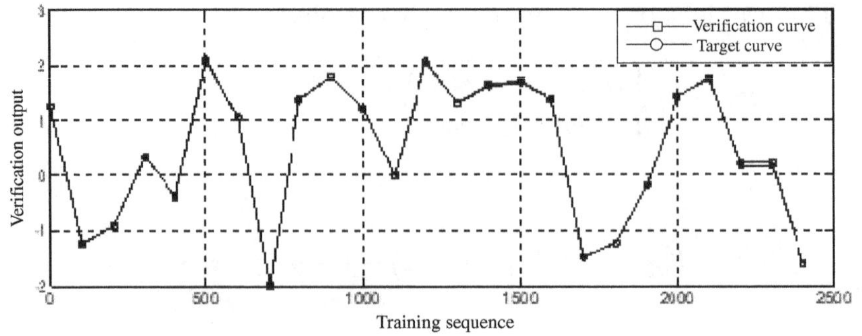

Figure 4.7 Verification curves of the Wiener model.

channel characteristics, so it is feasible to replace $y(n)$ with $\hat{y}(n)$ in the equalizer weight vector iteration.

Using the set of model parameters obtained by the convergence of the Wiener model channel estimator, several experiments were carried out, and the data with the group of data with better performance is selected as the final model parameter. A set of parameters from each of the 100 independent experiments was obtained, and the final estimated parameters of the Wiener model channel are obtained as follows:

$$\text{Linear part: } \boldsymbol{f}_{\text{lin}} = [1.1026 + \jmath\,0.1015, 0.2057 + \jmath\,0.0330, 0.1012 + \jmath\,0.0160],$$

$$\text{Nonlinear part: } \boldsymbol{f}_{\text{non}} = [1.5050 - \jmath\,0.0118, 0.0010 - \jmath\,0.0004, -0.0011 + \jmath\,0.0001].$$

In this way, the Wiener model modeling process of the nonlinear satellite channel is completed. Figure 4.8 shows the MSE curves of the predistortion and adaptive equalization, where the step sizes of the linear and nonlinear parts of the equalizer were set to 0.0008. It can be seen from Figure 4.8 that even if the channel noise superimposed on the predistorter error signal is not considered, the compensation performance using the model-equalization method is still better than the predistortion method. Therefore, in the actual channel environment, the superiority of the model-equalization method is more obvious than the predistortion method. However, the predistortion method converges faster than the model-equalization method. Figure 4.9 shows that the signal constellation after equalization is clear and compact, eliminating the phase rotation caused by nonlinear channels.

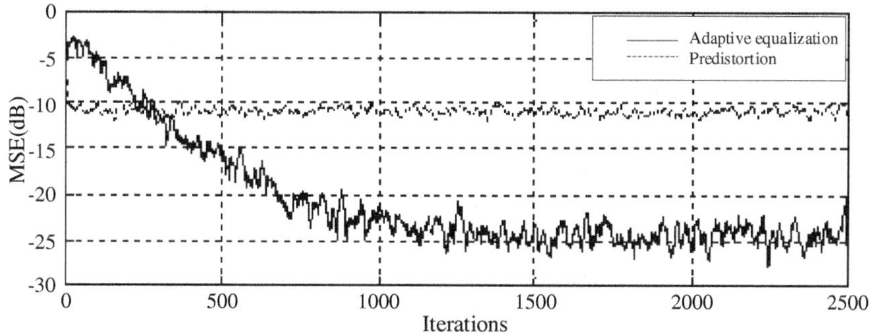

Figure 4.8 Curves of MSE.

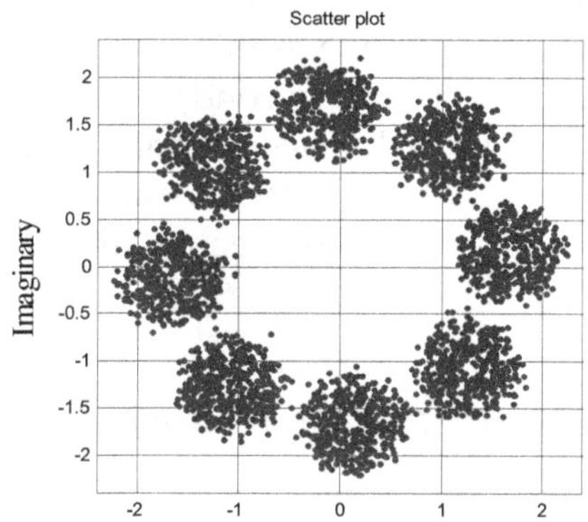

(a)

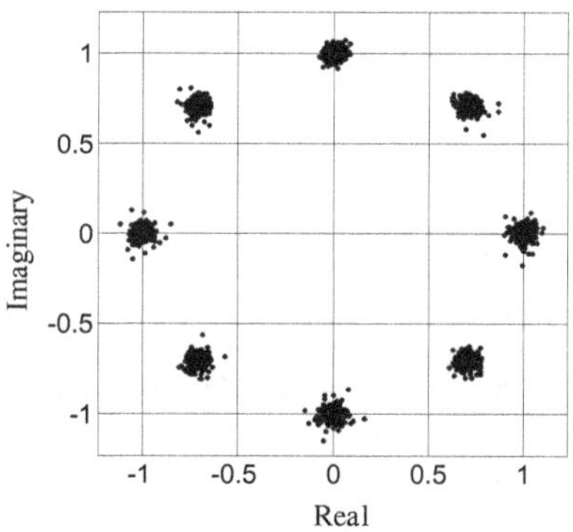

(b)

Figure 4.9 Input–output constellation of model-equalization method: (a) input of the equalizer; (b) output of the equalizer.

4.3 Nonlinear Satellite Channel Model and Equalization System under Alpha-Stable Distributed Noise Background

In existing research, the uplink noises are usually ignored or assumed to be Gaussian white noise. This was done in order to minimize the impact of the uplink noises on modeling to focus research on nonlinear channel modeling. However, the atmospheric noise is usually alpha-stable distributed noise, which is completely different from Gaussian white noise. Under alpha-stable distributed noise, the performance of the Wiener model will be greatly reduced, and it is impossible to approach the nonlinear satellite channel at all. This section first introduces the alpha-stable distributed noise model and describes the model parameters. Second, a simulation method generating alpha-stable distributed noise is introduced. Third, the ANFIS is used to model the nonlinear satellite channel in the alpha-stable distributed noise background, and the modeling results are compared with the Wiener model.

4.3.1 *Alpha-stable distribution model*

The alpha-stable distribution noise is usually expressed as the following characteristic function [15]:

$$\Phi(\omega) = \begin{cases} \exp\{j\delta t - \gamma|\omega|^{\alpha}[1 + j\beta \mathrm{sgn}(\omega)\tan(\pi\alpha/2)]\}, & \alpha \neq 1 \\ \exp\{j\delta t - \gamma|\omega|^{\alpha}[1 + j\beta \mathrm{sgn}(\omega)(2/\pi)\mathrm{lg}|\omega|]\}, & \alpha = 1, \end{cases} \tag{4.3.1}$$

$$\mathrm{sgn}(\omega) = \begin{cases} 1, & \omega > 0 \\ 0, & \omega = 0 \\ -1, & \omega < 0, \end{cases} \tag{4.3.2}$$

where $\alpha, \beta, \delta, \gamma$ are constants. These parameters are defined as follows:

(1) α is a characteristic index, and $\alpha \in (0, 2]$, which represents the trailing thickness of the probability density function of the alpha-stable distribution noise.
(2) β is the symmetric parameter and $\beta \in [-1, 1]$. If $\beta = 0$, the alpha-stable distribution noise has symmetry and is recorded as a

SαS distribution. There is no corresponding analytical expression for the probability density function $f(x)$ of SαS distribution. So, the $S\alpha S$ distribution is defined by its characteristic function $\Phi(\omega)$ as follows:

$$\Phi(\omega) = \int_{-\infty}^{\infty} f(x)e^{j\omega x}dx = \exp(j\delta\omega - \gamma|\omega|^{\alpha}). \tag{4.3.3}$$

(3) δ is a positional parameter, and $\delta \in (-\infty, \infty])$, which indicates the median or mean of the distribution.
(4) γ is a scatter factor, which is similar to the variance of the Gaussian distribution. It determines the deviation degree of alpha-stable distribution from δ.

For the SαS distribution, α determines its shape. If $\alpha \in (1, 2]$, δ denotes the mean of the SαS distribution. If $\alpha \in (0, 1]$, the mean of the SαS distribution does not exist and δ denotes the median of the SαS distribution. The dispersion coefficient γ determines the deviation degree of the SαS distribution from δ.

If $\alpha = 0$ and $\beta = 0$, the $S\alpha S$ distribution is called as the Gaussian distribution; its characteristic function is given by

$$\Phi(\omega) = \exp(j\delta\omega - \gamma|\omega|^2). \tag{4.3.4}$$

If $\delta = 0$ and $\gamma = 1$, the alpha-stable distribution is called as the standard alpha-stable distribution. The inverse of the Fourier transform is performed on the eigenfunction shown in Equation (4.3.1), and the probability density function of the standard alpha-stable distribution is written as

$$f(x;\alpha,\beta) = \begin{cases} \frac{1}{\pi}\int_0^x \exp(-\omega^{\alpha})\cos[x\omega + \beta\omega^{\alpha}\tan(\pi\alpha/2)]d\omega, & \alpha \neq 1 \\ \frac{1}{\pi}\int_0^x \exp(-\omega^{\alpha})\cos\left[x\omega + \beta\omega^{\alpha}(2/\pi)\lg|\omega|\right]d\omega, & \alpha = 1. \end{cases} \tag{4.3.5}$$

It can be seen from Equation (4.3.5) that when the parameter α is determined, the probability density function of the standard alpha-stable distribution meets the condition $f(x;\alpha,\beta) = f(-x;\alpha,-\beta)$ and is bounded and arbitrarily derivable. However, in addition to the Gaussian distribution, the Cauchy distribution, and Pearson distribution, probability distribution density function, and distribution function of the alpha-stable distribution have no closed expression. Of course, the probability density function of the alpha-stable distribution can be represented by series as

$$f(x;\alpha,\beta)=\begin{cases}\dfrac{1}{\pi x}\sum\limits_{k=1}^{x}\dfrac{(-1)^{k-1}}{k!}\Gamma(\alpha k+1)\left(\dfrac{x}{r}\right)^{-\alpha k}\\ \quad\sin\left[\dfrac{k\pi}{2}(\alpha+\xi)\right],0<\alpha<1\\ \dfrac{1}{\pi x}\sum\limits_{k=1}^{x}\dfrac{(-1)^{k-1}}{k!}\Gamma\left(\dfrac{k}{\alpha}+1\right)\left(\dfrac{x}{r}\right)^{k}\\ \quad\sin\left[\dfrac{k\pi}{2\alpha}(\alpha+\xi)\right],0<\alpha\le 2,\end{cases} \tag{4.3.6}$$

where

$x>0$ and

$$\Gamma(x)=\int_0^x t^{x-1}e^{-t}dt. \tag{4.3.7}$$

The computer simulation steps for the α stable distribution of random variables are as follows:

Step 1: Calculate the parameters of the alpha-stable distribution. The parameters of the alpha-stable distribution are calculated by

$$\begin{cases}\begin{cases}\beta_0=\dfrac{2\text{arc}\tan\left(\beta\tan\left(\dfrac{\pi\alpha}{2}\right)\right)}{\pi K(\alpha)}\\ \gamma_2=\gamma\left(1+\beta^2\tan^2\left(\dfrac{\pi\alpha}{2}\right)\right)^{1/(2\alpha)}\end{cases}, & \alpha\neq 1\\ \begin{cases}\beta_0=\beta\\ \gamma_2=\dfrac{2}{\pi}\gamma\end{cases}, & \alpha=1.\end{cases} \tag{4.3.8}$$

The parameter pair (γ,β) to be simulated in the standard parameter system $S_\alpha(\gamma,\beta,\delta)$ is transformed into (γ_2,β_0) in another parameter system.

Step 2: Calculate the random variable X.

The random variable X is calculated as follows:

$$X = \begin{cases} \dfrac{\sin\alpha(V-\gamma_0)}{(\cos\gamma)^{1/\alpha}}\left(\dfrac{\cos(V-\alpha(V-\gamma_0))}{\omega_1}\right)^{(1-\alpha)/\alpha}, & \alpha \neq 1 \\ \left(\dfrac{\pi}{2}+\beta_0\gamma\right)\tan\gamma - \beta_0\log\left(\dfrac{\omega_1\cos\gamma}{\dfrac{\pi}{2}+\beta\gamma}\right), & \alpha = 1, \end{cases} \tag{4.3.9}$$

where $X \sim S_\alpha(1,\beta_0,0)$, $\gamma_0 = -\frac{\pi\beta_0 K(\alpha)}{2\alpha}$, V obeys uniform distribution over $(-\pi/2,\pi/2)$, and ω_1 obeys an exponential distribution with unit mean.

Step 3: Calculate random variable Y.

According to $Y = \gamma_2 X$, $Y \sim S_\alpha(\gamma,\beta,0)$.

Step 4: Calculate the random variable U. Let $U = Y + \delta$. Here,

$$U \sim \begin{cases} S_\alpha(\gamma,\beta,\delta), & \alpha \neq 1 \\ S_1\left(\gamma,\beta,\delta - \dfrac{2}{\pi}\gamma\beta\ln\left(\dfrac{2}{\pi}\gamma\right)\right), & \alpha = 1, \end{cases} \tag{4.3.10}$$

where U is the final random variable to be obtained.

4.3.2 *ANFIS model for nonlinear satellite channels*

The nonlinear structure of the neural networks has a strong approximation ability to nonlinear functions. The ANFIS has the ability to approximate any nonlinear functions [16]. In this section, the ANFIS model of the nonlinear satellite channels and updating of the ANFIS model parameters using the backpropagation (BP) algorithm are discussed. Then, its modeling performance is compared with that of the Wiener model. The ANFIS model combines the advantages of fuzzy reasoning and neural networks, and uses the Takagi–Sugeno (TS) fuzzy model for system identification. Suppose that the three inputs of the TS model are $x(n)$, $x(n-1)$, $x(n-2)$, respectively. After each input is divided into M fuzzy sets, the first-order TS fuzzy model is

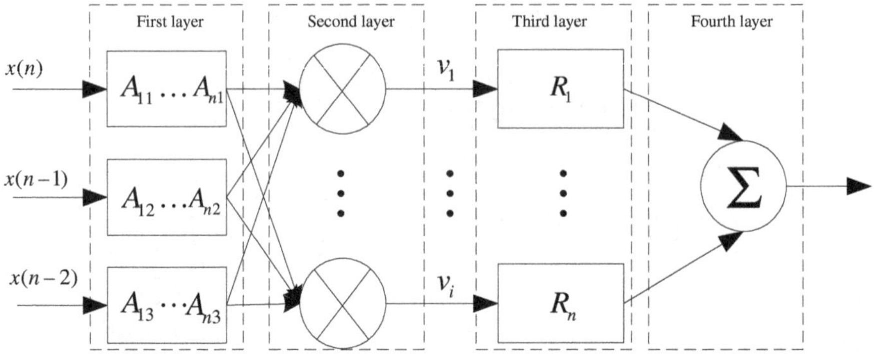

Figure 4.10 ANFIS model.

obtained:

$$\begin{cases} \text{if } x(n) \text{ is } A_{11} \text{ and } x(n-1) \text{ is } A_{12} \text{ and } x(n-2) \text{ is } A_{13} \\ \text{then } R_1 = r_{11}x(n) + r_{12}x(n-1) + r_{13}x(n-2) + r_{14} \\ \qquad\qquad\bullet \\ \qquad\qquad\bullet \\ \qquad\qquad\bullet \\ \text{if } x(n) \text{ is } A_{n1} \text{ and } x(n-1) \text{ is } A_{n2} \text{ and } x(n-2) \text{ is } A_{n3} \\ \text{then } R_n = r_{n1}x(n) + r_{n2}x(n-1) + r_{n3}(n-2) + r_{n4}, \end{cases} \tag{4.3.11}$$

where r_{ij} is the jth parameter of the ith linear function R_i, $i = 1, 2, \ldots, n$, and $j = 1, 2, 3, 4$. Its structure is shown in Figure 4.10.

In Figure 4.10, the forward part of the model from the first layer to the second layer can be regarded as the nonlinear processing of the signal. The latter part of the model from the third layer to the fourth layer are linear processes. In general, the ANFIS model is a series model of a nonlinear and a linear structure, which is similar to the Wiener model. The actual satellite channel can also be regarded as a structure in which the nonlinear TWTA and linear model FIR are connected in series. Therefore, the ANFIS model can be used to approximate the nonlinear satellite channel.

The steps for modeling with ANFIS are as follows:

Step 1: Fuzzify the input signal, that is, the signal is input into a system represented by a fuzzy membership function. Let $x_i(n-j+1)$ denote the jth input signal belonging to the ith fuzzy set. The fuzzy set to which it belongs is represented by

A_{ij}. $\mu_{ij}(x_{ij})$ represents the membership function belonging to $x_i(n-j+1)$.

$$\mu_{ij}(x_i(n-j+1)) = (1 + |(x_i(n-j+1) - c_{ij})/a_{ij}|^{2b_{ij}})^{-1}, \tag{4.3.12}$$

where a_{ij}, c_{ij}, b_{ij} are the parameters, $i = 1, 2, \ldots, n$, and $j = 1, 2, 3$.

Step 2: Calculate the normalized stimulation intensity of the ith fuzzy rule.

$$\nu_i = \frac{\mu_{i1}(x_i(n)) \times \mu_{i2}(x_i(n-1)) \times \mu_{i3}(x_i(n-2))}{\sum_j \mu_{j1}(x_j(n)) \times \mu_{j2}(x_j(n-1)) \times \mu_{j3}(x_j(n-2))}, \tag{4.3.13}$$

where $\mu_{i1}(x_{i1})$, $\mu_{i2}(x_{i2})$, and $\mu_{i3}(x_{i3})$ represent the membership functions of the first, second, and third signals belonging to the ith fuzzy set, respectively.

Step 3: Calculate the output of the linear function.

$$v_i R_i = v_i(r_{i1}x(n) + r_{i2}x(n-1) + r_{i3}x(n-2) + r_{i1}), \tag{4.3.14}$$

where $r_{i1}, r_{i2} \ldots$ denote the parameters of the linear function.

Step 4: Sum and output.

$$y = \frac{\sum_i v_i R_i}{\sum_i v_i}. \tag{4.3.15}$$

In parameter estimation, the antecedent parameters a_{ij}, b_{ij}, and c_{ij} (which are membership function parameters) are regarded as known. After fixing the antecedent parameters, the least square method is used to estimate them. The consequent parameters $r_{i1}, r_{i2}, \ldots,$ (which are linear function parameters) are first fitted by the least square method, and then the consequent parameters are fixed, finally BP [17] and least mean square (LMS) algorithms are used to identify the antecedent parameters a_{ij}, b_{ij}, and c_{ij}. After the antecedent parameters and consequent parameters are obtained, the channel modeling process is completed. The nonlinear satellite channel structure based on the ANFIS estimator is shown in Figure 4.11.

In Figure 4.11, $\mathbf{w}(n)$ is the uplink noise and z^{-1} is the signal delay. If the training sequence is a satellite receiving signal, it is delayed for a certain period of time and used as a training sequence of the ANFIS system.

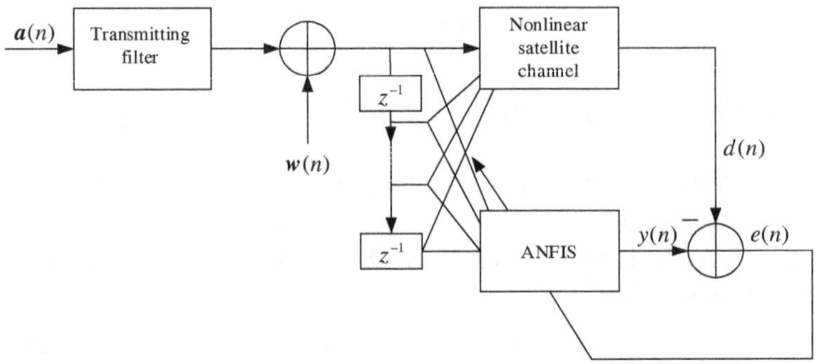

Figure 4.11 ANFIS estimation of nonlinear satellite channels.

4.3.3 *Simulation tests*

The transmitted signal was an 8PSK signal, and the high-power amplifier adopted an FIR filter in series with a memoryless nonlinear model structure and denoted as [0.7962 0.1538 0.07697]. If the uplink environmental noise was Gaussian white noise, the SNR was set to 5 dB. If the uplink environmental noise was the alpha-stable distribution noise, the SNR was set to 1 dB. The nonlinear part includes the amplitude–amplitude conversion and amplitude–phase conversion effects of the signal, as described in Equations (4.2.6) and (4.2.7). They are now approximated by the Wiener model channel estimator and the ANFIS channel estimator, respectively.

For the Wiener model channel estimator, the linear and nonlinear partial orders of the Wiener model were set to 3, the center tap was initialized, and the step sizes of the linear and nonlinear parts were set to 0.01 and 0.03, respectively.

For the ANFIS model channel estimator, the delay order of the satellite received signal is set to 3, that is, the training sequence is given by $[x(n), x(n-1), x(n-2)]$, the neural network is a three input–output structure, the membership function is a bell-shaped function, and each input signal has three membership functions. The number of training sequence was set to 40. Figure 4.12 shows the Wiener model channel estimator and the ANFIS model channel estimator validation graph when the uplink noise was Gaussian white noise and the input sequence length was 500. It can be seen from Figure 4.12(a) that the fitting of the model validation curve to the target curve is

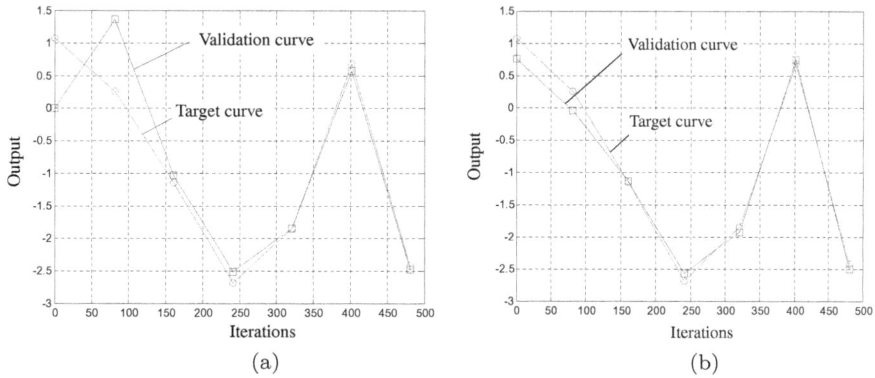

Figure 4.12 Validation results of two models under Gaussian white noise when the training sequence length was 500: (a) validation diagram of the Wiener model and (b) validation diagram of the ANFIS model.

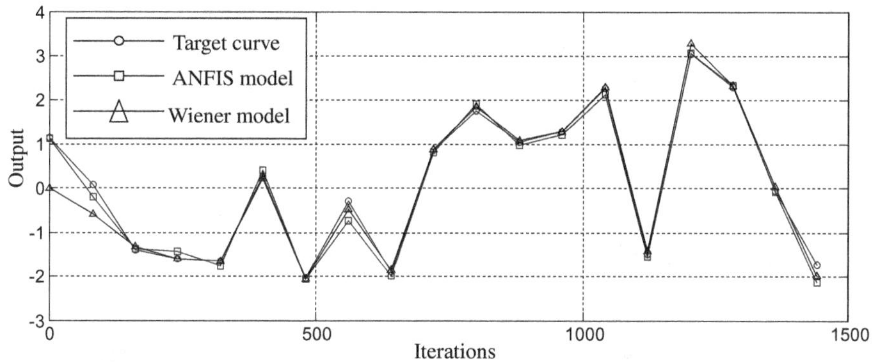

Figure 4.13 Validation results of two models in white Gaussian noise environment.

not good after 500 iterations and from Figure 4.12(b) that the fitting of the ANFIS validation chart is obviously better.

Figure 4.13 shows the validation diagram when the length of the transmitted signal was 1500 in the Gaussian white noise environment. It can be seen from Figure 4.13 that after the Wiener model converges, the fitting of the Wiener model is close to that of the ANFIS, but the ANFIS model can better fit the nonlinearity of the satellite channel from the beginning of the iteration.

Table 4.2 shows the root mean square error (RMSE) of the two model channel estimators for different training sequence lengths.

Table 4.2 Comparison of RMSEs.

Model Type	Sequence Length	RMSE
Wiener model	500	0.8447
	1500	0.2125
	3500	0.2034
ANFIS model	500	0.1307
	1500	0.1874
	3500	0.1877

When the signal points were 1500 and 3500, the first 400 signal points were eliminated, and the RMSE of the Wiener model was calculated. In other words, only the performance of the convergent Wiener model was considered. It can be seen from Table 4.2 that the performance of the ANFIS model channel estimator is significantly better than that of the Wiener model.

Figure 4.14 shows the validation diagram when the uplink signal was α-stable distributed noise environment and the transmitted signal length was 2000. The linear and nonlinear partial step sizes of the Wiener model were 0.0000000001 and 0.0000000003, respectively. It can be seen from Figure 4.14(b) that the ANFIS model channel estimator can fit the nonlinear channel well.

4.4 Nonlinear Satellite Channel Modeling Algorithm Based on TWTA and Group Delay

In addition to the nonlinear effects of the TWTA, the group delay effect of the satellite channels is also an important factor of the nonlinear distortion. The effects of the nonlinearity and group delay of the TWTA act together on the transmitted signal, which resulting in serious amplitude distortion and phase rotation of the received signal at the receiver. In the satellite channel modeling, these two effects should be considered at the same time. At present, if only one effect is considered, the main defects of the analytical model are as follows: the analytical model does not necessarily match the characteristics of the channel structure. Therefore, it is difficult to obtain the channel structure characteristics from the analytical model structure, which

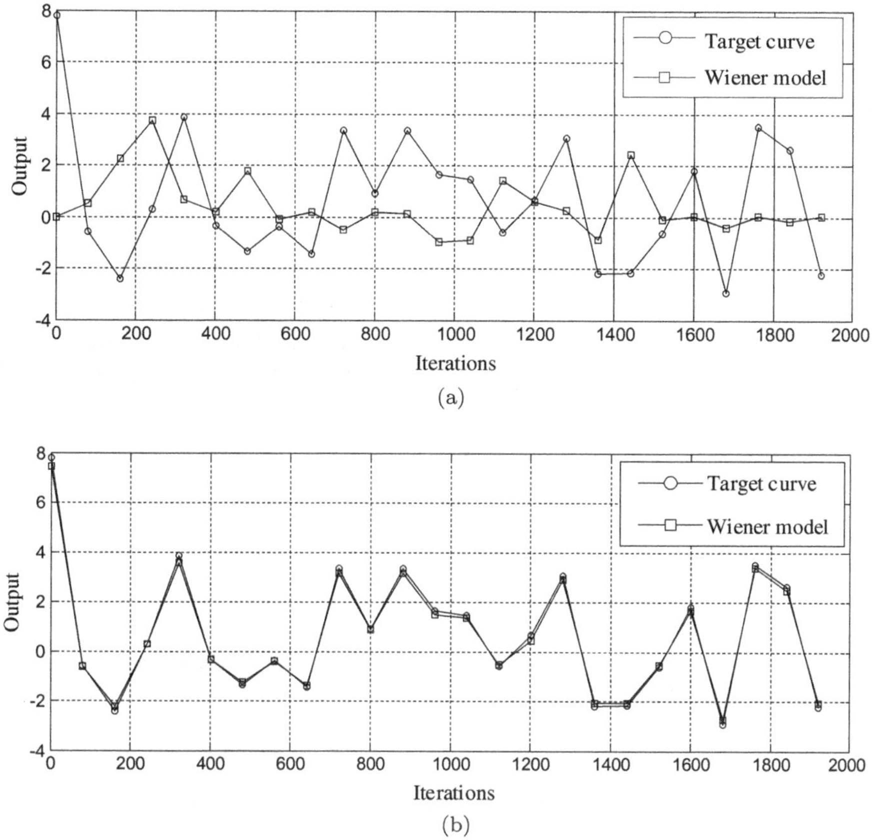

Figure 4.14 Validation diagrams of two models in the stable distributed noise environment: (a) validation diagram of Wiener model and (b) validation diagram of ANFIS model.

is not conducive to understanding and analysis. In addition, how to determine the appropriate analytical model is a difficult point of the research. In this section, the design method of the linear group delay channel is introduced.

The combined effects of both the nonlinearity of the TWTA and group delay are analyzed. The nonlinear satellite channel model, based on channel prior information, is constructed and verified through simulation.

4.4.1 *Design of linear group delay filter*

According to [5] and Equation (4.1.12), the following equation is obtained:

$$\frac{\omega d}{\omega_{\max}} = -\frac{d\theta(\omega)}{d\omega}, \tag{4.4.1}$$

where $\omega_{\max}$ is the maximum frequency of the signal, $\omega \in (-\omega_{\max}, \omega_{\max})$ denotes the frequency of the signal, d is the maximum group delay distortion, and $\theta(\omega)$ represents the phase–frequency response of the signal. On integrating both sides of Equation (4.4.1) with respect to angular frequency ω, $\theta(\omega)$ is obtained as follows:

$$\theta(\omega) = -\frac{\omega^2 d}{2\omega_{\max}} + \varphi_0, \tag{4.4.2}$$

where φ_0 is the initial phase; let $\varphi_0 = 0$.

In order to obtain the discrete impulse response $h(n)$ of the channel, the continuous impulse response $h(t)$ of the channel must first be obtained, then the samples were considered. Since the phase–frequency response of the signal is known, as long as the amplitude–frequency response is obtained, the frequency response expression of the system is constructed, and then inverse Fourier transform is used to obtain the continuous impulse response of the system.

The system's amplitude-frequency response uses a low-pass filter with cosine step-down characteristics [18]:

$$H(\omega) = \begin{cases} T_s & 0 \le \omega \le (1-\varepsilon)\pi/T_s \\ T_s/2 + (T_s/2) & \\ \sin\left[T_s/(2\varepsilon)\right]\left[(\pi/T_s) - \omega\right] & (1-\varepsilon)\pi/T_s \le \omega < (1+\varepsilon)\pi/T_s \\ 0 & \omega > (1+\varepsilon)\pi/T_s, \end{cases} \tag{4.4.3}$$

where T_s is the symbol duration and ε is the step-down ratio of the filter. The frequency response of the system is given by

$$H(e^{j\omega}) = H(\omega)e^{-j\frac{\omega^2 d}{2\omega_{\max}}}. \tag{4.4.4}$$

Inverse Fourier transform is performed on $H(\omega)$ and $\theta(\omega)$ to obtain $h(t)$, then $h(t)$ is sampled according to a limited time baud interval, and finally, the discrete-channel impulse response $h(n)$ is computed.

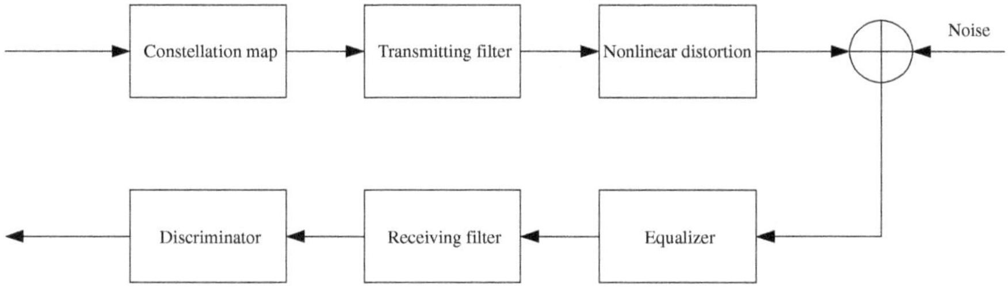

Figure 4.15 Equivalent baseband model for satellite channels.

4.4.2 *Combined effects of TWTA nonlinearity and group delay*

In digital satellite communication systems, the transmitted signal is processed by constellation mapping (i.e., modulation) as input of the transmitting filter, and then reaches the satellite receiver through the uplink of the satellite communication system. The internal TWTA in the satellite amplifies the received signals and then forwards the amplified signal. The satellite channel equivalent baseband model is shown in Figure 4.15 [19].

In Figure 4.15, the nonlinear channel is a joint channel considering both the nonlinearity of the TWTA and the group delay effect. It is known from Equations (4.2.6) and (4.2.7) that the nonlinearity of the TWTA is related to the instantaneous power of the input signal. Group delay and signal are related to the nonlinear phase caused by transmitter, wireless environment, high-power amplifier, etc. These two effects do not affect each other and can be considered independent of each other. Further analysis shows that the linear delay of the linear part of the TWTA and the wireless environment will cause a certain amplitude distortion of the input signal. The amplitude of the input signal in the nonlinear part of the group delay amplifier is different from that of the input signal without considering the group delay. It can be seen from Figure 4.15 that the group delay indirectly affects the amplitude conversion effect and the phase-modulation conversion effect of the nonlinear part of the amplifier.

In order to simplify the analysis, the linear part of the transmitting filter, uplink radio environment and high-power amplifier are combined into a linear combined channel when considering the nonlinearity of the TWTA and the group delay effect. This combined channel is depicted in Figure 4.16.

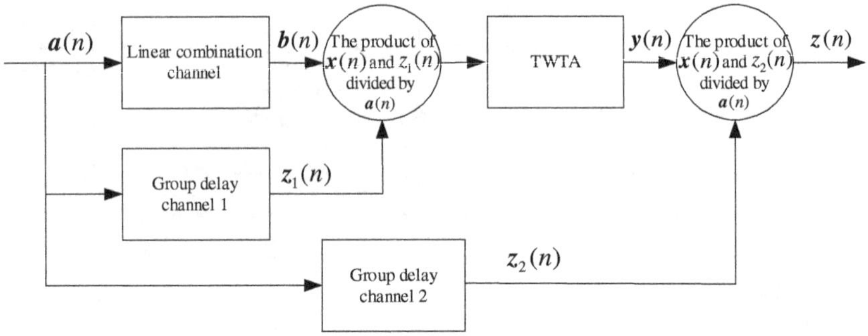

Figure 4.16 Satellite channel model considering both nonlinearity and group delay of the TWTA.

In Figure 4.16, $a(n)$ is the input signal of the system and $b(n)$ is the output of the linear combination channel when the input signal $a(n)$ passes through it. *Group delay channel 1* represents a group delay channel of a linear combined channel. $z_1(n)$ is the output signal of $a(n)$ through the group delay channel 1. $x(n)$ is recorded as the output signal of the linear combination channel when the input signal $a(n)$ passes through it. *The group delay channel 2* is supposed to represent the group delay channel of the whole system, which is the cascade of the *group delay channel 1* and the nonlinear amplifier group delay channel. However, in another branch, $a(n)$ has been multiplied by the group delay effect of the linear combination channel when passing through the combined channel, and if it is again multiplied by the group delay effect of the whole system, it will multiply once more the group delay effect of the linear filter. Therefore, the *group delay channel 2* here should represent the group delay channel of the nonlinear amplifier. $z_2(n)$ denotes the output signal of the *group delay channel 2* when the input signal $a(n)$ passes through it. $y(n)$ is regarded as the output signal of $x(n)$ affected by the amplitude conversion effect and the amplitude–phase conversion effect of the TWTA, and $z(n)$ is the output signal of the whole system.

Further analysis shows that the essence of the signal, which passes through the linear filter, nonlinear amplifier, and group delay channel, is to cause amplitude distortion and phase rotation in the input signal. Accordingly, the following equations are obtained:

$$b(n) = a(n)G_1(n)e^{j\phi_1(n)}, \tag{4.4.5}$$

$$z_1(n) = a(n)G_2(n)e^{j\phi_2(n)}, \tag{4.4.6}$$

$$x(n) = a(n)G(n)e^{j\phi(n)}, \tag{4.4.7}$$

$$y(n) = x(n)g_1(n)e^{j\varphi_1(n)}, \tag{4.4.8}$$

$$z_2(n) = a(n)g_2(n)e^{j\varphi_2(n)}, \tag{4.4.9}$$

$$z(n) = a(n)g(n)e^{j\varphi(n)}, \tag{4.4.10}$$

where $G_1(n)$, $G_2(n)$, and $G(n)$ are the amplitude gains of the linear combination channel, *group delay channel 1*, and the channel before the TWTA, respectively. $\phi_1(n), \phi_2(n)$, and $\phi(n)$ are the phases of the respective channels. $g_1(n), g_2(n)$, and $g(n)$ are the nonlinear channel of the TWTA with amplitude–amplitude conversion effect and amplitude–phase conversion effect, the *group delay channel 2*, and amplitude gain of the whole system, respectively. $\varphi_1(n), \varphi_2(n)$, and $\varphi(n)$ are the phases of the respective channels. Based on the independence of the systems in Figure 4.16, the corresponding group delay channels are written as

$$G(n) = G_1(n)G_2(n), \tag{4.4.11}$$

$$\phi(n) = \phi_1(n) + \phi_2(n), \tag{4.4.12}$$

$$g(n) = g_1(n)g_2(n), \tag{4.4.13}$$

$$\varphi(n) = \varphi_1(n) + \varphi_2(n). \tag{4.4.14}$$

Combining Equations (4.4.5) and (4.4.10), we get

$$x(n) = \frac{b(n)z_1(n)}{a(n)}, \tag{4.4.15}$$

$$z(n) = \frac{y(n)z_2(n)}{a(n)}. \tag{4.4.16}$$

4.4.3 *Nonlinear channel model based on channel prior information*

In this section, the construction method of the satellite channel analytical model considering both the nonlinearity of the TWTA and the group delay is discussed. Based on the Wiener model, the satellite channel nonlinear model is introduced to make full use of the known information of the channel to establish the channel prior information. Finally, the proposed model is verified [20–22].

4.4.3.1 *Prior information of nonlinear satellite channels*

According to the analyses in the previous section, the *a priori* information of the satellite channel has the following three types:

***A priori* information 1:** The linear combined channel in Figure 4.16 can be viewed as a linear filter with memory.

***A priori* information 2:** The group delay channel is a nonlinear channel.

***A priori* information 3:** It is known from Equations (4.4.15) and (4.4.16) that the linear combined channel and *group delay channel 1* are regarded as two independent filters, and the output of the system is before the nonlinear amplifier. The product of the output signals of the two filters is divided by the input signal of the system. The non-group delay channel of the whole system and *group delay channel 2* are regarded as two independent nonlinear filters, and the output of the whole system is the product of the output signal of the two filters, which is divided by the input signal.

In addition, the following instructions are required:

(1) For *a priori information 1*, the transmitting filter and the wireless environment cause a linear filtering effect and the linear portion of the TWTA causes a linear filtering effect and a memory effect, so the entire linear combined channel can be regarded as a linear filter with memory.
(2) Nonlinear channel includes Memoryless nonlinear channel and memory nonlinear channel. For *a priori information 2*, which kind of nonlinear group delay channel cannot be determined. If the group delay channel is a complex nonlinear channel with memory, the interaction between nonlinearity and memory effect in the channel makes the channel nonlinearity more complex. In this case, using the general nonlinear analytical model to identify the channel may lead to divergence.

4.4.3.2 *Modeling process*

The main differences between the modeling method given in this section and the traditional method are as follows: the traditional method directly uses a specific nonlinear function (analytical model) to approximate a nonlinear channel. In this section, the existing analytical expressions of the channel structure are divided into linear and nonlinear parts. Then, according to the *a priori* information of the known channel, the corresponding linear and nonlinear parts are used to represent the channel structure deduced from the *a priori* information and combined with the known channel information to complete the channel modeling. Finally, the new analytical model is used to approximate the nonlinear channel until the channel modeling process is completed.

Taking the Wiener model as an example, the expression of the Wiener model is written as

$$X(n) = \sum_{i=0}^{N} W_i Y(n-i), \tag{4.4.17}$$

$$Y(n) = \sum_{j=1}^{M} F_j (X(n))^j, \tag{4.4.18}$$

where $W_i (i = 1, 2, \ldots, N)$ represents the coefficients of the linear part of the Wiener model and $F_j (j = 1, 2, \ldots, M)$ represents the coefficients of the nonlinear part of the Wiener model. $X(n)$ and $Y(n)$ are the outputs of the linear and nonlinear parts, respectively. In this section, the linear and nonlinear partial expressions of the Wiener model are used to construct the linear and nonlinear parts of the new model.

Let the input signal of the whole system be $a(n)$; it can be shown from *a priori information 1* and *a priori information 3* that the input signal first passes through a cascaded filter with a memory linear combined channel and its corresponding group delay filter, which uses the linear portion of the Wiener model to represent the linear combined channel and the nonlinear portion to represent the group delay filter. So, the input–output relationship is written as

$$x(n) = \sum_{m_1=0}^{M_1} c_{1m_1} a(n-m_1) \frac{-\sum_{m_2=0}^{M_2} c_{2m_2} [a(n)]^{m_2}}{a(n)}, \tag{4.4.19}$$

where c_{1m_1} is the coefficient of the linear filter and $m_1 = 0, 1, \ldots, M_1$, c_{2m_2} is the coefficient of the group delay filter, and $m_2 = 0, 1, \ldots, M_2$. According to the amplitude–amplitude conversion and amplitude–phase conversion effects of the TWTA combined with the *a priori* information 1, the following equations are obtained:

$$\rho(n) = |x(n)|^2, \tag{4.4.20}$$

$$G(n) = \sum_{m_3=1}^{M_3} c_{3m_3} [\rho(n)]^{m_3}, \tag{4.4.21}$$

$$p(n) = \sum_{m_4=1}^{M_4} c_{4m_4} [\rho(n)]^{m_4}, \tag{4.4.22}$$

$$y(n) = x(n) G(n) \exp[jp(n)], \tag{4.4.23}$$

where $G(n)$ and $p(n)$ are the amplitude and phase changes of the nonlinear amplifier, respectively. $c_{3m_3}(m_3 = 0, 1, 2, \ldots, M_3)$ is the weight of the amplitude–amplitude conversion effect filter. $c_{4m_4}(m_4 = 0, 1, 2, \ldots, M_4)$ is the weight of the amplitude–phase conversion effect filter. c_{4m_5} is the weight of the *group delay channel 2* and $m_5 = 0, 1, \ldots, M_5$. $\rho(n)$ is the instantaneous power of x at time n. From *a priori* information 3, the output $z(n)$ of the entire system is obtained by

$$z(n) = y(n) \frac{\sum_{m_5=1}^{M_5} c_{5m_5} [a(n)]^{m_5}}{a(n)}. \tag{4.4.24}$$

According to the *a priori information 2*, in addition to knowing that the group delay channel is a nonlinear channel, it is impossible to obtain more information that is helpful for the construction of the analytical model. If the group delay channel is represented directly by the nonlinear part of the Wiener model, it will cause a problem: if the group delay channel is a complex memorized nonlinear channel and cannot be expressed as a cascaded structure with memory linear structure and memoryless nonlinear structure, the nonlinear part of the Wiener model cannot describe this nonlinearity well. If the nonlinear model of the group delay diverges in the process of approximating with the actual channel, the constructed entire analytical model in the process of approaching the channel may not even converge. In order to solve this problem, a nonlinear soft-decision function [22] is introduced as follows:

$$f(x) = \frac{1 - e^{-x}}{1 + e^{-x}}. \tag{4.4.25}$$

The function of this soft-decision function is to control the convergence of its independent variable function. If the analytical expression of the constructed group delay function can converge in the process of approaching the real channel, its performance is maintained. If it cannot converge, its output is limited to maintain the performance of the entire model.

The function of the soft-decision function is now analyzed. For Equation (4.4.25), if its independent variable x converges, i.e., $x \to 0$, then

$$\lim_{x\to 0} f(x) = \lim_{x\to 0} \frac{1 - \exp(-x)}{1 + \exp(-x)} = \frac{1 - \exp(0)}{1 + \exp(0)} = 0. \tag{4.4.26}$$

Equation (4.4.26) shows that, when the independent variable x of the soft-decision function converges, the soft-decision function also converges, that is, the soft-decision function does not improve the performance of the original independent variable function. If the independent variable x diverges, i.e., $x \to \infty$, then

$$\lim_{x\to \infty} f(x) = \lim_{x\to \infty} \frac{1 - \exp(-x)}{1 + \exp(-x)} = \frac{1 - \exp(-\infty)}{1 + \exp(-\infty)} = 1. \tag{4.4.27}$$

Equation (4.4.27) shows that, when the independent variable x of the soft-decision function diverges, the soft-decision function approaches one. That is, when the independent variable x diverges, the soft-decision function forces its own output to approach constant 1. Thereby, the influence of the divergent independent variable function on the whole system is eliminated. According to the analysis, the soft-decision function (4.4.25) can control the convergence of the independent function.

According to Equation (4.4.25), Equations (4.4.19) and (4.4.24) are rewritten as follows:

$$x(n) = \sum_{m_1=0}^{M_1} c_{1m_1} a(n-m_1) \frac{1-\exp\left(-\sum_{m_2=0}^{M_2} c_{2m_2}[a(n)]^{m_2}\right)}{a(n)\left\{1+\exp\left(-\sum_{m_2=0}^{M_2} c_{2m_2}[a(n)]^{m_2}\right)\right\}}, \tag{4.4.28}$$

$$z(n) = y(n) \frac{1-\exp\left(-\sum_{m_5=0}^{M_5} c_{5m_5}[a(n)]^{m_5}\right)}{a(n)\left\{1+\exp\left(-\sum_{m_5=0}^{M_5} c_{5m_5}[a(n)]^{m_5}\right)\right\}}. \tag{4.4.29}$$

It is obvious from Equations (4.4.19)–(4.4.27) that a nonlinear satellite channel analytical model based on the *a priori* information of the channel and both amplifier nonlinearity and group delay is obtained. The reality of the modeling process of the system is the process of identifying the channel using the corresponding analytical method. The cost function of the system identification is defined as

$$J(n) = E\{|e(n)|^2\}, \tag{4.4.30}$$

where $e(n) = z_{\text{out}}(n) - z(n)$ and $z_{\text{out}}(n)$ are actual outputs of the channel. System identification is to update the weight vectors of each part of the analytical model continuously by using the random gradient method to make the output of the analytical model approach the channel output as much as possible. That is, let $z(n)$ approach $z_{\text{out}}(n)$ as close as possible. The instantaneous MSE is used to replace the statistical MSE, and the weight vector of each part of the model is updated by the stochastic gradient method. The iterative formula of the weight vector is given by

$$\boldsymbol{C}_{1m_1}(n+1) = \boldsymbol{C}_{1m_1}(n) + \lambda e(n) \frac{\partial e^*(n)}{\partial \boldsymbol{C}_{1m_1}(n)}, \tag{4.4.31}$$

$$\boldsymbol{C}_{2m_2}(n+1) = \boldsymbol{C}_{2m_2}(n) + \lambda e(n) \frac{\partial e^*(n)}{\partial \boldsymbol{C}_{2m_2}(n)}, \tag{4.4.32}$$

$$\boldsymbol{C}_{3m_3}(n+1) = \boldsymbol{C}_{3m_3}(n) + \lambda e(n) \frac{\partial e^*(n)}{\partial \boldsymbol{C}_{3m_3}(n)}, \tag{4.4.33}$$

$$\boldsymbol{C}_{4m_4}(n+1) = \boldsymbol{C}_{4m_4}(n) + \lambda e(n)\frac{\partial e^*(n)}{\partial \boldsymbol{C}_{4m_4}(n)}, \tag{4.4.34}$$

$$\boldsymbol{C}_{5m_5}(n+1) = \boldsymbol{C}_{5m_5}(n) + \lambda e(n)\frac{\partial e^*(n)}{\partial \boldsymbol{C}_{5m_5}(n)}, \tag{4.4.35}$$

where $\boldsymbol{C}_{\boldsymbol{1m}_1} = \{c_{11}, c_{12}, \ldots, c_{1M_1}\}$, $\boldsymbol{C}_{\boldsymbol{2m}_2} = \{c_{21}, c_{22}, \ldots, c_{2M_2}\}$, $\boldsymbol{C}_{\boldsymbol{3m}_3} = \{c_{31}, c_{32}, \ldots, c_{3M_3}\}$, $\boldsymbol{C}_{\boldsymbol{4m}_4} = \{c_{41}, c_{42}, \ldots, c_{4M_5}\}$, and $\boldsymbol{C}_{\boldsymbol{5m}_5} = \{c_{51}, c_{52}, \ldots, c_{5M_5}\}$ are tap coefficients for each part of the model, respectively. Let

$$k_1(n) = f\left\{\sum_{m_2=1}^{M_2} c_{2m_2}[a(n)]^{m_2}\right\}, \tag{4.4.36}$$

$$k_2(n) = f\left\{\sum_{m_5=1}^{M_5} c_{2m_5}[a(n)]^{m_5}\right\}, \tag{4.4.37}$$

$$\begin{aligned} k_3(n) &= \frac{df\left(\sum_{m_2=1}^{M_2} c_{2m_2}[a(n)]^{m_2}\right)}{d\left(\sum_{m_2=1}^{M_2} c_{2m_2}[a(n)]^{m_2}\right)} \\ &= \frac{2\exp\left(-\sum_{m_2=1}^{M_2} c_{2m_2}[a(n)]^{m_2}\right)}{\left\{1+\exp\left(-\sum_{m_2=1}^{M_2} c_{2m_2}[a(n)]^{m_2}\right)\right\}^2}, \end{aligned} \tag{4.4.38}$$

$$\begin{aligned} k_4(n) &= \frac{df\left(\sum_{m_5=1}^{M_5} c_{5m_5}[a(n)]^{m_5}\right)}{d\left(\sum_{m_5=1}^{M_5} c_{5m_5}[a(n)]^{m_5}\right)} \\ &= \frac{2\exp\left(-\sum_{m_5=1}^{M_5} c_{5m_5}[a(n)]^{m_5}\right)}{\left\{1+\exp\left(-\sum_{m_5=1}^{M_5} c_{5m_5}[a(n)]^{m_5}\right)\right\}^2}. \end{aligned} \tag{4.4.39}$$

Symbols $\nabla_1(n), \nabla_2(n), \nabla_3(n), \nabla_4(n)$, and $\nabla_5(n)$ are used to denote $\frac{\partial e(n)}{\partial \boldsymbol{C}_{1m_1}(n)}$, $\frac{\partial e(n)}{\partial \boldsymbol{C}_{2m_2}(n)}$, $\frac{\partial e(n)}{\partial \boldsymbol{C}_{3m_3}(n)}$, $\frac{\partial e(n)}{\partial \boldsymbol{C}_{4m_4}(n)}$, and $\frac{\partial e(n)}{\partial \boldsymbol{C}_{5m_5}(n)}$, respectively.

The derivation process of $\nabla_1(n)$ is as shown from Equations (4.4.40)–(4.4.44):

$$\nabla_1(n) = \frac{\partial e(n)}{\partial \mathbf{C}_{1m_1}(n)} = \frac{\partial s(n)}{\partial y(n)} \times \left(G(n)\exp[jp(n)] + \frac{dy(n)}{dG(n)}\frac{dG(n)}{dx(n)} + \frac{dy(n)}{dp(n)}\frac{dp(n)}{dx(n)} \right) \times \frac{\partial x(n)}{\partial \boldsymbol{C}_{1m_1}(n)}, \tag{4.4.40}$$

where

$$\frac{\partial s(n)}{\partial y(n)} = \frac{k_2(n)}{a(n)}, \tag{4.4.41}$$

$$\frac{dy(n)}{dG(n)}\frac{dG(n)}{dx(n)} = \exp(jp(n))x^*(n)\sum_{m_3=1}^{M_3} m_3 c_{3m_3}(\rho(n))^{m_3-1}, \tag{4.4.42}$$

$$\frac{dy(n)}{dp(n)}\frac{dp(n)}{dx(n)} = j\exp(jp(n))x^*(n)\sum_{m_4=1}^{M_4} m_4 c_{4m_4}(\rho(n))^{m_4-1}, \tag{4.4.43}$$

$$\frac{\partial x(n)}{\partial \boldsymbol{C}_{1m_1}(n)} = \frac{k_1(n)}{a(n)}\{a(n), a(n-1), \ldots, a(n-M_1+1)\}, \tag{4.4.44}$$

where

$$\frac{\partial s(n)}{\partial y(n)} = \frac{k_2(n)}{a(n)}, \tag{4.4.41}$$

$$\frac{dy(n)}{dG(n)}\frac{dG(n)}{dx(n)} = \exp(jp(n))x^*(n)\sum_{m_3=1}^{M_3} m_3 c_{3m_3}(\rho(n))^{m_3-1}, \tag{4.4.42}$$

$$\frac{dy(n)}{dp(n)}\frac{dp(n)}{dx(n)} = j\exp(jp(n))x^*(n)\sum_{m_4=1}^{M_4} m_4 c_{4m_4}(\rho(n))^{m_4-1}, \tag{4.4.43}$$

$$\frac{\partial x(n)}{\partial \boldsymbol{C}_{1m_1}(n)} = \frac{k_1(n)}{a(n)}\{a(n), a(n-1), \ldots, a(n-M_1+1)\}. \tag{4.4.44}$$

The derivation process of $\nabla_2(n)$ is as shown in Equations (4.4.45) and (4.4.46).

$$\begin{aligned}\nabla_2(n) &= \frac{\partial e(n)}{\partial \boldsymbol{C}_{2m_2}(n)}\\ &= \frac{\partial z(n)}{\partial y(n)}\left(G(n)\exp[jp(n)] + \frac{dy(n)}{dG(n)}\frac{dG(n)}{dx(n)} + \frac{dy(n)}{dp(n)}\frac{dp(n)}{dx(n)}\right)\\ &\quad \times \frac{\partial x(n)}{\partial \boldsymbol{C}_{2m_2}(n)},\end{aligned} \tag{4.4.45}$$

where

$$\begin{aligned}\frac{\partial x(n)}{\partial \boldsymbol{C}_{2m_2}(n)} &= k_3(n)\frac{\sum_{m_1=0}^{M_1} c_{1m_1}a(n-m_1)}{a(n)}\\ &\quad \times \left\{a(n), a^2(n), \ldots, a^{M_2}(n)\right\}.\end{aligned} \tag{4.4.46}$$

The derivation process of $\nabla_3(n)$ is as shown from Equations (4.4.47)–(4.4.49).

$$\nabla_3(n) = \frac{\partial e(n)}{\partial \boldsymbol{C}_{3m_3}(n)} = \frac{\partial z(n)}{\partial y(n)}\frac{dy(n)}{dG(n)}\frac{\partial G(n)}{\partial \boldsymbol{C}_{3m_3}(n)}, \tag{4.4.47}$$

where

$$\frac{dy(n)}{dG(n)} = x(n)\exp[jp(n)], \tag{4.4.48}$$

$$\frac{\partial G(n)}{\partial \boldsymbol{C}_{3m_3}(n)} = \{\rho(n), \rho^2(n), \ldots, \rho^{M_3}(n)\}. \tag{4.4.49}$$

The derivation process of $\nabla_4(n)$ is written as

$$\nabla_4(n) = \frac{\partial e(n)}{\partial \boldsymbol{C}_{4m_4}(n)} = \frac{\partial s(n)}{\partial y(n)}\frac{dy(n)}{dp(n)}\frac{\partial p(n)}{\partial \boldsymbol{C}_{4m_4}(n)}, \tag{4.4.50}$$

where

$$\frac{dy(n)}{dp(n)} = jx(n)\exp[jp(n)]G(n), \tag{4.4.51}$$

$$\frac{\partial p(n)}{\partial \boldsymbol{C}_{4m_4}(n)} = \{\rho(n), \rho^2(n), \ldots, \rho^{M_4}(n)\}. \tag{4.4.52}$$

The derivation process of $\nabla_5(n)$ is written as

$$\begin{aligned}\nabla_5(n) &= \frac{\partial e(n)}{\partial \boldsymbol{C}_{5m_5}(n)} = \frac{\partial s(n)}{\partial \boldsymbol{C}_{5m_5}(n)} \\ &= \frac{y(n)}{a(n)} k_4(n)\{a(n), a^2(n), \ldots, a^{M_5}(n)\}. \end{aligned} \tag{4.4.53}$$

It can be shown from Equations (4.4.39)–(4.4.52) that the cost of fully considering the channel prior information is the complex gradient derivation process. On this basis, the whole derivation process must be repeated again once for considering one more channel prior information each time.

4.4.3.3 *Simulation*

In order to verify the effectiveness of the proposed method, the 8PSK signal was used as the training signal of the model and the satellite channel shown in Figure 4.15. The parameters of the linear combined channel were [0.7962 0.1538 0.0769]. The nonlinear effect of the TWTA is described by the deformation model of the Salah model. As shown in Equations(4.2.5) and (4.2.7), $\alpha_A = 2.1587$, $\beta_A = 1.1517$, $\alpha_\phi = 4.0033$, $\beta_\phi = 9.1040$. The group delay channel parameters of the whole system were given by

$$\begin{aligned} h = [&0.0091 + j0.0014 - 0.0247 + j0.0044\ 0.0549 - j0.0056 - 0.1246 \\ &- j0.0149\ 0.2381 + j0.2518\ 0.7079 - j0.4688 - 0.1246 \\ &- j0.0149\, 0.0549 - j0.0056 - 0.0247 + j0.0044\, 0.0091 + j0.0014]. \end{aligned}$$

The group delay channel of the whole system is a cascade of the *group delay channel 1* and *group delay channel 2*, i.e.,

$$h_1 \otimes h_1 = h, \tag{4.4.54}$$

where $\otimes$ denotes convolution. Discrete Fourier transform (DFT) on both sides of Equation (4.5.54) is performed to get the following

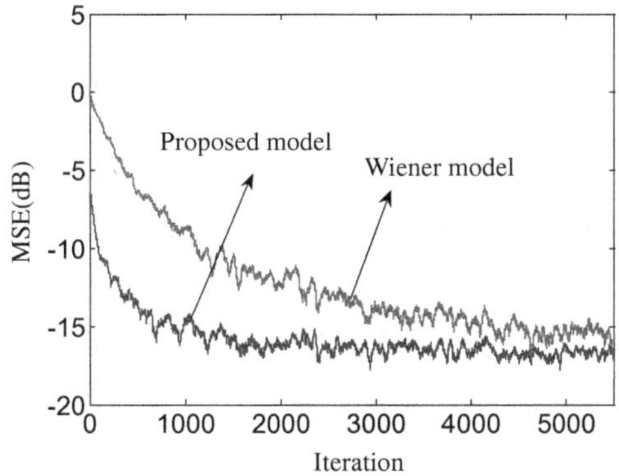

Figure 4.17 Training curves of the two models for input power backoff at 3 dB.

equation:

$$H_1 H_1 = H \tag{4.4.55}$$

where H_1 is the Fourier transform of h_1 and H is the Fourier transform of h. h_1 can be obtained from Equation (4.5.55) by

$$h_1 = \mathsf{F}^{-1}(\sqrt{H}). \tag{4.4.56}$$

The simulation results of the two models (the Wiener model and the proposed model) were compared through the simulation experiment. The weight order of the two models was set to 3, the center tap was initialized to 1, the iterative step size of the linear and nonlinear tap coefficients of the Wiener model was set to 0.0004, and the iteration step size of each part of the proposed model was set to 0.001. Figures 4.17 and 4.18 show that the MSE curves of the two models approach those of the actual channel when the input power of the nonlinear amplifier was backed off to 3 dB and 6 dB, respectively. In order to better observe the convergence position of the curves, each curve was smoothed to make the curves jitter less at the convergence position. The input power backoff is the difference between the input signal power and the actual input signal power when the amplifier operates at the saturation point, which reflects the degree of the nonlinear distortion caused by the amplitude–amplitude conversion effect of the nonlinear amplifier and the amplitude–phase conversion effect.

It can be seen from Figure 4.17 that when the number of iterations was set to 0, the MSE of the proposed model was –6 dB and the MSE

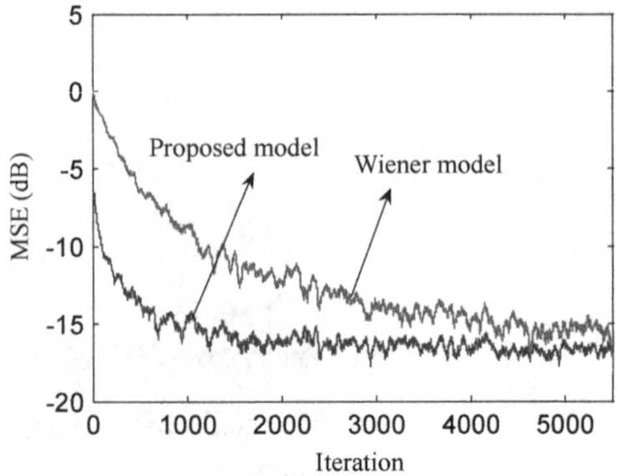

Figure 4.18 Training curves of the two models for input power backoff at 6 dB.

of the Wiener model was 0 dB. This is because the proposed model makes full use of the prior information of the channel. The fitting degree of the proposed structure to the actual channel structure is higher than that of the Wiener model. When the iteration has not started, the proposed model has been fitted to the actual channel to a certain extent, and the MSE is relatively small. At the beginning of the iteration, the MSE of the proposed model is smaller than that of the Wiener model. In the iterative process, the parameters of the proposed model can be matched with those of the actual channel with only minor adjustments, so the convergence speed is faster. In Figure 4.17, the MSE curve of the proposed model was approximately 1,500 steps around the convergence. The MSE curve of the Wiener model converged at about 4,800 steps. It can be seen from Figure 4.17 that the steady-state MSE of the proposed model is slightly lower than that of the Wiener model.

Compared with Figure 4.17, Figure 4.18 shows that the weaker the nonlinear effect, the closer the approximation of the two models to the channel. In other words, the nonlinear effect has a great influence on the approximation effect of the proposed model.

References

[1] Benedetto S., Biglieri E., and Daffara R. Modelling and performance evaluation of nonlinear satellite links-a Volterra series approach. *IEEE Transactions on Aerospace and Electronic Systems*, 1979, AES-15(4): 494–507.

[2] Li Z. C. The current status and developmental trends of space travelling wave tube amplifier. *Space Electronic Technology*, 2012, 9(4): 28–34 (in Chinese).
[3] Tseng C. H., Lin C. F., Lu F. S., *et al.* Frequency-domain channel estimation for nonlinear multicarrier underwater communication systems. *Oceans 2010 IEEE-Sydney*, Sydney, NSW, Australia, 2010, pp. 1–5.
[4] Zhu X. W., Li H. L., and Yong S. W. A new definition measurement method of group delay and its application. *Acta Electronic Sinica*, 2008, 36(9): 1819–1823 (in Chinese).
[5] Zhang X. J. Dissertation for the master degree in engineering. Yanshan University, 2010 (in Chinese).
[6] Qiu T. S., Yang Z. C., and Li X. B. A weighted average least p-Norm algorithm under alpha stable noise conditions. *Journal of Electronics & Information Technology*, 2007, 29(2): 410–413 (in Chinese).
[7] Guo Y. C. *Swarm Intelligence and Computational Intelligence Optimization Based Blind Equalization Algorithm.* Tsinghua University Press, Beijing, China, 2018 (in Chinese).
[8] Chen Z. Y., Zhou S. H., and Feng S. M. Removal of impulsive atmosphere noise based on gaussian filter. *Journal of Data Acquisition and Processing*, 2013, 28(6): 784–789 (in Chinese).
[9] Ma K. Q. and Ge L. D. Algorithm and simulation of atmospheric noise generation based on Alpha stable distribution. *Journal of Information Engineering University*, 2010, 11(2): 214–217 (in Chinese).
[10] Xu F. *Wavelet Transform Multi-Modulus Blind Equalization Algorithms Based on Fractional Lower Order Statistic.* Nanjing: Nanjing University of Information Science and Technology; 2012 (in Chinese).
[11] Guo J. *Multi-Modulus Frequency Domain Blind Equalization Algorithms Based on Lower Order Statistics.* Nanjing: Nanjing University of Information Science and Technology, 2013 (in Chinese).
[12] Guo Y. C., Xu F., and Gong X. Orthogonal wavelet blind equalization algorithm and simulation based on fraction lower order statistics. *Journal of System Simulation*, 2012, 24 (11): 2344–2349 (in Chinese).
[13] Wang X. L. and Qiang Q. Identification of nonlinear channel based on ANFIS. *Computer Simulation*, 2008, 05: 129–131, 221 (in Chinese).
[14] Kang H. W., Cho Y. S., and Youn D. H. An efficient adaptive predistorter for nonlinear high power amplifier in satellite communication. *IEEE International Symposium on Circuits and Systems*, 1997, 4(9–12): 2288–2291.
[15] Li X. T. *Research on Alpha Stable Distribution Model and its Applications.* Huazhong University of Science & Technology Wuhan, Wuhan, China, 2006 (in Chinese).

[16] Ma R. *Principles of Artificial Neural Networks*. Beijing: Machinery Industry Press, 2010 (in Chinese).
[17] Wang L. Based on the BP algorithm of artificial neural network modeling research. *Equipment Man-ufacturing Technology*, 2014, (1): 162–164, 169 (in Chinese).
[18] Guo Y. C. *Communication Signal Analysis and Processing*. Hefei: Hefei University of Technology Press, 2009 (in Chinese).
[19] Benammar B., Thomas N., Poulliat C., *et al.* On linear MMSE based turbo-equalization of nonlinear Volterra channels. *IEEE International Conference on Military Communications*, 2013, 4703–4707.
[20] Guo Y. C., Xu R., and Wang N. N. Modeling of satellite channel by introduction of priori information of channel. *Journal of Henan University of Science and Technology: Natural Science*, 2014, 35(3): 43–51 (in Chinese).
[21] Xu R. *Modeling and Equalization Technique of Nonlinear Satellite Channel.* Nanjing: Nanjing University of Information Science and Technology, 2014 (in Chinese).
[22] Zhang J. S. and Xiao X. C. A reduced parameter second-order Volterra filter with application to nonlinear adaptive prediction of chaotic time series. *Acta Physica Sinica*, 2001, 50(7): 1248–1254 (in Chinese).

Chapter 5

Nonlinear Channel Blind Equalization Algorithm Based on Multiwavelet Double Transform

Summary

Based on the influence of the nonlinearity of the TWTA on the modulation signal, the multi-wavelet basis function is used to transform the input signal of the equalizer and the output signal of the decision device, and a nonlinear blind channel equalization algorithm based on the multi-wavelet double transform is proposed. The nonlinear blind equalization algorithm based on the multiwavelet neural network is studied. By combining support vector machine, diversity technology, and neural network, the nonlinear blind equalization algorithm based on spatial diversity support vector machine multiwavelet neural network is studied. The nonlinear blind equalization algorithm based on double neural networks is optimized by chaos search algorithm. With the characteristics of feedback loop in echo state network (ESN) dynamic library, simplified training process, and strong processing capacity, Volterra filter was used as the ESN output layer and principal component analysis (PCA) was utilized to study the equalization algorithm based on Volterra filter ESN and PCA.

In the satellite communication system, the nonlinearity of the satellite channel seriously affects the performance of the satellite communication system, which can cause in-band distortion and out-of-band spectrum regeneration, and these influences cannot be ignored [1, 2]. Such nonlinearity mainly includes the nonlinearity of the satellite communication system radio frequency (RF) terminal, the satellite transponder traveling wave tube amplifier (TWTA), and the group delay caused by the channel distortion fading [3, 4]. Generally, the transmitter pre-distortion or receiver equalization or higher-order multimodulus modulation can be used to reduce the influence of the nonlinearity [5]. Although the transmitter pre-distortion treatment can effectively improve the spectrum and power utilization, it will result in serious distortion and affect the communication quality [6]. Although the receiver equalization technology can effectively overcome the nonlinear influences, the weight vector length of the equalizer affects the computation complexity. For example, Volterra equalizer can effectively compensate the nonlinearity of the satellite channel, but the length of the weight vector is quite large, the computation complexity is quite high, and the convergence speed is quite slow [7, 8]. Researches have shown that the nonlinear channel could have been compensated by using decision feedback equalizer without considering channel memory, but the effect was not ideal [9]. The Hammerstein–Wiener model presented in Ref. [10] can reduce the length of the weight vector, the computation complexity, and the running time, but when the ratio of the maximum eigenvalue to the minimum eigenvalue of the input signal autocorrelation matrix (also known as the condition number) is larger, the convergence speed is reduced and the tracking performance becomes worse. Researches also show that the autocorrelation matrix of the equalizer is more orderly and sparse after the signals are transformed by multi-wavelet, which not only greatly reduces the computational complexity but also improves the convergence performance [11].

In recent years, the intelligence computation, which is formed by a combination of artificial neural network (ANN) model, fuzzy system, genetic algorithm, evolutionary mechanism, etc., gradually becomes an important field of artificial intelligence (AI). Among them, after neural network has been applied in nonlinear channel blind equalization, some achievements have been made [12, 13]. The support vector machine (SVM) can minimize the actual risk by seeking the minimization of structural risk and pursue the optimal result under the condition of limited information. Therefore, some achievements have been made in the research of SVM equalization technology [14, 15]. Chaos refers to a seemingly irregular complex movement in the real world. It is a kind of motion state exhibited by nonlinear dynamic systems under certain conditions. Chaotic searching based on chaotic motion can effectively avoid local minimum values.

With the combination of chaos and neural network, the application of chaotic neural network in global optimization and noise control of nonlinear systems has become a new hotspot [16–18].

In view of the dynamic, real-time, and nonlinear characteristics of the actual problems to be solved in the field of signal processing, the established model should be able to perform nonlinear mapping, access the time history and internal state of input and output signals, and also have good memory. Recurrent neural network (RNN) is a relatively ideal model because it allows feedback connections between different layers of neurons and have good memory. However, during the training phase, the feedback loop makes the cost function's derivative of the connection weight difficult to calculate, and the instability of the network may also cause various inherent network functions to fail. In this case, the echo state network (ESN) originally proposed in Ref. [19] retains the processing potential of the RNN. In the training stage, the simplified processing is realized by setting the connection weight of the recursive inner layer (that is, establishing the so-called dynamic library), and the desired signals are not used; the weight vector of the output linear combinator is adjusted by the least mean square method and the echo state network is applied to the channel equalization problem [20].

Aiming at the influence of satellite channel nonlinearity on communication quality, in this chapter, the following contents are studied: (1) the nonlinear satellite channel blind equalization algorithm based on multi-wavelet double transform; (2) the nonlinear blind equalization algorithm based on multi-wavelet neural network; (3) the nonlinear blind equalization algorithm based on support vector machine and neural network; (4) the double neural network blind equalization algorithm based on chaos optimization algorithm; (5) the equalization algorithm based on Volterra filter ESN and PCA.

5.1 Volterra Blind Equalization System for Nonlinear Satellite Channel

The Volterra blind equalization system for nonlinear satellite communication is shown in Figure 5.1. The nonlinear satellite channel model in Figure 5.1 is composed of the transmitting filter, the TWTA, and the receiving filter. Among them, the TWTA is nonlinear and the Volterra equalizer is used.

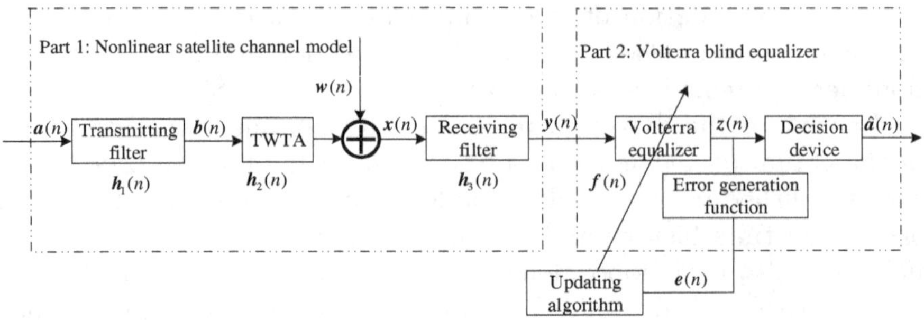

Figure 5.1 Volterra blind equalization system for nonlinear satellite channel.

In Figure 5.1, $\boldsymbol{a}(n)$ represents the input signal vector, $\boldsymbol{b}(n)$ is the output signal vector of the transmitting filter, $\boldsymbol{w}(n)$ denotes white noise vector, $\boldsymbol{x}(n)$ is the output vector of the TWTA with additive noise, $\boldsymbol{y}(n)$ is the output vector of the receiving filter, $\boldsymbol{h}_1(n)$, $\boldsymbol{h}_2(n)$, and $\boldsymbol{h}_3(n)$ are the weight vectors of the transmitting filter, the TWTA, and the receiving filter, respectively, $\boldsymbol{f}(n)$ is the weight vector of the Volterra equalizer, $\boldsymbol{z}(n)$ represents the output vector to equalizer, $e(n)$ is an error signals generated by the error generation function, and $\hat{\boldsymbol{a}}(n)$ is regarded as the output vector of judgment device.

5.1.1 *Influence of nonlinearity of TWTA on modulation signals*

When the output vector $\boldsymbol{b}(n)$ of the transmitting filter passes through the TWTA as shown in Figure 5.1, amplitude-to-amplitude conversion effect (AM/AM) and amplitude-to-phase conversion effect (AM/PM) will be generated, which will seriously affect the communication quality. According to Saleh model of the TWTA [21,22], the input and output functions of AM/AM and AM/PM are given as follows:

$$A(r) = \alpha_a r/(1 + \beta_a r^2) \tag{5.1.1}$$

$$\Phi(r) = \alpha_p r^2/(1 + \beta_p r^2) \tag{5.1.2}$$

where $\alpha_a = 2$, $\beta_a = 1$, $\alpha_p = \pi/3$, $\beta_p = 1$. The characteristic curves of the TWTA are shown in Figure 5.2.

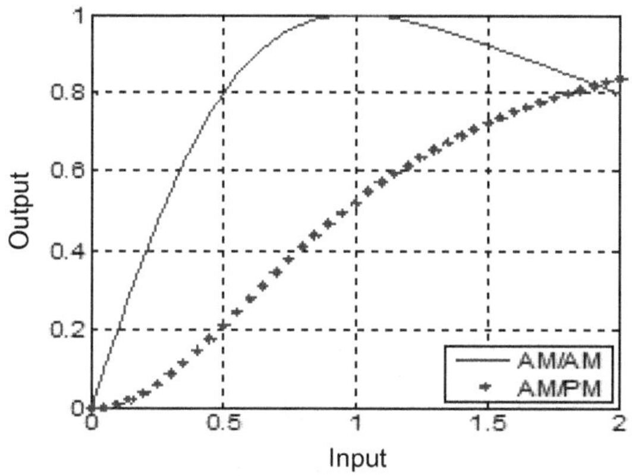

Figure 5.2 Characteristic curves of AM/AM and AM/PM functions.

It has been seen from Figure 5.2 that after the signal passes through the TWTA, the amplitude and phase produce significant distortion. In satellite communications, the quadrature phase shift keying (QPSK) signals perform better in terms of bandwidth and power tradeoffs than binary phase shift keying (BPSK) and 16QAM signals. Although BPSK and QPSK have the same bit error rate (BER), the energy efficiency of the latter is only half that of the former. Although 16QAM can improve the spectrum utilization, the TWTA must have high transmission power under the same BER as other modulation methods [23]. In satellite communication, frequency resources are relatively abundant and transmission bandwidth is wide, but the communication distance is long, transmission loss is large, and power resources are limited. Therefore, satellite communication generally adopts a modulation mode with high power utilization rate. At present, most communication satellites and TV relay satellites use QPSK modulation to transmit service data. APSK signal, as a method combining orthogonal amplitude modulation with phase shift keying (PSK) modulation, has relatively few signal modulus values and small fluctuations of the signal envelope, which not only reduces the nonlinear sensitivity to the TWTA but also improves the spectrum utilization, and is an appropriate modulation method in satellite communication. In order to analyze the effect of the TWTA on APSK signal, 16QAM and 16PSK signals were compared and

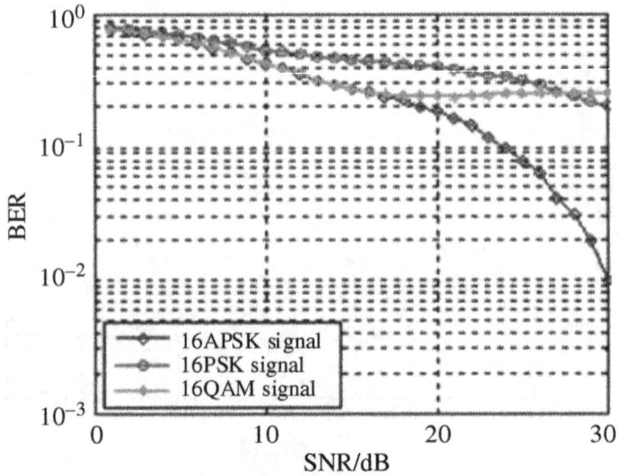

Figure 5.3 Influence of the amplifier on the three modulated signals.

16APSK was used as the input signal of Saleh TWTA to analyze the effect of the TWTA on APSK signals, as shown in Figure 5.3.

Figure 5.3 shows that after 16APSK passes the TWTA, the BER is the minimum. In other words, the nonlinearity of the TWTA has the largest effect on 16APSK distortion.

5.1.2 *Blind equalization algorithm based on nonlinear filter*

This section introduces two basic nonlinear filters: decision feedback filter and Volterra filter.

5.1.2.1 *Decision feedback filter*

Decision feedback equalization (DFE) consists of a feedback loop and is a nonlinear structure, whose feedforward filter, like the linear equalizer, takes the output $\boldsymbol{y}(n)$ of the channel as its input and whose feedback filter takes the decision signal $\hat{a}(n)$ of the equalizer as its input [24]. The output of the feedback is used to offset the partial distortion caused by the feedforward filter.

Suppose that the length of the feedforward filter of the equalizer is N_f, that of the feedback filter is N_b, and the weight vectors of the feedforward part and the feedback part are $\boldsymbol{f}(k)$ and $\boldsymbol{b}(k)$, respectively, then the decision input of the feedback equalizer is

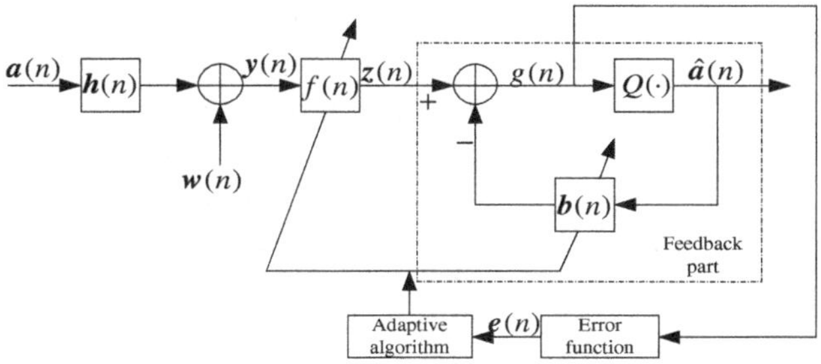

Figure 5.4 Decision feedback equalization system.

written as (Figure 5.4):

$$g(n) = \sum_{i=0}^{N_f-1} f^*(n,i)y(n-i) - \sum_{i=0}^{N_b-1} b^*(n,i)\hat{a}(n-i)$$
$$= \boldsymbol{f}^H(n)\boldsymbol{y}(n) - \boldsymbol{b}^H(n)\hat{\boldsymbol{a}}(n) \tag{5.1.3}$$

In the training mode of the DFE, $\hat{a}(n) = a(n)$. In data mode, $\hat{a}(n) = \mathrm{dec}(a(n))$.

5.1.2.2 *Volterra filter*

Volterra series is a kind of functional series [25]. As a memory extension of Taylor series, Volterra series is widely used in the analysis of nonlinear systems and has strong system representation ability. The relationship between the input $y(n)$ and output $z(n)$ of nonlinear time-invariant system is written as:

$$\begin{aligned} z(n) &= h_0 + \int_{-\infty}^{\infty} h_1(\tau_1)y(n-\tau_1)d\tau_1 \\ &+ \int_{-\infty}^{\infty}\int_{-\infty}^{\infty} h_2(\tau_1,\tau_2)y(n-\tau_1)y(n-\tau_2)d\tau_1\,d\tau_2 \\ &+ \int_{-\infty}^{\infty}\int_{-\infty}^{\infty}\cdots\int_{-\infty}^{\infty} h_k(\tau_1,\tau_2,\ldots,\tau_k)y(n-\tau_1)y(n-\tau_2)\cdots \\ &\times y(n-\tau_k)d\tau_1\,d\tau_2\cdots d\tau_k \end{aligned} \tag{5.1.4}$$

where $(h_0, h_1(\tau_1), h_2(\tau_1,\tau_2), \ldots, h_k(\tau_1,\tau_2,\ldots,\tau_k))$ are called as the core of the system, h_0 represents the direct-current (DC) response of the system, h_1 represents the linear response, and the second-order and higher-order kernel represent the nonlinear response.

The equalizer with the transverse filter structure is essentially a finite impulse response filter, and the Volterra series has numerous items, so it is necessary to intercept the Volterra series according to the actual requirements of the system. Based on Equation (5.1.4), the output of the n-order Volterra nonlinear equalizer can be expressed as:

$$\begin{aligned} z(n) = &\sum_{m_1=0}^{N-1} f(m_1)y(n-m_1) \\ &+ \sum_{m_1=0}^{N-1}\sum_{m_2=0}^{N-1}\sum_{m_3=0}^{N-1} f(m_1,m_2,m_3)y(n-m_1)y(n-m_2) \\ &\times y^*(n-m_3) \end{aligned} \tag{5.1.5}$$

where $f(m_1)$ and $f(m_1, m_2, m_3)$ are the weights of the equalizer and N is the memory length of the filter. In satellite communication, nonlinear channel is a bandpass filter, and the frequency component generated by the even-order term is far away from the carrier frequency and is filtered by bandpass filter, so only the odd-order term is considered. In Equation (5.1.5), the first term is a linear term, which is used to achieve linear equalization of the satellite channel, and the third term is nonlinear and is used to realize the nonlinear equalization of the satellite channel. According to Equation (5.1.5), the weight vector length of the n-order Volterra nonlinear equalizer is computed by:

$$L = \sum_{n=1}^{(N+1)/2} N^{2n-1} \tag{5.1.6}$$

For convenience, the input $y(n)$ of the equalizer or the output of the receiving filter and the weight coefficient of the Volterra equalizer may be expressed in matrix form.

Let the weight vector of the Volterra equalizer be written as:

$$\boldsymbol{f}(n) = [f(n), f(n-1), \ldots, f(n-L+1)]^{\mathrm{T}} \tag{5.1.7}$$

The input to the equalizer is denoted as:

$$\boldsymbol{y}(n) = [y(n), y(n-1), \ldots, y(n-L+1), y^2(n)y^*(n), \ldots, \\ y^2(n-L+1)y^*(n-L+1)]^{\mathrm{T}} \tag{5.1.8}$$

At this point, the output to the equalizer is written as:

$$z(n) = \boldsymbol{f}^{\mathrm{T}}(n)\boldsymbol{y}(n) \tag{5.1.9}$$

5.1.3 *Volterra blind equalization algorithm*

In Figure 5.1, the weight vector of the Volterra equalizer may be updated according to constant modulus blind equalization algorithm (CMA). In the CMA, its error function is defined as:

$$e(n) = R^2 - |z(n)|^2 \tag{5.1.10}$$

where R^2 is the modulus value of the CMA and

$$R^2 = \frac{E\{|a(n)|^4\}}{E\{|a(n)|^2\}} \tag{5.1.11}$$

The cost function of the CMA is defined as:

$$J[e(n)] = E\{e^2(n)\} \tag{5.1.12}$$

Then, the weight vector of the equalizer is updated as follows:

$$\boldsymbol{f}(n+1) = \boldsymbol{f}(n) - \mu\hat{\nabla}_{\boldsymbol{f}} J[e(n)] \tag{5.1.13}$$

where μ is the step-size, and the choice of step-size μ is very important. The larger μ is, the quicker the convergence rate is, but the bigger the steady-state MSE is. The smaller μ is, the smaller the steady-state MSE is, but the slower the convergence rate is.

$\hat{\nabla}_{\boldsymbol{f}} J[e(n)]$ means the instantaneous value of the partial derivative of $J[e(n)]$ to $\boldsymbol{f}(n)$. The gradient vector should be found by minimizing the cost function, namely,

$$\hat{\nabla}_f J[e(n)] = \frac{\partial}{\partial \boldsymbol{f}(n)}\{e^2(n)\} = 2e(n)\frac{\partial e(n)}{\partial \boldsymbol{f}(n)} \tag{5.1.14}$$

While

$$\begin{aligned}\frac{\partial e(n)}{\partial \boldsymbol{f}(n)} &= \frac{\partial[z(n)z^*(n)]}{\partial \boldsymbol{f}}(n) = \frac{\partial[\boldsymbol{f}^{\mathrm{T}}(n)\boldsymbol{y}(n)(\boldsymbol{y}^{\mathrm{T}}(n)\boldsymbol{f}(n))^*]}{\partial \boldsymbol{f}(n)} \\ &= 2\boldsymbol{y}^*(n)\boldsymbol{y}^{\mathrm{T}}(n)\boldsymbol{f}(n) = 2\boldsymbol{y}(n)z^*(n)\end{aligned} \tag{5.1.15}$$

Therefore, the iterative equation of the weight vector of the Volterra equalizer is written as:

$$\boldsymbol{f}(n+1) = \boldsymbol{f}(n) - 4\mu e(n)z^*(n)\boldsymbol{y}(n) \tag{5.1.16}$$

The Volterra equalizer gives consideration to both linear and nonlinear structure of the system and is suitable for various nonlinear system models. However, it has a large number of weight coefficients, and the computational complexity increases rapidly with the increase of memory depth. On the other hand, the dimension of the input vector to the nonlinear equalizer is much larger than that of the linear equalizer, which will increase the ratio of the maximum eigenvalue to the minimum eigenvalue of the autocorrelation matrix of the input vector to result in a rather slow convergence rate. In view of the nonlinear characteristics of the satellite channel and the defects of the Volterra equalizer, this section is intended to improve on the following aspects:

(1) The Volterra equalizer will be replaced with the Wiener equalizer to reduce the complexity of the equalizer structure.
(2) The input vector of the Wiener equalizer is transformed by the balanced orthogonal multiwavelet basis function to reduce the autocorrelation of the input signals.
(3) A first-order decision feedback filter is added to the output end of the Wiener equalizer, and the inputs of the decision feedback filter are used as the decision output of the Wiener equalizer.

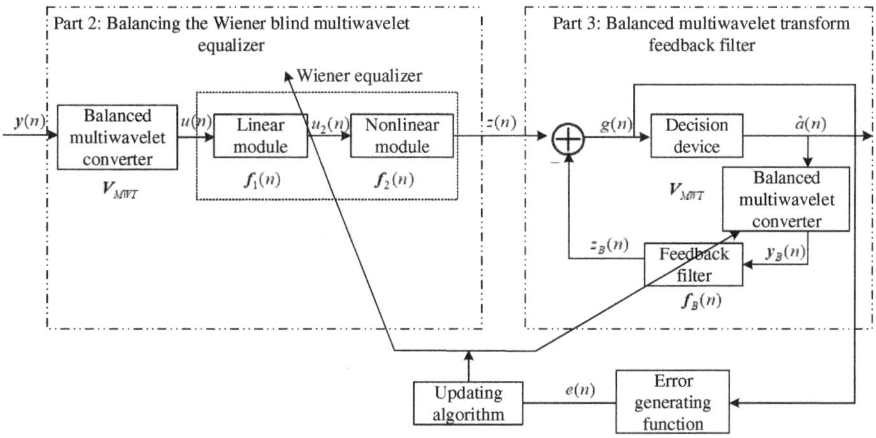

Figure 5.5 Structure of improved nonlinear satellite channel blind equalizer.

5.1.4 *Nonlinear blind equalization algorithm based on balanced orthogonal multiwavelet double transform*

Aiming at the defects of the Volterra equalizer of the nonlinear satellite channel, part 2 in Figure 5.1 is modified into Figure 5.5 according to the improvement discussed in the previous section [26].

In Figure 5.5, the balanced multiwavelet converter (BMC) and Wiener blind equalization algorithm constitute the balanced multiwavelet transform Wiener blind equalization algorithm. The BMC and decision feedback filter (DFF) constitute the balanced multiwavelet transform feedback filter. This blind equalization algorithm with double balanced multiwavelet transform is called as the blind equalization algorithm based on balanced orthogonal multiwavelet double transform.

5.1.4.1 *Multiwavelet representation of the equalizer*

Multiwavelets are composed of multiple scale functions, which are supported on any scale without heavy barriers. The correlation among different frequencies in multiwavelets is smaller, and the decorrelation is better. The ratio of the maximum and minimum eigenvalues of the autocorrelation matrix after multiwavelet transform is smaller. At the same time, it has orthogonality, symmetry, and short support. Therefore, the performance of the equalizer constructed by multiwavelet is better than that constructed by single wavelet [26, 27].

In Figure 5.5, the impulse response $f_1(n)$ of the linear module of the Wiener equalizer is a finite number of values and is denoted by multiwavelet as:

$$f_{1i}(n) = \sum_{l=1}^{r}\sum_{j=1}^{J}\sum_{k=0}^{k_j-1} w_{j,k,l}(n)\psi_{j,k,l}(i) + \sum_{l=1}^{r}\sum_{k=0}^{k_j-1} v_{J,k,l}(n)\varphi_{J,k,l}(i) \tag{5.1.17}$$

When the multiwavelet function and scale function are determined, the characteristics of the equalizer are determined by $w_{j,k,l}(n)$ and $v_{J,k,l}(n)$ and then the weight coefficients of the equalizer consist of $w_{j,k,l}(n)$ and $v_{J,k,l}(n)$. The output of the equalizer is expressed as:

$$\begin{aligned} u(n) &= \sum_{i=0}^{N-1} f_{1i}(n)y(n-i) \\ &= \sum_{l=1}^{r}\left(\sum_{j=1}^{l}\sum_{k=0}^{k_j-1} w_{j,k,l}(n)r^{w}_{j,k,l}(n) + \sum_{k=0}^{k_J-1} v_{J,k,l}(n)r^{v}_{J,k,l}(n)\right) \end{aligned} \tag{5.1.18}$$

where

$$r^{w}_{j,k,l}(n) = \sum_{i=0}^{N-1} \psi_{j,k,l}(i)y(n-i) \tag{5.1.19a}$$

$$r^{v}_{J,k,l}(n) = \sum_{i=0}^{N-1} \varphi_{J,k,l}(i)y(n-i) \tag{5.1.19b}$$

where $\psi_{j,k,l}(n)$ represents the ith wavelet basis function with scale parameter j and translation parameter k, $\varphi_{J,k,l}(n)$ represents the lth scale function with scale parameter J and translation parameter k. Equation (5.1.8) shows that the discrete orthogonal multiwavelet transform of the input $y(n)$ is equivalent to the convolution of the input $y(n)$ with the multiwavelet basis function on each scale.

From Equation (5.1.17) to Equation (5.1.19), the essence of using multiwavelet to represent equalizer is to carry out balanced orthogonal multiwavelet transform on the input signal of the equalizer, but in practical application, a prefiltering process must be carried out first, and the prefiltering process will destroy the symmetry and orthogonality of the multiwavelet. In order to avoid these problems, Lebrun, Vetterli, *et al.* proposed the balanced orthogonal multiwavelet theory, and its basic idea is to make the known unbalanced multiwavelet rotate at a certain equilibrium degree to turn it into the balanced multiwavelet. In general, the orthogonal unitary matrix is used for the first- or multi-order balance processing of the orthogonal multiwavelet, but the first-order balance processing is helpful in reducing the computational complexity. Therefore, the orthogonal unitary matrix selected for first-order equilibrium processing is defined as:

$$\boldsymbol{U} = \frac{1}{\sqrt{2}} \begin{pmatrix} 1 & -1 \\ 1 & 1 \end{pmatrix} \tag{5.1.20}$$

The balanced multi-low pass filter and multi-high pass filter are denoted as, respectively,

$$\begin{cases} \tilde{\boldsymbol{H}}(\omega) = \boldsymbol{U}^T \boldsymbol{H}(\omega) \boldsymbol{U} \\ \tilde{\boldsymbol{G}}(\omega) = \boldsymbol{G}(\omega) \boldsymbol{U} \end{cases} \tag{5.1.21}$$

where $H(\omega)$, $G(\omega)$ are multi-low pass filter and multi-high pass filter before balancing, respectively.

Based on the Mallat algorithm, the input vector $\boldsymbol{y}(n)$ of the equalizer is decomposed into a balanced multiwavelet. In order to decompose the input vector $\boldsymbol{y}(n) = [y(n), y(n-1), \ldots, y(n-m+1)]^{\mathrm{T}}$ of the equalizer, it is necessary to define block matrices $\boldsymbol{P}_j$ and $\boldsymbol{Q}_j$, where $\boldsymbol{P}_j$ and $\boldsymbol{Q}_j$ are composed of the coefficients of the new multi-low-pass filter $\tilde{\boldsymbol{H}}(\omega)$ and multi-high-pass filter $\tilde{\boldsymbol{G}}(\omega)$ after balancing, respectively, and they are written as:

$$\boldsymbol{P}_j = \begin{bmatrix} \tilde{\boldsymbol{H}}_0 & \tilde{\boldsymbol{H}}_1 & \tilde{\boldsymbol{H}}_2 & \ldots & 0 & \ldots & 0 \\ 0 & 0 & \tilde{\boldsymbol{H}}_0 & \tilde{\boldsymbol{H}}_1 & \tilde{\boldsymbol{H}}_2 & 0 & \ldots \\ 0 & \ldots & & & \ldots & 0 & \ldots \\ 0 & 0 & 0 & 0 & \tilde{\boldsymbol{H}}_0 & \tilde{\boldsymbol{H}}_1 & \ldots \end{bmatrix}_{\frac{l}{2} \times l} \tag{5.1.22a}$$

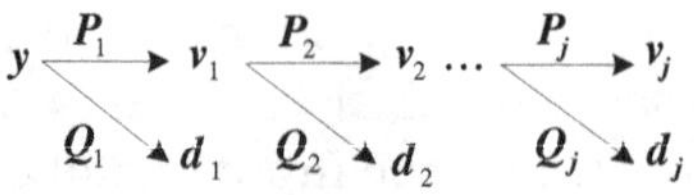

Figure 5.6 Mallat decomposition structure.

$$Q_j = \begin{bmatrix} \tilde{G}_0 & \tilde{G}_1 & \tilde{G}_2 & \ldots & 0 & \ldots & 0 \\ 0 & 0 & \tilde{G}_0 & \tilde{G}_1 & \tilde{G}_2 & 0 & \ldots \\ 0 & \ldots & & & \ldots & 0 & \ldots \\ 0 & 0 & 0 & 0 & \tilde{G}_0 & \tilde{G}_1 & \ldots \end{bmatrix}_{\frac{l}{2}\times l} \tag{5.1.22b}$$

where $j = 1 \sim J, l = m/2^{j-1}$, $J, l \in \mathbb{Z}$, so the Mallat decomposition equation is written as:

$$v_J = P_J P_{J-1} \cdots P_1 y, \quad w_J = Q_J P_{J-1} \cdots P_1 y \tag{5.1.23}$$

The Mallat decomposition structure is shown in Figure 5.6.

In Figure 5.6, v_j and w_j are the low-pass coefficients and high-pass coefficients of the input $y(n)$ after the j-level decomposition, respectively, and the signal vector after decomposition of balanced orthogonal multi-wavelet is given by:

$$\begin{aligned} u = [d_1; d_2; \cdots; d_J; v_J] = [&Q_1; Q_2 P_1; Q_2 P_1 P_0; \cdots; \\ &Q_J P_{J-1} \cdots P_2 P_1; P_J P_{J-1} \cdots P_2 P_1] y \end{aligned} \tag{5.1.24}$$

Therefore, the balanced orthogonal multiwavelet transformation matrix is written as:

$$\begin{aligned} V_{\mathrm{MWT}} = [&Q_1; Q_2 P_1; Q_2 P_1 P_0; \cdots; \\ &Q_J P_{J-1} \cdots P_2 P_1; P_J P_{J-1} \cdots P_2 P_1] \end{aligned} \tag{5.1.25}$$

Then, Equation (5.1.24) becomes

$$u = V_{\mathrm{MWT}} y \tag{5.1.26}$$

5.1.4.2 *Balanced orthogonal multiwavelet Wiener equalization algorithm*

In Figure 5.5, the input vector of the linear module of the Wiener equalizer is denoted as $\boldsymbol{u}$ as shown in Equation (5.1.26), so the output of the linear module of the Wiener equalizer is given as follows:

$$\begin{aligned} u_2(n) &= f_{10}(n)u(n) + f_{11}(n)u(n-1) + \cdots + f_{1M-1}(n)u(n-M+1) \\ &= f_1^{\mathrm{T}}(n)u(n) = f_1^{\mathrm{T}}(n)V_{\mathrm{MWT}}y(n) \end{aligned} \tag{5.1.27}$$

where $\boldsymbol{u} = [u(n), u(n-1), \ldots, u(n-M+1)]^{\mathrm{T}}$, and

$$\boldsymbol{f}_1(n) = [f_{10}(n), f_{11}(n), \ldots, f_{1M-1}(n)]^{\mathrm{T}}.$$

The output of the nonlinear module of the Wiener equalizer is given as follows:

$$\begin{aligned} z(n) &= f_{21}(n)u_2(n) + f_{21}(n)u_2^2(n) + \cdots + f_{2M_2}(n)u_2^{M_2}(n) \\ &= \boldsymbol{f}_2^{\mathrm{T}}(n)\boldsymbol{u}_2(n) \end{aligned} \tag{5.1.28}$$

where $\boldsymbol{f}_2(n) = [f_{21}(n), f_{22}(n), \ldots, f_{2M_2}(n)]^{\mathrm{T}}$, $\boldsymbol{u}_2(n) = [u_2(n), u_2^2(n), \ldots, u_2^{M_2}(n)]^{\mathrm{T}}$.

The constant modulus algorithm (CMA) is used to adjust the weight vectors $\boldsymbol{f}_1(n)$ and $\boldsymbol{f}_2(n)$, respectively, and their updated equations are given by:

$$\begin{cases} \boldsymbol{f}_1(n+1) = \boldsymbol{f}_1(n) + \mu_{f_1}\hat{\boldsymbol{R}}^{-1}(n)e(n)z^*(n)\dfrac{\partial z(n)}{\partial \boldsymbol{f}_1(n)} \\ \boldsymbol{f}_2(n+1) = \boldsymbol{f}_2(n) + \mu_{f_2}\hat{\boldsymbol{R}}^{-1}(n)e(n)z^*(n)\dfrac{\partial z(n)}{\partial \boldsymbol{f}_2(n)} \end{cases} \tag{5.1.29}$$

where $\hat{\boldsymbol{R}}^{-1}(n) = \mathrm{diag}[\sigma_{1,0,1}^{2\omega}(n), \ldots, \boldsymbol{\sigma}_{J,k,l}^{2\omega}(n), \ldots, \boldsymbol{\sigma}_{J,0,1}^{2v}(n), \ldots, \boldsymbol{\sigma}_{J,k,l}^{2v}(n)]$, $\boldsymbol{\sigma}_{J,k,l}^{2\omega}(n)$ and $\boldsymbol{\sigma}_{J,k,l}^{2v}(n)$ are the average power estimation of the lth wavelet coefficient $r_{J,k,l}^{\omega}(n)$ and the lth scale coefficient

$r^{v}_{J,k,l}(n)$ with scale parameter J and translation parameter k, respectively, and

$$\begin{cases}\sigma^{2\omega}_{J,k,l}(n+1) = \beta_\sigma \sigma^{2\omega}_{J,k,l}(n) + (1-\beta_\sigma)|r^{2\omega}_{J,k,l}(n)|^2 \\ \sigma^{2v}_{J,k,l}(n+1) = \beta_\sigma \sigma^{2v}_{J,k,l}(n) + (1-\beta_\sigma)|r^{2v}_{J,k,l}(n)|^2\end{cases} \tag{5.1.30}$$

where β_σ is the forgetting factor and $0 < \beta_\sigma < 1$.

$$\begin{aligned}\frac{\partial z(n)}{\partial \boldsymbol{f}_1(n)} &= \frac{\partial(\boldsymbol{f}_2^{\mathrm{T}}(n)\boldsymbol{u}_2(n))}{\partial \boldsymbol{f}_1(n)} \\ &= \boldsymbol{f}_2^{\mathrm{T}}(n)\left(\left(\frac{\partial u_2(n)}{\partial \boldsymbol{f}_1(n)}\right)^{\mathrm{T}}, \left(\frac{\partial u_2^2(n)}{\partial \boldsymbol{f}_1(n)}\right)^{\mathrm{T}}, \ldots, \left(\frac{\partial u_2^{M_2}(n)}{\partial \boldsymbol{f}_1(n)}\right)^{\mathrm{T}}\right) \\ &= \boldsymbol{f}_2^{T}(n)\Bigg(\left(\frac{\partial u_2(n)}{\partial \boldsymbol{f}_1(n)}\right)^{\mathrm{T}}, 2u_2(n)\left(\frac{\partial u_2(n)}{\partial \boldsymbol{f}_1(n)}\right)^{\mathrm{T}}, \ldots, \\ &\qquad M_2 u_2^{M_2-1}(n)\left(\frac{\partial u_2(n)}{\partial \boldsymbol{f}_1(n)}\right)^{\mathrm{T}}\Bigg) \\ &= \begin{pmatrix} \boldsymbol{u}(n)[\boldsymbol{f}_2^{\mathrm{T}}(n)\boldsymbol{U}(n)] \\ \boldsymbol{u}(n-1)[\boldsymbol{f}_2^{\mathrm{T}}(n)\boldsymbol{U}(n)] \\ \vdots \\ \boldsymbol{u}(n-M_1+1)[\boldsymbol{f}_2^{T}(n)\boldsymbol{U}(n)] \end{pmatrix}\end{aligned} \tag{5.1.31}$$

where $\boldsymbol{U}(n) = [1, 2[\boldsymbol{f}_1^{\mathrm{T}}(n)\boldsymbol{u}(n)], \ldots, M_2[\boldsymbol{f}_1^{\mathrm{T}}(n)\boldsymbol{u}(n)]^{M_2-1}]]$

$$\frac{\partial z(n)}{\partial \boldsymbol{f}_2(n)} = \frac{\partial(\boldsymbol{f}_2^{\mathrm{T}}(n)\boldsymbol{u}_2(n))}{\partial \boldsymbol{f}_2(n)} = \boldsymbol{u}_2(n) = \begin{pmatrix} [\boldsymbol{f}_1^{T}(n)\boldsymbol{u}(n)] \\ [\boldsymbol{f}_1^{T}(n)\boldsymbol{u}(n)]^2 \\ \cdots \\ [\boldsymbol{f}_1^{\mathrm{T}}(n)\boldsymbol{u}(n)]^{M_2} \end{pmatrix} \tag{5.1.32}$$

From Equation (5.1.24)–(5.1.32), a Wiener blind equalization algorithm based on balanced orthogonal multiwavelet transform is proposed. In this proposed algorithm, the balanced orthogonal multiwavelet transform is put before the Wiener equalizer, and the input signal is transformed by orthogonal multiwavelet transform to reduce the signal autocorrelation so as to achieve the purpose of accelerating convergence.

5.1.4.3 *Balanced orthogonal multiwavelet double transform decision feedback filter*

According to the idea of [28] and Figure 5.5, after the input of the decision feedback filter (DFF) $\boldsymbol{f}_B(n)$, namely, the decision output $\hat{a}(n)$ of the Wiener equalizer, is transformed by balanced orthogonal multiwavelet, its input vector is written as:

$$\boldsymbol{y}_B(n) = \boldsymbol{V}_{\mathrm{MWT}}\hat{\boldsymbol{a}}(n) \tag{5.1.33}$$

At this point, the input of the decision device is given by:

$$g(n) = z(n) - z_B(n) = z(n) - \boldsymbol{f}_B(n)\boldsymbol{y}_B(n) \tag{5.1.34}$$

where $z_B(n)$ is the output of the DFF $\boldsymbol{f}_B(n)$ and

$$z_B(n) = \boldsymbol{f}_B(n)\boldsymbol{y}_B(n) \tag{5.1.35}$$

Since the input of the DFF is transformed by the balanced orthogonal multiwavelets, the DFF is called as the balanced orthogonal multiwavelet DFF. The weight vector of the balanced orthogonal multiwavelet DFF is updated as follows:

$$\boldsymbol{f}_B(n+1) = \boldsymbol{f}_B(n) - \mu_B\hat{\boldsymbol{R}}_B^{-1}(n)e(n)\boldsymbol{y}_b(n)g^*(n) \tag{5.1.36}$$

where $\hat{\boldsymbol{R}}_B^{-1}(n) = \mathrm{diag}[\sigma^{2\omega}_{B(1,0,1)}(n), \ldots, \sigma^{2\omega}_{B(J,k,l)}(n), \ldots, \sigma^{2v}_{B(J,0,1)}(n), \ldots, \sigma^{2v}_{B(J,k,l)}(n)]$,$\sigma^{2\omega}_{B(J,k,l)}(n)$ and $\sigma^{2v}_{B(J,k,l)}(n)$ are the average power estimation of the lth wavelet coefficient $\boldsymbol{r}^{\omega}_{B(J,k,l)}(n)$ and the lth scale coefficient $\boldsymbol{r}^{v}_{B(J,k,l)}(n)$ with scale parameter k and translation parameter J, respectively, and

$$\begin{cases} \boldsymbol{\sigma}^{2\omega}_{B(J,k,l)}(n+1) = \boldsymbol{\beta}_B\boldsymbol{\sigma}^{2\omega}_{B(J,k,l)}(n) + (1-\boldsymbol{\beta}_B)|\boldsymbol{r}^{2\omega}_{B(J,k,l)}(n)|^2 \\ \boldsymbol{\sigma}^{2v}_{B(J,k,l)}(n+1) = \boldsymbol{\beta}_B\boldsymbol{\sigma}^{2v}_{B(J,k,l)}(n) + (1-\boldsymbol{\beta}_B)|\boldsymbol{r}^{2v}_{B(J,k,l)}(n)|^2 \end{cases} \tag{5.1.37}$$

where $\boldsymbol{\beta}_B$ is the forgetting factor and $0 < \boldsymbol{\beta}_B < 1$.

After the inputs of the feedback filter and the Wiener equalizer are transformed by the balanced orthogonal multiwavelets, the structure of the Wiener equalizer and the feedback filter is improved, and the weight vectors of the improved Wiener equalizer and the improved feedback filter are still updated by CMA. This improved algorithm is called as double multi-wavelet transform-based nonlinear constant modulus blind equalization algorithm and decision feedback equalizer (DMWT-NCMA-DFE).

5.1.4.4 *Computational complexity*

The computational complexity of the Volterra blind equalization algorithm and the DMWT-NCMA-DFE in this section are analyzed. Since the amount of multiplication is much larger than that of addition, the frequency of multiplication required by the DMWT-NCMA-DFE algorithm to run once is taken as the measure of the amount of calculation.

According to Equation (5.1.5), there are a large number of multiplication operations in the third-order nonlinear term of the Volterra equalizer, and $3N^3 + N$ multiplication operations are required when the signal passes through the equalizer. According to Equation (5.1.16), $N^3 + N + 1$ multiplication operations are required for each update of the weight vector, so the multiplication time required for the weight vector update of the Volterra equalizer is $4N^3 + 2N + 1$ for each time. Assume that the length of input real signal is $\boldsymbol{H}$, the orthogonal multiwavelet transform of the input signal $\boldsymbol{y}(n)$ (note that $\boldsymbol{V}_{\text{MWT}}$ is $\boldsymbol{H} \times \boldsymbol{H}$ orthogonal matrix) according to Equation (5.1.26) needs to multiply $\boldsymbol{H}^2$ times since the DMWT-NCMA-DFE needs to run H point balanced orthogonal multiwavelet transform in each weight vector iteration. Consider that the actual signal length is very long and the filter length is very short, that is, $\boldsymbol{V}_{\text{MWT}}$ is a sparse matrix. Assume that the number of non-zero elements in each row of $\boldsymbol{V}_{\text{MWT}}$ is $M_1(M_1 \ll H)$, the multiplication time of Equation (5.1.22) is M_1H, when the signal passes through the Wiener equalizer, and the multiplication time of Equations (5.1.27) and (5.1.28) is $(1+2+\cdots+N+M_1)$. The computation loads brought by Equations (5.1.31) and (5.1.32) are $(1+2+\cdots+N+M_1)$. After the decision feedback module is calculated from Equation (5.1.33) to Equation (5.1.37), the required computation load is $M_1H + 7M_1$, and the required multiplication time of the DMWT-NCMA-DFE is $N^2+N+2M_1H+7M_1$. Obviously, it is lower than the Volterra blind equalizer.

5.1.5 *Algorithm simulation*

To verify the performance of the DMWT-NCMA-DFE, the Volterra blind equalization algorithm, nonlinear constant modulus blind equalization algorithm and decision feedback equalizer (NCMA-DFE), multi-wavelet transform-based nonlinear constant modulus blind equalization algorithm and decision feedback equalizer (MWT-NCMA-DFE) were compared. In the simulations, the memory length of the Volterra series model was set to 3, the parameters $M_1 = 16$ and $M_2 = 3$ of the linear and nonlinear modules of the Wiener equalizer, and the initialization weight vector $\boldsymbol{f}_1(0) = [1, 0, \ldots, 0]^{\mathrm{T}}$ and $\boldsymbol{f}_2(0) = [1, 0, \ldots, 0]^{\mathrm{T}}$. The weight length of the feedback filter was set to $M_B = 5$, and the weight vector was initialized to $\boldsymbol{f}_B(0) = [0, 0, \ldots, 0]^{\mathrm{T}}$. The outer and inner radius ratio of the APSK constellation was set to 2.73. Both the transmitting filter and the receiving filter adopted the square root ascending cosine filter, whose roll down factor was 0.35 and sampling rate was 8. The SNR was 20, $J = 2$, and the average power was initialized to 25. The results with 200 Monte Carlo simulations are shown in Figure 5.7.

In Figure 5.7(a), it has been seen that the convergence speed of the DMWT-NCMA-DFE has an improvement of about 4000 steps in comparison with that of the MWT-NCMA-DFE, and the convergence speed of the Volterra blind equalization algorithm and the NCMA-DFE is very low and shows almost no convergence. In 2000 iterations, the MSE of the DMWT-NCMA-DFE (converged) has a drop of about 3 dB in comparison with that of the MWT-NCMA-DFE, while the NCMA-DFE (unconverged) has a drop of about 5 dB in comparison with that of the Volterra equalization algorithm. It has been seen from Figure 5.7(c) to Figure 5.7(f) that serious distortion occurs when 16APSK signals pass through the nonlinear satellite channel (as shown in Figure 5.7(b)), and that when the DMWT-NCMA-DFE, the MWT-NCMA-DFE, the MWT-NVMA-DFE and the Volterra blind equalization algorithms are used to restore the signal constellation, the output constellation of the DMWT-NCMA-DFE is the clearest, most concentrated, and compact.

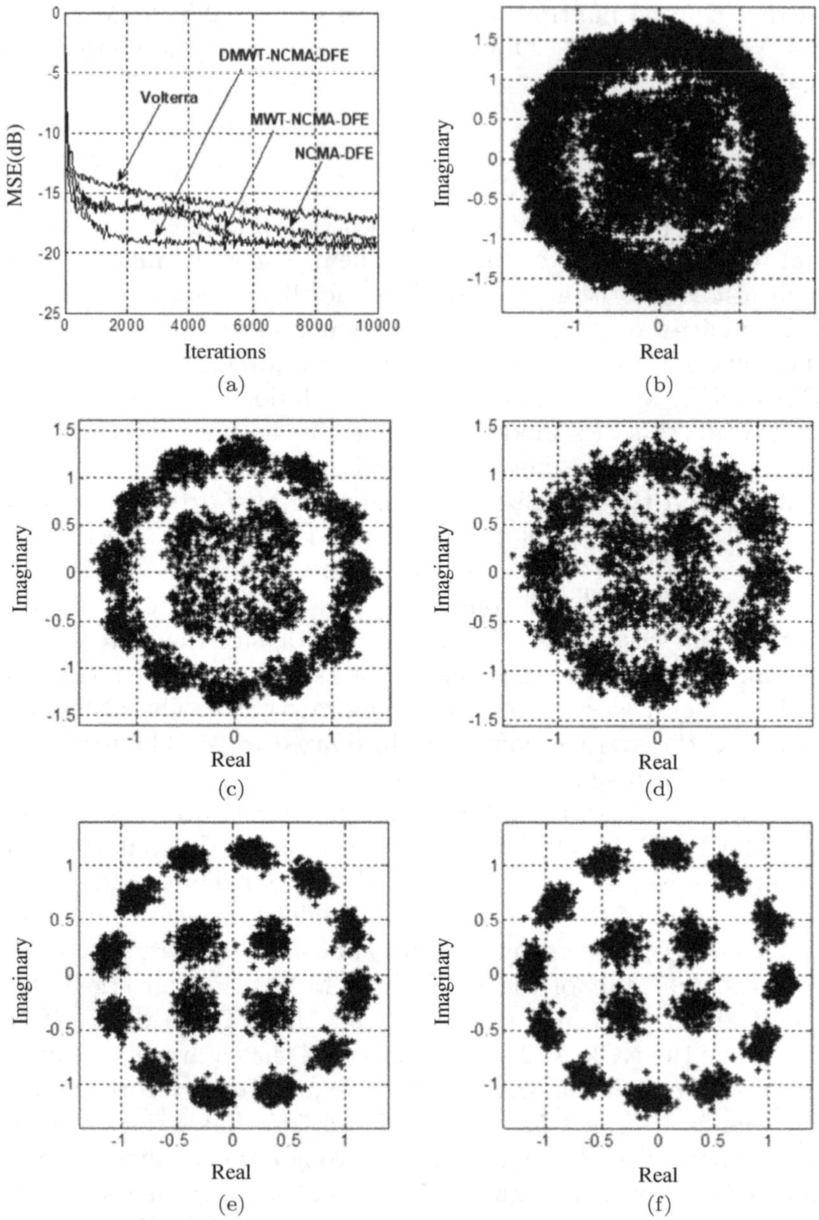

Figure 5.7 Simulation results: (a) MSE curves of each algorithm; (b) Input of equalizer; (c) Output of Volterra equalizer; (d) Output of NCMA-DFE; (e) Output of MWT-NCMA-DFE and (f) Output of DMWT-NCMA-DFE.

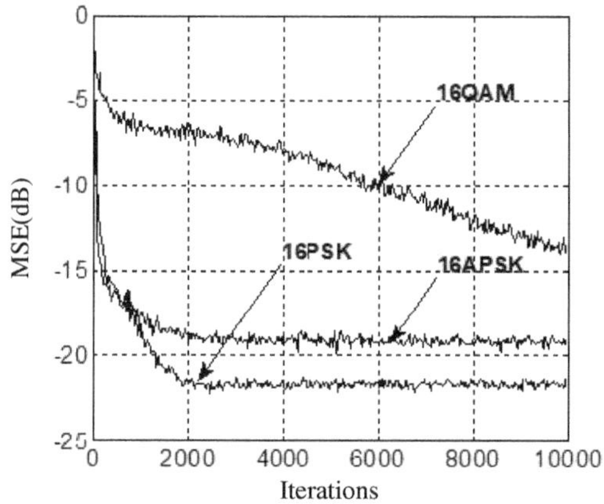

Figure 5.8 MSE curves of the DMWT-NCMA-DFE for three modulated signals.

In order to test the recovery ability of the DMWT-NCMA-DFE, 16QAM, 16PSK, and 16APSK signals were selected as the input of the nonlinear satellite channel. The simulation parameters are the same as before and the simulation results are shown in Figure 5.8.

It has been seen from Figure 5.8 that for 16QAM signal, the convergence rate of the DMWT-NCMA-DFE is extremely slow, and it has not yet converged after 10,000 iterations, and for 16PSK signal, the MSE of the DMWT-NCMA-DFE after convergence has a drop of about 3 dB in comparison with that of 16APSK, but the convergence speed has a drop of about 500–1000 steps in comparison with that of 16APSK.

5.2 Nonlinear Blind Equalization Algorithm Based on Multiwavelet Neural Network

5.2.1 *Neural network model*

5.2.1.1 *Neuron model*

Neuron is the basic unit of a neural network, which mainly contains three basic elements: synapse or connection chain, adder, and activation function, as shown in Figure 5.9.

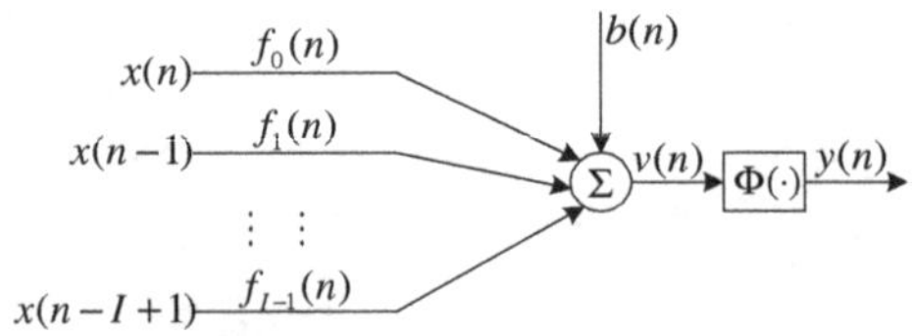

Figure 5.9 Neuron model.

The output of the neuron is defined as:

$$y(n) = \mathbf{\Phi}(v(n)) = \mathbf{\Phi}\left(\sum_{i=0}^{I-1} x(n-i)f_i(n) + \boldsymbol{b}(n)\right)$$
$$= \mathbf{\Phi}\left(\boldsymbol{x}^{\mathrm{T}}(n)\boldsymbol{f}(n) + \boldsymbol{b}(n)\right) \tag{5.2.1}$$

where $\boldsymbol{x}(n) = \{x(n), x(n-1), \ldots, x(n-I+1)\}$ is the output of the neuron, $\boldsymbol{f}(n) = \{f_0(n), f_1(n), \ldots, f_{I-1}(n)\}$ is the weight vector of the neuron connection, and $\boldsymbol{b}(n)$ is the external bias. If the value of the $b(n)$ is positive, the network input of the activation function will be increased. If it is negative, it decreases correspondingly. $y(n)$ is the output of the neurons, and $\mathbf{\Phi}(\cdot)$ is an activation function and generally has three basic structures, which are described as follows:

(1) *Threshold function*: It is also called as Heaviside function and defined as:

$$\mathbf{\Phi}(x) = \begin{cases} 1, & x \geq 0 \\ 0, & x < 0 \end{cases}, \quad \text{or} \quad \mathbf{\Phi}(x) = \begin{cases} 1, & x \geq 0 \\ -1, & x < 0 \end{cases} \tag{5.2.2}$$

(2) *Gaussian function*: It is defined as:

$$\mathbf{\Phi}(x) = \exp(-\frac{1}{2\sigma_i^2}\sum_j (x_j - w_{ji})^2) \tag{5.2.3}$$

(3) *Sigmoid function*: It is a s-shaped and the most general activation function in artificial neural network with many features such as smoothness, asymptotic, and monotonicity and shows good balance performance between linearity and nonlinearity.

The most general function form is written as:

$$\mathbf{\Phi}(x) = (1 + e^{-\alpha x}) - 1 \tag{5.2.4}$$

where parameter α controls the slope of the function.

5.2.1.2 *Neural network model*

Neural network is a directed graph composed of interconnected synaptic nodes and activated connections. Among them, the construction of neuron is closely related to the learning algorithm of the training network. From the connection mode, the neural network is mainly divided into the feedforward neural network and feedback neural network, which are described as follows:

(1) Feedforward neural network, as shown in Figure 5.10.
In a feedforward neural network, the output of the neurons in the previous layer serves as the input of the neurons in the next layer, and the input and output layers are connected with the outside world. As a hidden layer, other middle layers, called as hidden layers, exist between external input and network output in some way. Adding one or more hidden layers can improve network performance.
(2) Feedback neural network, as shown in Figure 5.11.
Figure 5.11 shows that there is at least one feedback loop in the feedback neural network. The output of each neuron in the

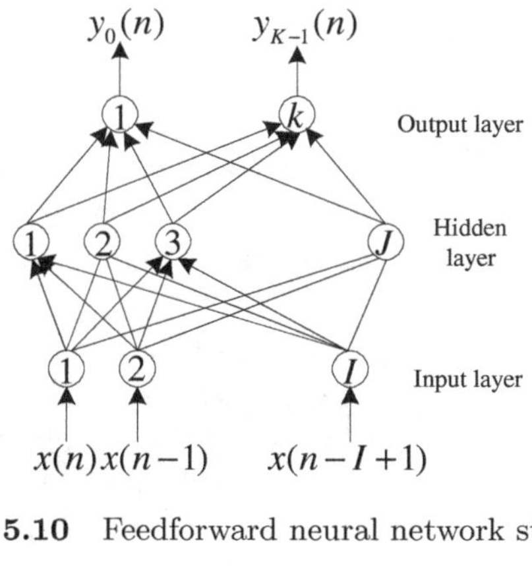

Figure 5.10 Feedforward neural network structure.

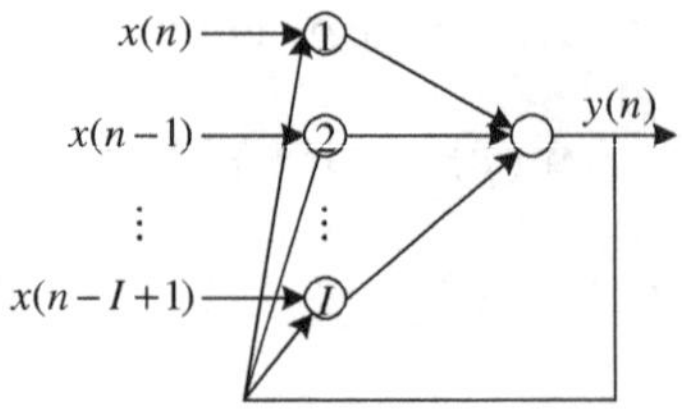

Figure 5.11 Feedback neural network.

single-layer network is fed back to other neurons as input, and the input of the neuron is affected by the output of the neuron.

5.2.2 *Nonlinear blind equalization algorithm based on multiwavelet neural network*

The feedforward neural network (FNN), structurally speaking, is hierarchical, and the signal from the input layer to its next layer and any unit of each layer has a connection, and also its traditional training method is the back propagation (BP) algorithm. In terms of system theory, the feedforward neural network is a steady-state nonlinear system with good nonlinear classification function.

5.2.2.1 *Neural network blind equalization system model*

The principle of neural network blind equalization algorithm is shown in Figure 5.12. $\boldsymbol{a}(n)$ is the transmitted signal vector, $\boldsymbol{x}(n)$ is the output vector to the channel, $\boldsymbol{w}(n)$ represents additive Gaussian white noise vector, $\boldsymbol{y}(n)$ denotes the input vector to the equalizer, $z(n)$ represents the output to the equalizer, and $\hat{a}(n)$ represents the output of the decision device.

Cybenc has proved that the neural network structure can approximate a continuous function with arbitrary accuracy. In this section, the three-layer feedforward neural network structure is shown in Figure 5.13.

In Figure 5.13, $f_{ij}(n)$ is the connection weight coefficient between the ith neuron in the input layer and the jth neuron in the hidden layer, $f_{jp}(n)$ is the connection weight coefficient between the jth neuron in the hidden layer and the pth neuron in the output layer, and the input $\boldsymbol{y}(n) = \{y(n-1), y(n-2), \ldots, y(n-I)\}^{\mathrm{T}}$ denotes the

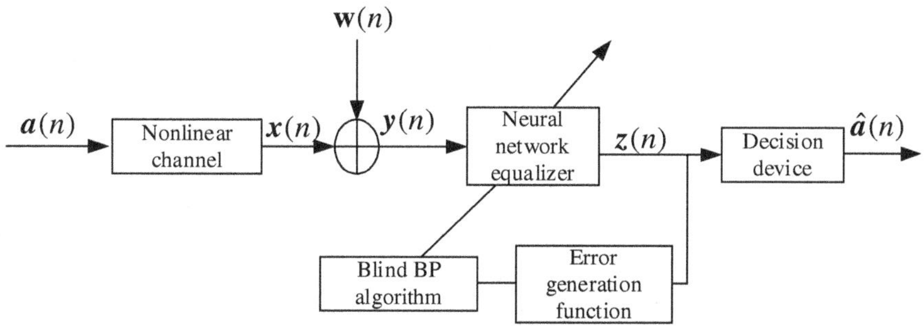

Figure 5.12 Blind neural network equalization system.

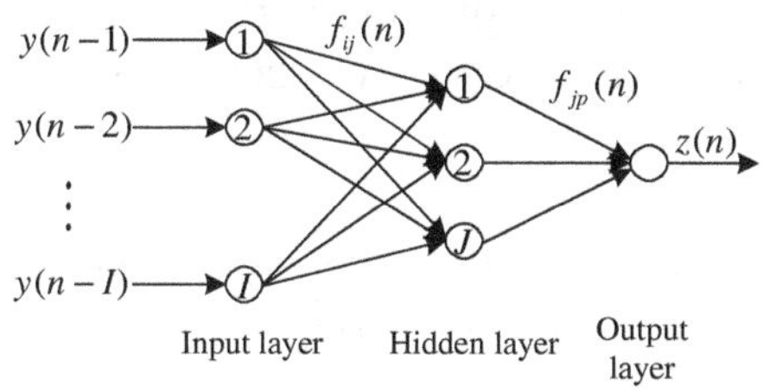

Figure 5.13 Basic structure of the feedforward neural network.

input of the input layer. The input of the jth neuron in the hidden layer is $u_j^J(n)$. The output of the jth neuron in the hidden layer is $v_j^J(n)$ and the output of the pth neuron in the output layer is $u_p^P(n)$. The output of the neural network is $z(n)$. The activation function of the neuron is selected as:

$$F(x) = x + a\sin(\pi x) \tag{5.2.5}$$

where $-\infty < x < \infty$, $0 < \alpha < 1 \cdot x$ represents $u_j^J(n)$ or $u_p^P(n)$. Equation (5.2.5) has good identification ability in the input of the neural network. The state equations of the neural network are defined or written as:

$$u_j^J(n) = \sum_{i=1}^{I} f_{ij}(n)y(n-i) \tag{5.2.6}$$

$$v_j^J(n) = F\left(\sum_{i=1}^{I} f_{ij}(n)y(n-i)\right) \tag{5.2.7}$$

$$u_p^P(n) = \sum_{j=1}^{J} f_{jp}(n)v_j^J(n) \tag{5.2.8}$$

$$z(n) = F\left(\sum_{j=1}^{J} f_{jp}(n)v_j^J(n)\right) \tag{5.2.9}$$

In the neural network blind equalization algorithm, the determination of the connection weight coefficient between neurons and the selection of activation function play a key role in improving the performance algorithm.

5.2.2.2 *Nonlinear blind equalization algorithm based on multiwavelet neural network*

The idea of introducing balanced orthogonal multiwavelet transform into the neural network nonlinear blind equalization algorithm is to carry out orthogonal multiwavelet transform on the input signal of the neural network equalizer at the same decomposition scale and adjust the weight vector from the input layer to the hidden layer of the neural network with the signals transformed by the multiwavelet. At this point, the principle of the multiwavelet transform feedback neural network nonlinear blind equalization algorithm (MWT-FNN) [29] is as shown in Figure 5.14.

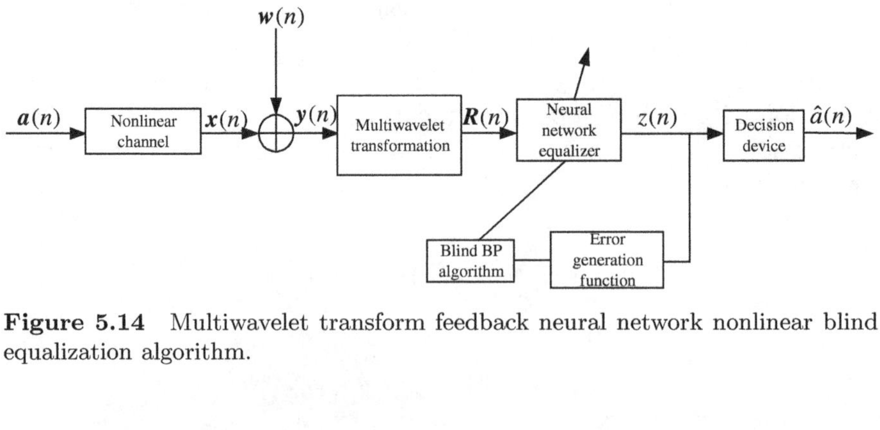

Figure 5.14 Multiwavelet transform feedback neural network nonlinear blind equalization algorithm.

In Figure 5.14, the input $\boldsymbol{R}(n)$ of the neural network equalizer can be obtained after the channel output signal $\boldsymbol{y}(n)$ passes through the multiwavelet transform, namely,

$$\boldsymbol{R}(n) = \boldsymbol{V}_{\mathrm{MWT}}\boldsymbol{y}(n) \tag{5.2.10}$$

where $\boldsymbol{V}_{\mathrm{MWT}}$ represents the matrix of the balanced orthogonal multiwavelet transform. According to Equation (5.2.10), the input and output of the hidden layer are written as:

$$u_j^J(n) = \sum_{j}^{I} f_{ij}(n)R_i(n) \tag{5.2.11}$$

$$v_j^J(n) = F\left(\sum_{i=1}^{I} f_{ij}(n)R_i(n)\right) \tag{5.2.12}$$

$$u_p^P(n) = \sum_{j=1}^{J} f_{jp}(n)v_j^J(n) \tag{5.2.13}$$

$$z(n) = F\left(\sum_{j=1}^{J} f_{jp}(n)v_j^J(n)\right) \tag{5.2.14}$$

The iteration equation of the weight vector for the input layer is obtained as follows:

$$\begin{aligned} \boldsymbol{f}_{ij}(n+1) &= \boldsymbol{f}_{ij}(n) - \mu_{ij}\frac{\partial J(n)}{\partial \boldsymbol{f}_{ij}(n)} \\ &= \boldsymbol{f}_{ij}(n) - \mu_{ij}\frac{\partial J(n)}{\partial z(n)} \cdot \frac{\partial z(n)}{\partial u_p^P(n)} \cdot \frac{\partial u_p^P(n)}{\partial v_j^J(n)} \cdot \frac{\partial v_j^J(n)}{\partial u_j^J(n)} \cdot \frac{\partial u_j^J(n)}{\partial \boldsymbol{f}_{ij}(n)} \end{aligned} \tag{5.2.15}$$

where μ_{ij} is the iteration step-size of the weight vector.

$$\frac{\partial J(n)}{\partial z(n)} = 2|z(n)|(|z(n)|^2 - R^2) \tag{5.2.16}$$

$$\frac{\partial z(n)}{\partial u_p^P(n)} = F'(u_p^P(n)) \tag{5.2.17}$$

$$\frac{\partial u_p^P(n)}{\partial v_j^J(n)} = \boldsymbol{f}_{jp}(n) \tag{5.2.18}$$

$$\frac{\partial v_j^J(n)}{\partial u_j^J(n)} = F'(u_j^J(n)) \tag{5.2.19}$$

$$\frac{\partial v_j^J(n)}{\partial \boldsymbol{f}_{ij}(n)} = \boldsymbol{R}_i(n) \tag{5.2.20}$$

Similarly, the iterative equation of the weight vector from the hidden layer to the output layer is updated by:

$$\boldsymbol{f}_{jp}(n+1) = \boldsymbol{f}_{jp}(n) - \mu_{jp}\frac{\partial J(n)}{\partial \boldsymbol{f}_{jp}(n)} \tag{5.2.21}$$

where μ_{jp} is the iteration step-size. In the same way, there is also the following relation.

$$\frac{\partial J(n)}{\partial \boldsymbol{f}_{jp}(n)} = \frac{\partial J(n)}{\partial z(n)} \cdot \frac{\partial z(n)}{\partial u_p^P(n)} \cdot \frac{\partial u_p^P(n)}{\partial \boldsymbol{f}_{jp}(n)} \tag{5.2.22}$$

Thus,

$$\begin{aligned}\boldsymbol{f}_{jp}(n+1) &= \boldsymbol{f}_{jp}(n) - \mu_{jp}\frac{\partial J(n)}{\partial \boldsymbol{f}_{jp}(n)} \\ &= \boldsymbol{f}_{jp}(n) - \mu_{jp}\frac{\partial J(n)}{\partial z(n)} \cdot \frac{\partial z(n)}{\partial u_p^P(n)} \cdot \frac{\partial u_p^P(n)}{\partial \boldsymbol{f}_{jp}(n)} \end{aligned} \tag{5.2.23}$$

where μ_{jp} is the step-size.

$$\frac{\partial u_p^P(n)}{\partial \boldsymbol{f}_{jp}(n)} = v_j^J(n) \tag{5.2.24}$$

Equations (5.2.10)–(5.2.24) is called as the multiwavelet transform feedback neural network nonlinear blind equalization algorithm (MWT-FNN).

5.2.2.3 *Computational complexity*

In the MWT-FNN algorithm, it is necessary to make sure the signal $y(n)$ is transformed by orthogonal wavelet in the process of updating the weight vector. Compared with the conventional FNN algorithm, the computational load increases to some extent. Now, multiplications are taken as the measure of computation complexity. Let the length of input signal be N and the weight length of the neural network equalizer of the input layer be L, then the multiplications of Equation (5.2.10) are LN, and the multiplications from Equation (5.2.11) to Equation (5.2.14) are $(2I + 2J)N$, and the multiplications required to update the weight vector of the input layer are $8N$. Similarly, the multiplications required to update the weight vector between the hidden layer and the output layer are $6N$, then the multiplications of the MWT-FNN are $(14 + 2I + 2J + L)N$.

5.2.3 *Algorithm simulation*

In order to verify the performance of the nonlinear blind equalization algorithm based on multi-wavelet neural network, the input signal was selected as 2PAM, the weight vector length of the neural network equalizer was set to 16, the fourth tap was initialized to 1, the step-size was given by $\mu = 1 \times 10^{-5}$, and the SNR was set to 20 dB. The weight vector length from the input layer to the hidden layer of the multi-wavelet neural network was set to 16, DB2 wavelet was selected, and the step-size was given by $\mu = 8 \times 10^{-4}$. The input–output relationship based on the nonlinear Wiener channel model is shown in Figure 5.15.

In Figure 5.15, $H = [0.3122 - 0.1040\ 0.8908\ 0.3134]$ represents the linear part of the nonlinear channel. In the nonlinear part, $D = 0.1$ is the quadratic coefficient and $D_1 = 0.05$ is the cubic coefficient. Simulation results with 500 Monte Carlo simulations are shown in Figure 5.16.

Figure 5.16(a) shows that the convergence speed of the MWT-N-FNN algorithm based on the multi-wavelet neural network has an improvement of about 2000 steps and the steady-state MSE is reduced by 6 dB in comparison with those of the N-FNN algorithm.

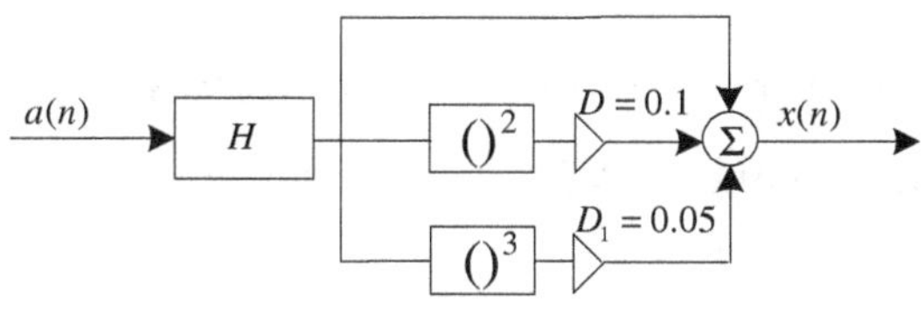

Figure 5.15 Wiener model of nonlinear channel.

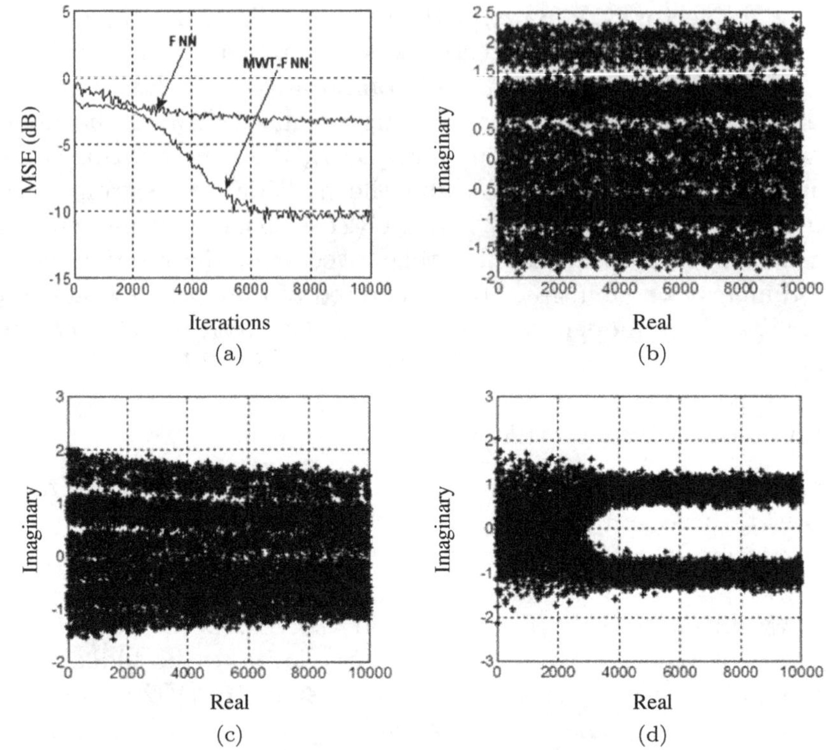

Figure 5.16 Simulation results: (a) MSE curves of all algorithm; (b) input of equalizer; (c) output of FNN and (d) output of MWT-FNN.

The constellations of the output signals are clearer and more compact. This may be due to the good decorrelation of the orthogonal multiwavelets, the nonlinearity of the neural networks, and the ability to process signals in a large parallel distribution.

5.3 Nonlinear Blind Equalization Algorithm Based on Support Vector Machine and Neural Network

5.3.1 *Support vector machine foundation*

Support vector machine (SVM) [30] is developed from the optimal classification surface in the case of linear separability, and its idea is shown in Figure 5.17.

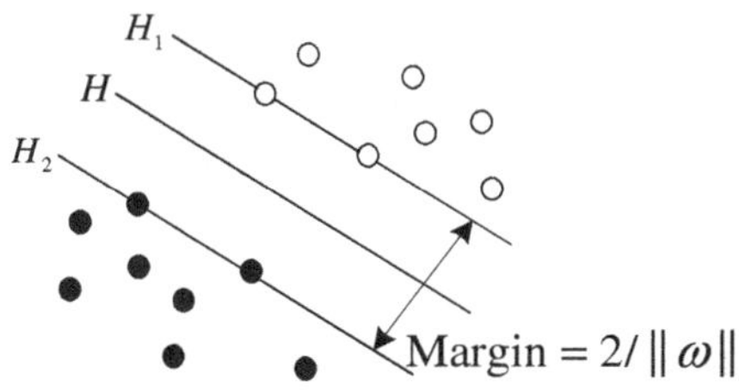

Figure 5.17 Schematic diagram of the optimal classification plane.

In Figure 5.17, black and white dots represent two different types of data samples, H_1 and H_2 which refer to the line nearest and parallel to the classification line and pass through various data samples as much as possible. The distance Δ between them is called as the classification interval (margin). The optimal classification line not only requires that the classification line can separate the two kinds of data sets but also that it can maximize the classification distance between them. Among them, the former ensures the empirical risk minimum, whereas the latter ensures the minimum confidence range of generalization, i.e., the real risk minimum. If two-dimensional classification is extended to multi-dimensional space, the optimal classification surface can be obtained.

5.3.1.1 *Optimal classification surface*

Let the linear separable sample set be (x_i, y_i), and $i = 1, \ldots, n$, $x_i \in R^d$ and $y_i \in \{+1, -1\}$ is the category symbol. The general linear discriminant function is given by $f(x) = \boldsymbol{\omega} \cdot \boldsymbol{x} + \boldsymbol{b}$, and the corresponding classification surface equation is:

$$\boldsymbol{\omega} \cdot \boldsymbol{x} + \boldsymbol{b} = 0 \tag{5.3.1}$$

To normalize the discriminant function, it is required that all samples can be classified correctly and $|f(x)| \geq 1$ must be satisfied. At this point, the sample closest to the classification surface meets the conditions as follows:

$$y_i[(\boldsymbol{\omega} \cdot \boldsymbol{x}) + \boldsymbol{b}] - 1 \geq 0, \quad i = 1, 2, \ldots, n \tag{5.3.2}$$

At this point, the maximization of the classification interval $2/||\boldsymbol{\omega}||$ is equivalent to the minimum of the classification interval

$||\boldsymbol{\omega}||^2$. The optimal classification surface H in Figure 5.17 is the classification surface which satisfies Equation (5.4.2) and makes $\frac{1}{2}||\boldsymbol{\omega}||^2$ the minimum. The data samples on hyperplanes H_1 and H_2 parallel to classification plane H are those data samples that make the equal sign hold in Equation (5.4.2). These data samples are support vectors (SV) that support the whole optimal classification surface.

Therefore, the optimal classification surface problem can be transformed into the constrained optimization problem of Equation (5.4.2), namely, to find the minimum value of the function $J(\boldsymbol{\omega})$, i.e.,

$$J(\boldsymbol{\omega}) = \frac{1}{2}||\boldsymbol{\omega}||^2 \tag{5.3.3}$$

This constrained optimization problem is called as the original problem, which has two characteristics: (1) cost function $J(\boldsymbol{\omega})$ is convex function of $\boldsymbol{\omega}$; (2) the constraint is linear with respect to $\boldsymbol{\omega}$. Now, the Lagrange multiplier method is used to solve the constrained optimal problem. The Lagrange function is defined as:

$$\boldsymbol{L}(\boldsymbol{\omega}, \boldsymbol{b}, a) = \frac{1}{2}||\boldsymbol{\omega}||^2 - \sum_{i=1}^{n} \alpha_i[y_i(\boldsymbol{\omega} \cdot \boldsymbol{x}_i + \boldsymbol{b}) - 1] \tag{5.3.4}$$

where $\alpha_i \geq 0$ is the multiplier of Lagrange. In order to find the minimum value of Equation (5.3.4), let the partial differential of Equation (5.4.4) with respect to $\boldsymbol{\omega}$, $\boldsymbol{b}$ and α_i, respectively, be zero i.e.,

$$\begin{cases} \dfrac{\partial L}{\partial \omega} = 0 \Rightarrow \omega = \sum_{i=1}^{n} \alpha_i y_i x_i \\ \dfrac{\partial L}{\partial b} = 0 \Rightarrow \sum_{i=1}^{n} \alpha_i y_i = 0 \\ \dfrac{\partial L}{\partial \alpha_i} = 0 \Rightarrow \alpha_i[y_i(\omega \cdot x_i + b) - 1] = 0 \end{cases} \tag{5.3.5}$$

By using the constraint conditions of Equations (5.3.2) and (5.3.5), the problem of solving the optimal classification surface can be transformed into a dual problem of finding the optimal solution by the convex quadratic programming method, that is,

$$\begin{cases} \max \sum_{i=1}^{n} \alpha_i - \frac{1}{2} \sum_{i=1}^{n} \sum_{j=1}^{n} \alpha_i \alpha_j y_i y_j (x_i \cdot x_j) \\ s.t. \quad \alpha_i \geq 0, \quad i = 1, \dots, n \\ \sum_{i=1}^{n} \alpha_i y_i = 0 \end{cases} \tag{5.3.6}$$

where α_i is the multiplier of Lagrange. Equation (5.3.6) is a quadratic function optimization problem with a unique solution. If α_i^* is the optimal solution, then

$$\omega^* = \sum_{i=1}^{n} \boldsymbol{\alpha}_i^* y_i \boldsymbol{x}_i \tag{5.3.7}$$

where $\boldsymbol{\alpha}_i^*$ is a nonzero sample and the support vector.

The classification threshold $\boldsymbol{b}^*$ can be solved by the constraint condition $\alpha_i[y_i(\boldsymbol{\omega} \cdot \boldsymbol{x}_i + \boldsymbol{b}) - 1] = 0$. The optimal classification surface function is obtained as follows:

$$f(x) = \operatorname{sgn}\{(\boldsymbol{\omega} \cdot \boldsymbol{x}) + \boldsymbol{b}\} = \operatorname{sgn}\left\{\sum_{i=1}^{n} \boldsymbol{\alpha}_i^* y_i (\boldsymbol{x}_i \cdot \boldsymbol{x}) + \boldsymbol{b}^*\right\} \tag{5.3.8}$$

The above is the general expression of support vector machine.

5.3.1.2 *Generalized optimal classification surface*

The above method is feasible only on the premise of ensuring that all training samples are correctly classified, and the best generalization performance can be obtained by maximizing the classification interval. When the optimal classification surface cannot correctly separate the two kinds of data points, in order to achieve a balance between empirical risk and generalization performance, it can be considered to introduce the relaxation factor ξ and acquiesce to the existence of misclassification samples. At this point, the classification surface $\boldsymbol{\omega} \cdot \boldsymbol{x} + \boldsymbol{b} = 0$ satisfies the following condition:

$$y_i[(\boldsymbol{\omega} \cdot \boldsymbol{x}_i) + \boldsymbol{b}] \geq 1 - \xi_i, \quad i = 1, 2, \dots, n \tag{5.3.9}$$

When $0 < \xi_i < 1$, the sample point x_i is classified correctly. When $\xi_i \geq 1$, the sample point x_i is wrongly classified. If the penalty term

$C\sum_{i=1}^{n}\xi_i$ is added to minimize the cost function $\frac{1}{2}||\boldsymbol{\omega}||^2$, then the cost function is written as:

$$J(\boldsymbol{\omega}, \xi) = \frac{1}{2}||\boldsymbol{\omega}||^2 + C\sum_{i=1}^{n}\xi_i \tag{5.3.10}$$

where C is the penalty factor, which is a normal number. Similar to the case of linear separability, Equation (5.4.10) can be obtained by convex quadratic programming.

$$\begin{cases} \max \sum_{i=1}^{n}\alpha_i - \frac{1}{2}\sum_{i=1}^{n}\sum_{j=1}^{n}\alpha_i\alpha_j y_i y_j (x_i \cdot x_j) \\ \text{s.t.} \quad 0 \leq \alpha_i \leq C, \quad i = 1, \ldots, n \\ \sum_{i=1}^{n}\alpha_i y_i = 0 \end{cases} \tag{5.3.11}$$

For the nonlinear classification problem, if the training samples are not classified correctly in the original space, it can be transformed into the linear problem in the high-dimensional characteristic space by some nonlinear mapping. Support vector machine (SVM) is an effective tool to solve this problem by kernel transformation.

5.3.1.3 *Kernel function*

Kernel function is one of the key techniques of support vector machine (SVM). Kernel function transformation is a very convenient way, which uses kernel function transformation $\boldsymbol{\Phi}$ to map the random vector $\boldsymbol{x}$ from the n-dimensional vector space to the high-dimensional vector space, and turns the nonlinear inseparable into linear separable, and designs the learning algorithm in the high-dimensional vector space. The principle is shown in Figure 5.18.

There are three common kernel functions of support vector machines: polynomial function, radial basis function, and sigmoid function. These are described as follows:

5.3.1.3.1 q-order polynomial function

The kernel function is selected as:

$$K(\boldsymbol{x}, \boldsymbol{x}_i) = [(\boldsymbol{x}, \boldsymbol{x}_i) + 1]^q \tag{5.3.12}$$

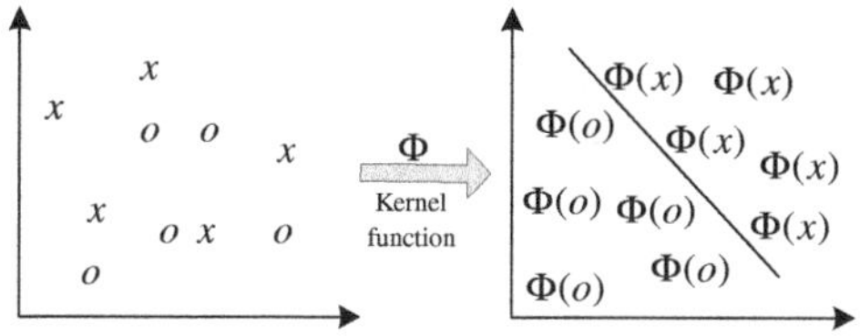

Figure 5.18 SVM kernel function transformation diagram.

Then, the discriminant function of the SVM is defined as:

$$f(\boldsymbol{x},\alpha) = \operatorname{sgn}\left\{\sum_{i=1}^{s} \alpha_i y_i[(\boldsymbol{x}_i \cdot \boldsymbol{x}) + 1]^q - \boldsymbol{b}\right\} \tag{5.3.13}$$

where s is the number of support vectors. The order q of polynomial can control the VC dimension of the system.

5.3.1.3.2 Radial basis function

The kernel function is chosen as:

$$K(\boldsymbol{x},\boldsymbol{x}_i) = \exp\left\{-\frac{|\boldsymbol{x}-\boldsymbol{x}_i|^2}{\sigma^2}\right\} \tag{5.3.14}$$

Then, the discriminant function of the SVM is written as:

$$f(\boldsymbol{x}) = \operatorname{sgn}\left\{\sum_{i=1}^{N_C} \alpha_i \exp\left\{-\frac{|\boldsymbol{x}-\boldsymbol{x}_i|^2}{\sigma^2}\right\} - \boldsymbol{b}\right\} \tag{5.3.15}$$

In Equation (5.4.15), the center position of the radial basis function is determined by the support vector $\boldsymbol{x}_i$, and N_C is the number of centers. Assume that the data sample is hypersphere distribution in the input space, the value of the scale parameter σ is the radius of the hypersphere.

5.3.1.3.3 Sigmoid function

The sigmoid kernel function is as follows:

$$K(\boldsymbol{x},\boldsymbol{x}_i) = \tan\,\mathrm{h}[\upsilon(\boldsymbol{x}\cdot\boldsymbol{x}_i) + a] \tag{5.3.16}$$

In Equation (5.3.16), sigmoid is a hyperbolic tangent function. If and only if the values of υ and a meet the Mercer condition, the possible cases are $\upsilon = 2$ and $a = 1$. At this point, the discriminant function of the SVM is given as follows:

$$f(\boldsymbol{x}) = \operatorname{sgn}\left\{\sum_{i=1}^{s} \boldsymbol{\alpha}_i \tan\text{h}[\upsilon(\boldsymbol{x} \cdot \boldsymbol{x}_i) + a] + \boldsymbol{b}\right\} \tag{5.3.17}$$

Equation (5.3.17) is the discriminant function of the three-layer neural network model, whose hidden layer node corresponds to the support vector.

5.3.2 *Regression principle of support vector machine*

The regression problem is to set up a new sample of the input data $\boldsymbol{x}$ and infer the corresponding output value y_i (which can be any real number) according to the given sample data. Set the given data sample set as $\{(x_i, y_i), \ldots, (x_l, y_l)\}$, $x_i \in R^n$, $i = 1, 2, 3, \ldots, l$ and $y_i \in R$, the regression problem is to find a function $y = f(x)$ on R^n, and infer the corresponding y value from any input $\boldsymbol{x}$, as shown in Figure 5.19.

In Figure 5.19, "×" represents the sample points in the set of the data samples, and the curve represents the regression objective function $f(x)$.

When the SVM is applied to the regression problem, the main features of the maximum interval algorithm can be retained and the nonlinear function can be obtained from the linear learning machine in the kernel feature space.

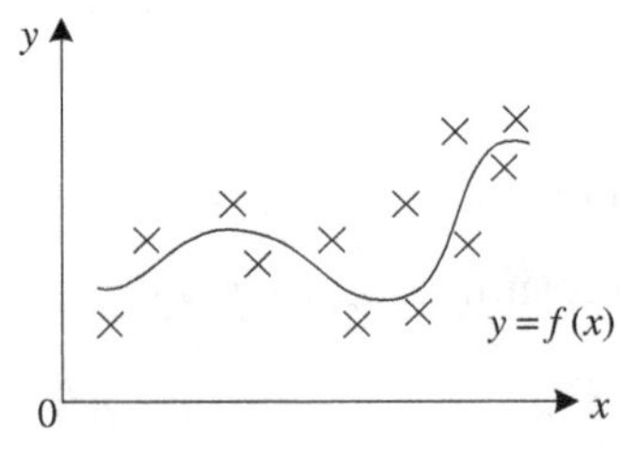

Figure 5.19 Geometric significance of regression algorithm.

5.3.2.1 *Linear support vector machine regression*

Let's say there is an n-dimensional sample vector, and the data samples in the space are $(x_1, y_1), (x_2, y_2), \ldots, (x_l, y_l) \in R^n \times R$. Let the linear function be defined as:

$$f(\boldsymbol{x}) = \boldsymbol{\omega} \cdot \boldsymbol{x} + \boldsymbol{b} \tag{5.3.18}$$

This optimization problem is expressed as:

$$R(\boldsymbol{\omega}, \xi, \xi^*) = \frac{1}{2}\boldsymbol{\omega} \cdot \boldsymbol{\omega} + C\sum_{i=1}^{l}(\xi + \xi^*) \tag{5.3.19}$$

where $\frac{1}{2}\boldsymbol{\omega} \cdot \boldsymbol{\omega}$ can make the function more flat and improve the generalization ability. In order to make the regression problem more precise, the "softening method" is introduced here. Assume that C is the penalty parameter and ξ and ξ^* represent the relaxation variables, then the constraint condition is written as:

$$f(x_i) - y_i \leq \xi_i^* + \varepsilon \tag{5.3.20}$$

$$f(x_i) - y_i \leq \xi_i + \varepsilon \tag{5.3.21}$$

where $\xi_i^*, \xi_i \geq 0, \quad i = 1, \ldots, l$, ε is a normal number.

After the Lagrange function is introduced, the convex quadratic optimization problem required to be solved is represented as:

$$\begin{aligned} &L(\boldsymbol{\omega}, b, \xi, \xi^*, \boldsymbol{\alpha}, \boldsymbol{\alpha}^*, \gamma, \gamma^*) \\ &= \frac{1}{2}\boldsymbol{\omega} \cdot \boldsymbol{\omega} + C\sum_{i=1}^{l}(\xi_i + \xi_i^*) - \sum_{i=1}^{l}\boldsymbol{\alpha}_i[\xi_i + \varepsilon - y_i + f(x_i)] \\ &\quad - \sum_{i=1}^{l}\boldsymbol{\alpha}_i^*[\xi_i + \varepsilon + y_i - f(x_i)] - \sum_{i=1}^{l}(\xi_i\gamma_i - \xi_i^*\gamma_i^*) \end{aligned} \tag{5.3.22}$$

where $\boldsymbol{\alpha} = (\alpha_1, \ldots, \alpha_N)$ and $\boldsymbol{\alpha}^* = (\alpha_1^*, \ldots, \alpha_N^*)$ are the Lagrange multipliers. The minimum value of Lagrange function is obtained by partial derivative of Equation (5.3.22) with respect to $\boldsymbol{\omega}$, $\boldsymbol{b}$, ξ_i, and ξ_i^*.

The maximum value is obtained by partial derivative of Equation (5.3.22) with respect to α_i, α_i^*, γ_i, and γ_i^*. The extreme value of Equation (5.3.22) should satisfy the following conditions:

$$\frac{\partial L}{\partial \omega} = 0, \quad \frac{\partial L}{\partial b} = 0, \quad \frac{\partial L}{\partial \xi_i} = 0, \quad \frac{\partial L}{\partial \xi_i^*} = 0 \tag{5.3.23}$$

Then,

$$\sum_{i=1}^{l} (\alpha_i - \alpha_i^*) = 0 \tag{5.3.24}$$

$$\boldsymbol{\omega} = \sum_{i=1}^{l} (\alpha_i - \alpha_i^*)\boldsymbol{x}_i \tag{5.3.25}$$

$$C - \alpha_i - \gamma_i = 0, \tag{5.3.26}$$

$$C - \alpha_i^* - \gamma_i^* = 0, \tag{5.3.27}$$

where $i = 1, 2, \ldots, l$.

From Equation (5.3.18) to Equation (5.3.27), the dual form of the optimization problem can be written as:

$$W(\boldsymbol{\alpha}, \boldsymbol{\alpha}^*) = -\frac{1}{2}\sum_{i,j=1}^{l} (\alpha_i - \alpha_i^*)(\alpha_j - \alpha_j^*)(x_i \cdot x_j) + \sum_{i=1}^{l} (\alpha_i - \alpha_i^*)y_i - \sum_{i=1}^{l} (\alpha_i + \alpha_i^*)\varepsilon \tag{5.3.28}$$

The constraint conditions are given as:

$$\boldsymbol{\omega} = \sum_{i=1}^{l} (\alpha_i - \alpha_i^*)\boldsymbol{x}_i, \quad 0 \le \alpha_i, \quad \alpha_i^* \le C \tag{5.3.29}$$

This is also a quadratic optimization problem.

5.3.2.2 *Regression principle of nonlinear SVM*

Nonlinear regression and nonlinear classification problems are similar. The data samples can be transformed from regression problems in low-dimensional characteristic space to linear regression problems in high-dimensional characteristic space by nonlinear mapping.

The key to solve the regression problem is the choice of kernel function $K(x, y)$. Here, the optimization problem is to maximize the optimization function under the constraint of Equation (5.3.29), that is,

$$W(\boldsymbol{\alpha}, \boldsymbol{\alpha}^*) = -\frac{1}{2}\sum_{i,j=1}^{l}(\boldsymbol{\alpha}_i - \boldsymbol{\alpha}_i^*)(\boldsymbol{\alpha}_j - \boldsymbol{\alpha}_j^*)K(x_i, x_j) + \sum_{i=1}^{l}(\alpha_i - \alpha_i^*)y_i - \sum_{i=1}^{l}(\alpha_i + \alpha_i^*)\varepsilon \tag{5.3.30}$$

Here,

$$\boldsymbol{\omega} = \sum_{i=1}^{l}(\boldsymbol{\alpha}_i - \boldsymbol{\alpha}_i^*)\phi(x_i) \tag{5.3.31}$$

The function $f(x)$ can be directly expressed as:

$$f(\boldsymbol{x}) = \sum_{i=1}^{l}(\boldsymbol{\alpha}_i - \boldsymbol{\alpha}_i^*)K(\boldsymbol{x},\mathbf{x}_i) + \boldsymbol{b} \tag{5.3.32}$$

According to the Kuhn–Tucker theorem, one has

$$\alpha_i[\varepsilon + \xi_i - y_i + f(x_i)] = 0 \tag{5.3.33}$$

$$\alpha_i[\varepsilon + \xi_i^* - y_i + f(x_i)] = 0 \tag{5.3.34}$$

$$\xi_i\gamma_i = 0 \tag{5.3.35}$$

$$\xi_i^*\gamma_i^* = 0 \tag{5.3.36}$$

According to Equations (5.3.35) and (5.3.36), that is, $\alpha_i\alpha_i^* = 0$, any set of α_i and α_i^* will not be 0 at the same time. According to Equations (5.3.26), (5.3.29), (5.3.35), and (5.3.36), one gets

$$(C - \alpha_i)\xi_i = 0 \tag{5.3.37}$$

$$(C - \alpha_i)\xi_l^* = 0 \tag{5.3.38}$$

To sum up, the steps of the SVM regression are given as follows:

Step 1: Give training set $T = \{(x_1, y_1), (x_2, y_2), \ldots, (x_N, y_N)\} \in (R^n \times Y^n)$, where $x_i \in R^n$, $y_i \in R$, $\quad i = 1, \ldots, N$.
Step 2: Select the appropriate parameter $\varepsilon > 0$.
Step 3: Establish and solve the optimization problem of the convex quadratic programming to get the vectors $\bar{\boldsymbol{\alpha}} = (\bar{\alpha}_1, \bar{\alpha}_2, \ldots, \bar{\alpha}_N)$ and $\bar{\boldsymbol{\alpha}}^* = (\bar{\alpha}_1^*, \bar{\alpha}_2^*, \ldots, \bar{\alpha}_N^*)$.
Step 4: Calculate $\bar{\omega} = \sum_{i=1}^{N}(\bar{\alpha}_i^* - \bar{\alpha}_i)x_i$ and $\bar{b} = y_j - (\bar{\omega} \cdot x_j) + \varepsilon$.
Step 5: Construct the decision function, i.e., the regression function of the training set.

$$y = f(\boldsymbol{x}) = (\bar{\boldsymbol{\omega}} \cdot \boldsymbol{x}) + \bar{b} = \sum_{i=1}^{N} (\bar{\alpha}_i^* - \bar{\alpha}_i)(\boldsymbol{x}_i \cdot \boldsymbol{x}) + \bar{b}$$

5.3.3 *Multiwavelet neural network blind equalization algorithm based on spatial diversity SVM*

5.3.3.1 *SVM multi-wavelet neural network blind equalization algorithm*

In view of the lack of global searching ability of the multi-wavelet neural network nonlinear blind equalization algorithm, in this section, the SVM is used to search the optimal initial connection weight vector between the input layer and the hidden layer of the multi-wavelet neural network, so as to obtain the principle of the SVM multi-wavelet transform feedback neural network nonlinear blind equalization algorithm (SVM-MWT-FNN) [29], as shown in Figure 5.20.

In Figure 5.20, $\boldsymbol{a}(n)$ is the transmitted signal vector, $\boldsymbol{x}(n)$ is the output vector of the Wiener model of the nonlinear channel, $\boldsymbol{w}(n)$

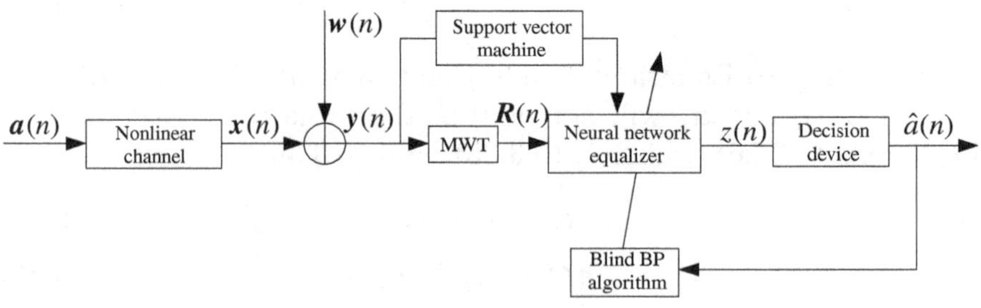

Figure 5.20 Principle of SVM-MWT-FNN.

denotes an additive white Gaussian noise vector, $\boldsymbol{R}(n)$ represents the input vector of the neural network equalizer transformed by multi-wavelet, $z(n)$ is called as the output of the neural network equalizer, and $\hat{a}(n)$ is the decision output of $a(n)$.

When the transmitted signal vector $\boldsymbol{a}(n)$ is a constant modulus vector, the top N-group vectors of the $\boldsymbol{y}(n)$, which are received by the equalizer, can be equalized by the SVM. According to the statistical characteristics of the transmitted signals and the principle of the structural risk minimization, the initial connection weight vector $\boldsymbol{f}_{ij}(n)$ between the input layer and the hidden layer of the MWT-FNN is estimated with precision ε, then the SVM regression problem can be described as [30–34]:

$$\min E_{\text{svm}}(\boldsymbol{f}_{ij}(n)) = \frac{1}{2}||\boldsymbol{f}_{ij}(n)||^2 \tag{5.3.39}$$

The constraint conditions are written as follows:

$$\begin{cases} R^2 - ([\boldsymbol{f}_{ij}(n)]^{\mathrm{T}}\boldsymbol{y}(n))^2 \le \varepsilon \\ ([\boldsymbol{f}_{ij}(n)]^{\mathrm{T}}\boldsymbol{y}(n))^2 - R \le \varepsilon \end{cases} \tag{5.3.40}$$

where

$$R^2 = E\{|a(n)|^4\}/E\{|a(n)|^2\}, \varepsilon = 3\sigma_n^2\frac{\ln N}{N} \tag{5.3.41}$$

where σ_n^2 denotes noise variance. Parameter ε determines the width of the insensitive region and the number of support vectors.

In order to relax constraints, relaxation variables $\xi(n)$ and $\tilde{\xi}(n)$ and penalty coefficient C are introduced, and the optimization problem of Equations (5.3.39) and (5.3.40) can be transformed into the problem of solving constraint optimization, that is,

$$\min E_{\text{svm}}(\boldsymbol{f}_{ij}(n)) = \frac{1}{2}||\boldsymbol{f}_{ij}(n)||^2 + C\sum_{n=1}^{N}(\xi(n) + \tilde{\xi}(n)) \tag{5.3.42}$$

The constraint conditions are as follows:

$$\begin{cases} R^2 - ([\boldsymbol{f}_{ij}(n)]^{\mathrm{T}}\boldsymbol{y}(n))^2 \le \varepsilon + \xi(n) \\ ([\mathbf{f}_{ij}(n)]^{\mathrm{T}}\boldsymbol{y}(n))^2 - R^2 \le \varepsilon + \tilde{\xi}(n) \\ \xi(n), \tilde{\xi}(n) \ge 0 \end{cases} \tag{5.3.43}$$

In Equations (5.3.42) and (5.3.43), $\xi(n)$ and $\tilde{\xi}(n)$ are used to measure the distance of sample outliers, and the penalty variable C reflects the importance of outlier sample points, and

$$C = E[g_n] + 3\sigma_\varepsilon \tag{5.3.44}$$

where $g_n = |y(n)|^2$, $E[g_n]$ denotes the mean of g_n, and σ_ε represents the variance of the received signals.

In the above original optimization problem of constructing the SVM multi-wavelet neural network nonlinear blind equalization algorithm, the weight vector $\boldsymbol{f}_{ij}$ of the equalizer in the constraint condition is quadratic, and the linear programming algorithm adopted by SVM cannot solve the quadratic original optimization problem. Therefore, an iterative weight quadratic programming algorithm is used to solve this problem. Suppose that the input of the hidden layer of the neural network is $u^J_{j,\text{SVM}}(n)$, and after obtaining $u^J_{j,svm}(n)$, the quadratic constraint in Equation (5.3.43) can be changed to the linear constraint, namely,

$$\begin{cases} R - ([\boldsymbol{f}_{ij,\text{svm}}(n)]^{\text{T}}\boldsymbol{x}(n))\boldsymbol{u}^J_{j,\text{svm}}(n) \le \varepsilon + \xi(n) \\ ([\boldsymbol{f}_{ij,\text{svm}}(n)]^{\text{T}}\boldsymbol{x}(n))\boldsymbol{u}^J_{j,\text{svm}}(n) - R \le \varepsilon + \tilde{\xi}(n) \end{cases} \tag{5.3.45}$$

After transforming the original optimization problem from Equation (5.4.41) to Equation (5.4.44) into a convex quadratic programming problem, one gets

$$\begin{aligned} \max E'_{\text{svm}} = & -\frac{1}{2}\sum_{i,n=1}^{N}(\tilde{\alpha}(n) - \alpha(n))(\tilde{\alpha}(i) - \alpha(i)) \\ & \times (\boldsymbol{u}^J_{j,\text{svm}}(n)\boldsymbol{u}^J_{j,\text{svm}}(i)) < \boldsymbol{x}(n), \boldsymbol{x}(i) > \\ & -\varepsilon\sum_{n=1}^{N}(\tilde{\alpha}(n) + \alpha(n)) + \sum_{n=1}^{N}(\tilde{\alpha}(n) - \alpha(n)) \end{aligned} \tag{5.3.46}$$

where $\tilde{\alpha}(n)$ and $\alpha(n)$ are the Lagrange multiplier.

The constraint conditions are written as follows:

$$\begin{cases} \sum_{n=1}^{N}(\tilde{\alpha}(n) - \alpha(n)) = 0 \\ 0 \le \tilde{\alpha}(n), \alpha(n) \le C, \quad n = 1, \ldots, N \end{cases} \tag{5.3.47}$$

where $< \boldsymbol{x}(n), \boldsymbol{x}(i) >$ is the inner product.

By comparing the original problem with the dual problem, the initial weight vector of connecting the input layer and hidden layer of the neural network equalizer can be expressed as:

$$\boldsymbol{f}_{j,\mathrm{svm}}^{J}(n) = \sum_{n=1}^{N} (\tilde{\alpha}(n) - \alpha(n)) u_{j,\mathrm{svm}}^{J}(n)\mathbf{x}(n) \tag{5.3.48}$$

According to the above steps, the initial vector $\boldsymbol{f}_{j,\mathrm{svm}}^{J}(0)$ of the connection weight vector $\boldsymbol{f}_{j,\mathrm{svm}}^{J}(n)$ between the input layer and the hidden layer of the neural network can be calculated, and $\boldsymbol{f}_{j,\mathrm{svm}}^{J}(n)$ can be recirculated and updated until the switching condition is satisfied. The updating equation of $\boldsymbol{f}_{j,\mathrm{svm}}^{J}(n)$ is given by:

$$\boldsymbol{f}_{j,\mathrm{svm}}^{J}(n) = \lambda \boldsymbol{f}_{j,\mathrm{svm}}^{J}(n-1) + (1-\lambda)\boldsymbol{f}_{j,\mathrm{svm}}^{J}(n-2) \tag{5.3.49}$$

where λ represents the missing factor.

When the switching conditions are satisfied as follows:

$$\begin{cases} \mathrm{MSE}(n) = \dfrac{1}{N}\displaystyle\sum_{n=1}^{N} (|\boldsymbol{u}_{j,\mathrm{svm}}^{J}(n)|^2 - R) \\ |\mathrm{MSE}(n) - \mathrm{MSE}(n-1)| \le \eta \end{cases}, \tag{5.3.50}$$

the initial vector of the connection weight vector $\boldsymbol{f}_{j,\mathrm{svm}}^{J}(n)$ can be obtained, and η is the switching threshold.

5.3.3.2 *Nonlinear blind equalization algorithm based on spatial diversity SVM and multiwavelet neural network*

The basic principle of the diversity technology is to combine multiple statistical independent fading channels with unified information to reduce the influence of the fading. It is widely used in mobile communication because it can effectively improve the reliability of communication via tracking each path signal and processing more signal components in time. According to different diversity resources, the diversity technology can be divided into time diversity, space diversity, and angle diversity. In this section, the spatial diversity [35,36], also known as antenna diversity, will be selected, which uses multiple transmitting or receiving antennas. In order to obtain a satisfactory diversity effect, the distance between antennas at the receiving end

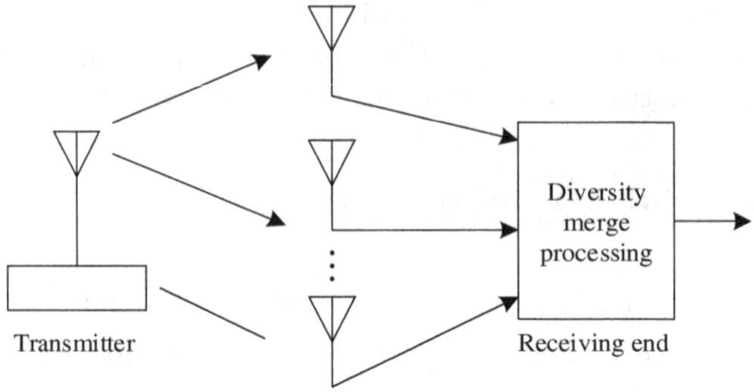

Figure 5.21 Structure of spatial diversity.

$d \geq 3\lambda$ and λ is the wavelength of working frequency. Usually, the more diversity antennas there are, the more obvious the fading elimination effect will be. However, when the diversity multiplicity is increased to a certain amount, the improvement of the diversity performance is obviously weakened. Therefore, it is advisable to have a diversity multiplicity belonging to the range from 2 to 4. Since the fading of signals in diversity branches is independent of each other and the probability of the simultaneous fading is very low, the application of spatial diversity to blind equalization system can effectively eliminate channel fading and inter-symbol interference (ISI). Monsen was the first to combine spatial diversity equalizer (SDE) with blind equalizer. The spatial diversity requires at least two antennas at a certain interval, and the diversity structure is shown in Figure 5.21.

When the spatial diversity technique is applied to the SVM multiwavelet neural network nonlinear blind equalization algorithm, each branch of the diversity contains a SVM multiwavelet neural network blind equalizer [37–40], and the outputs of all branches are weighted and combined. The spatial diversity blind equalizer based on the SVM multi-wavelet transform feedback neural network (SDE-SVM-MWT-FNN) is proposed, as shown in Figure 5.22.

According to the principle of the wavelet neural network equalizer optimized by support vector machine, the input and output signals of the dth branch equalizer are written as follows:

$$\boldsymbol{R}^{(d)}(n) = \boldsymbol{V}_{\mathrm{MWT}}\boldsymbol{y}^{(d)}(n) \tag{5.3.51}$$

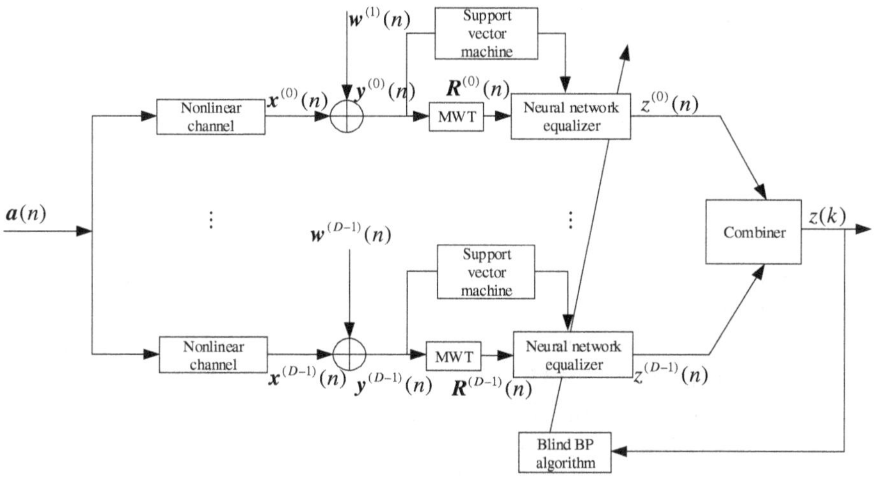

Figure 5.22 Structure of SDE-SVM-MWT-FNN.

$$u_j^{J(d)}(n) = \sum_{i=1}^{I} \boldsymbol{f}_{ij,\text{svm}}^{(d)}(n)\boldsymbol{R}^{(d)}(n) \tag{5.3.52}$$

$$v_j^{J(d)}(n) = F\left(\sum_{i=1}^{I} \boldsymbol{f}_{ij,\text{svm}}^{(d)}(n)\boldsymbol{R}^{(d)}(n)\right) \tag{5.3.53}$$

$$z^{(d)}(n) = F\left(\sum_{j=1}^{J} \boldsymbol{f}_{jp}^{(d)}(n)\boldsymbol{v}_j^{J(d)}(n)\right) \tag{5.3.54}$$

where $d = 0, 1, \ldots, D-1$, $f_{ij,\text{svm}}^{(d)}(n)$ is the connection weight coefficient between the ith input layer and the hidden layer of the dth branch equalizer, and $f_{jp}^{(d)}(n)$ is the connection weight coefficient between the ith hidden layer and the pth output layer of the dth branch equalizer. After all branch signals are weighted and combined in the combinator, the output signal is obtained:

$$z(n) = \sum_{d=0}^{D} g_d z_d(n) \tag{5.3.55}$$

where g_d is the weighting coefficient. If the maximum ratio combination is adopted, $g_d = A_d/\sigma_d^2$. A_d is the signal amplitude of the dth branch, and σ_d^2 is the noise power of the dth branch. If the equal gain combination is adopted, $g_d = 1$.

5.3.3.3 *Computational complexity*

Since there are the SVM, $D - 1$ branches, and a weighted combinator in the structure of the blind equalization algorithm, the computational complexity of this blind equalization algorithm will be increased to some extent. According to literature [30], let n be the number of training samples. For a test sample, the computational complexity of the SVM is $O(n^3)$, while the computational complexity of the MWT-FNN is $O(MN)$, then it can been known from Equation (5.3.51) to Equation (5.3.56) that the computational complexity of the SDE-SVM-MWT-FNN is $O(Dn^3 + DMN + D) \approx O(Dn^3)$. Obviously, it is the same order of magnitude as the complexity of the SVM.

5.3.4 *Algorithm simulation*

In order to test the performance of the SDE-SVM-MWT-FNN algorithm, the FNN, the SVM-FNN, and the SVM-MWT-FNN are compared. In the simulation experiment and Figure 5.22, the spatial diversity multiplicity d is 2. The SDE-SVM-MWT-FNN adopts two nonlinear channel hybrid Wiener models.

Channel 1:

$$\begin{aligned} a_2^{(1)}(n) &= (1.1026 + j0.1015)a(n) \\ &\quad + (0.2057 + j0.0330)a(n-1) \\ &\quad + (0.1012 + j0.0160)a(n-2) \\ x^{(1)}(n) &= (1.5050 - j0.0118)a_2^{(1)}(n) \\ &\quad + (0.0010 - j0.0004)(a_2^{(1)}(n))^2 \\ &\quad + (-0.0011 + j0.0001)(a_2^{(1)}(n))^3 \end{aligned}$$

Channel 2:

$$\begin{aligned} a_2^{(2)}(n) &= (1.22 + j0.646)a(n) \\ &\quad + (0.063 - j0.001)a(n-1) \\ &\quad + (-0.024 - j0.014)a(n-2) \end{aligned}$$

$$
\begin{aligned}
x^{(2)}(n) = &(0.039 - j0.022)a_2^{(2)}(n) \\
&+ (0.035 - j0.035)(a_2^{(2)}(n))^2 \\
&+ (-0.01 - j0.017)(a_2^{(2)}(n))^3
\end{aligned}
$$

The FNN, the SVM-FNN, and the SVM-MWT-FNN were simulated by *channel 1*. In the simulation, the transmitted signal was 8PSK, the SNR was set to 20 dB, the three-layer multi-wavelet neural network was selected, the weight length of the input layer of the equalizer was set to 16, and the center tap of the equalizer was initialized to 1. For the N-FNN, the step-size was set to $\mu_1 = 1.2 \times 10^{-5}$. For the SVM-N-FNN, the step-size was set to $\mu_2 = 0.0003$. For the SVM-MWT-FNN, the step-size was set to $\mu_3 = 0.0003$. For *channel 1*, in the SDE-SVM-MWT-FNN, the step-size was set to $\mu_4^{(1)} = 0.0012$. For *channel 2*, the step-size was set to $\mu_4^{(2)} = 0.0003$. DB2 multi-wavelet was used, the training sample number of the SVM blind equalization algorithm was set to $N = 100$, $\lambda = 0.9$, and the switching threshold was $\eta = 10^{-4}$. The results of 500 Monte Carlo simulations are shown in Figure 5.23.

It can be seen from Figure 5.23(a) that the SDE-SVM-MWT-FNN has an improvement of about 1500 steps, 2500 steps, and 7500 steps in comparison with that of the SVM-MWT-FNN, the SVM-FNN, and the FNN. In terms of steady-state MSE, the SDE-SVM-MWT-FNN has a drop of about 5 dB, 8 Db, and 13 dB IN comparison with that of the SVM-MWT-FNN, the SVM-MWT-FNN, the SVM-FNN, and the FNN. It can be seen from Figure 5.23(b) to Figure 5.23(f) that the constellations of the SDE-SVM-MWT-FNN are clearer and more compact than those of the SVM-MWT-FNN, the SVM- FNN, and the FNN, and the ISI is significantly reduced.

5.4 Blind Equalization Algorithm Based on Chaos Algorithm

Chaos is a kind of complex motion form which seems irregular in the real world, and it is a kind of motion state of nonlinear dynamic system under certain conditions. Chaos has its own characteristics that are as follows: randomness, ergodicity, and regularity. When the space is stable, the chaotic motion track can traverse every state

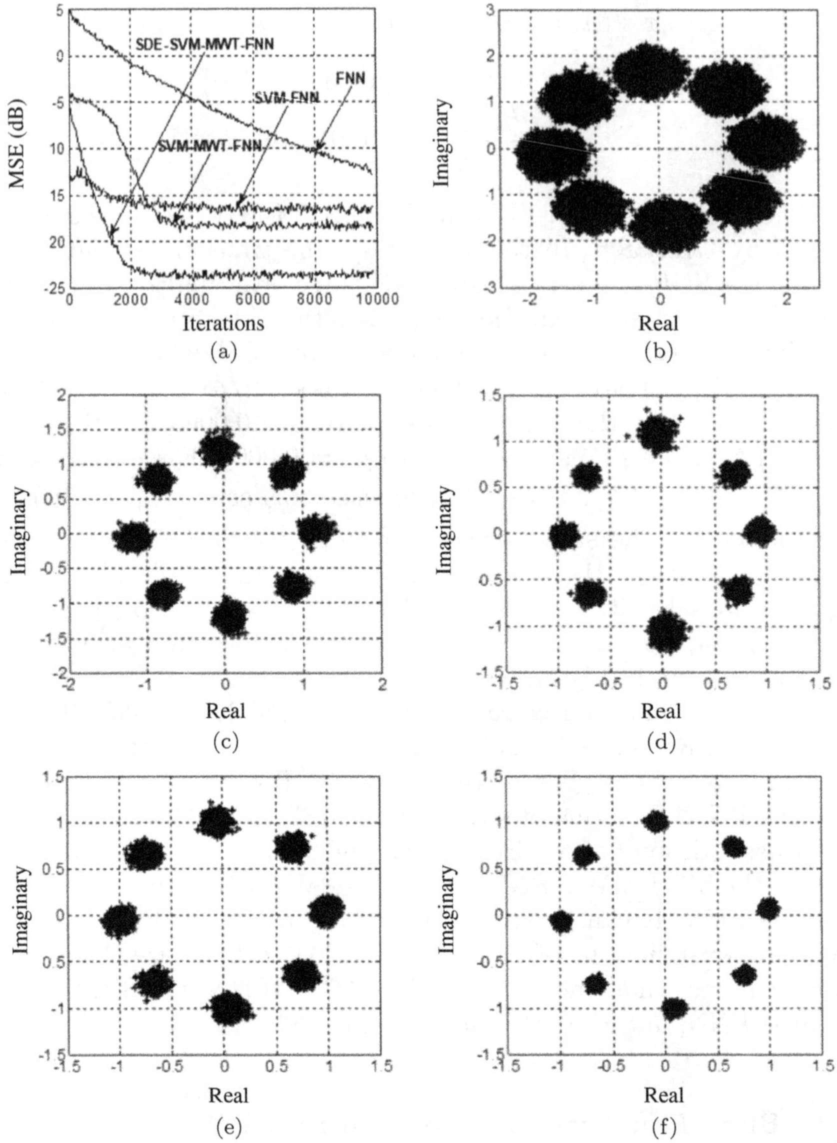

Figure 5.23 Simulation results: (a) MSE curves of all algorithms; (b) input of equalizer I; (c) output of FNN; (d) output of SVM-FNN; (e) output of SVM-MWT-FNN and (f) output of SDE-SVM-MWT-FNN.

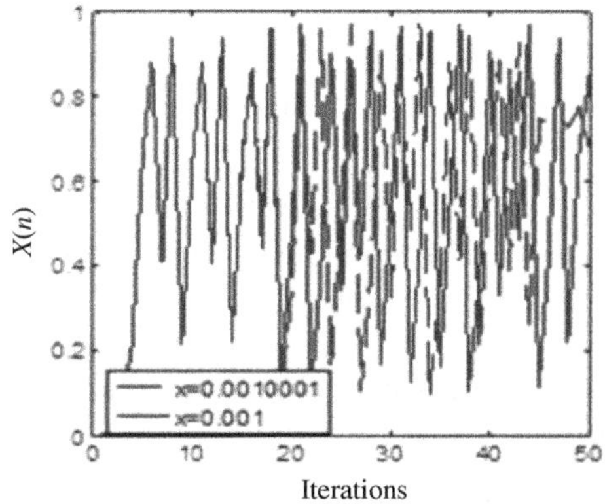

Figure 5.24 Initial value sensitivity of chaos.

point in the space, without repetition and disorder. This feature can avoid local minimum effectively in the process of chaos searching.

5.4.1 *Basis of the chaos algorithm*

5.4.1.1 *Chaos theory*

Chaotic signals have the following characteristics:

5.4.1.1.1 Sensitive dependency of initial value

The initial value sensitive dependence of chaos is also known as the butterfly effect. If the initial value is slightly different, the subsequent system will have great change. Therefore, the long-term moving trajectory of the chaos system is unpredictable. The characteristics can be explained by logistic mapping of the chaotic system. The logistic mapping system can be expressed as:

$$x(n+1) = \mu x(n)(1 - x(n)), \quad x(n) \in [0, 1], \quad \mu \in [0, 4] \tag{5.4.1}$$

When the parameter $\mu \in [3.7, 4]$, the mapping is chaotic, as shown in Figure 5.24.

5.4.1.1.2 Elongation and folding characteristic

Chaos has the characteristics of elongation and folding. When the system is unstable, the distance between the points will be enlarged, which is called as the elongation characteristic. But the whole system is stable and the stable system will limit the distance between points, which is called as folding property. Chaos is produced by the repeated stretching and folding of the system.

5.4.1.1.3 Fractal and self-similarity

All kinds of strange attractors in chaos have a fractal structure, which is described by fractal dimension. At the same time, fractal in chaos has self-similarity, which can be strict or traditional. Fractal and chaos are different manifestations of the same law. They are fractal in space and chaotic in time.

5.4.1.1.4 Boundedness and inner randomness

Chaos has inherent randomness, which comes from the sensitive dependence of the system on the initial value. The system does not need any additional random factors and is completely described by deterministic equations, but it will behave like randomness. The trajectories shown in Figure 5.24 remain within [0,1], so the chaotic system is stable and bounded on the whole.

5.4.1.2 *Chaos algorithm*

The chaos algorithm [36, 37] based on the carrier transformation method takes the chaos variable as the carrier, whereas chaos variable is mapped to the optimization variable region by linear mapping. In the chaos algorithm, the searching optimization variable is replaced with the updated chaos sequence, the chaos variable is used to traverse all the states in the space one by one during searching, and all the experienced states will not be searched repeatedly, which greatly saves the search time.

Logistic model is a special case of the chaos dynamic model. Its dynamic characteristic equation is simple and mature, so it is often used in the chaos optimization method. After using the model to generate chaotic variables $x(n)$, the chaotic variable $y(n)$ may be expressed as:

$$y(n) = a + x(n) \times (b - a) \tag{5.4.2}$$

where b and a are the upper and lower limits of the value range of the optimized variable $y(n)$. During the whole optimization process, $x(n)$

keeps searching for the updated value in (0,1), and $y(n)$ is mapped by $x(n)$ and simultaneously traverses the entire optimization variable space with $x(n)$. After n times of the traversal optimization, the optimal solution of the problem can always be found.

5.4.2 *Chaotic optimization process*

Compared with other optimization algorithms, in terms of cost function, the chaos optimization algorithm does not need many restrictions. It only needs to repeatedly calculate the value of the cost function and then compare and update, which simplifies the intermediate operation and improves the efficiency. Moreover, according to the ergodicity of chaotic searching, the searching process avoids falling into the local optimal solution problem. In this section, the chaos optimization method in literature [41–44] is adopted. At the current point of the disturbance weight vector of the chaos variable, the time-varying parameter is introduced, so that the amplitude of the disturbance changes at any time, and the update of the parameter are gradually reduced. Accordingly, the problem of the initial value sensitivity of the classical chaos algorithm is solved. After getting the optimal solution, the local fine search is carried out, so as to obtain the required global optimal weight vector.

In the chaotic optimization process, similar carriers are introduced into the chaotic variable to search, the time-varying parameter $\Psi(t)$ is brought in the second carrier process, and $f(K) = f_{\text{opt}} + \Psi(t)(C(K) - 0.5)$ is introduced into the optimization variable. Among them, $f(K)$ is the optimization variable, K and t are the iteration times, f_{opt} represents the current optimal variable, $\boldsymbol{C}(K) = \{C_1(K), C_2(K), \ldots, C_n(K)\}$ represents the chaotic variable vector generated by chaotic mapping, n is the dimension of the optimization variable, and $C_i(K)$ denotes the ith dimension value of $C(K)$. Let the iteration equation of the time-varying parameters be $\Psi(t+1) = \lambda\Psi(t)$, λ is the attenuation factor of the time-varying parameters, and $\lambda \in [0.95, 0.999]$. The initial value of $\Psi(t)$ is denoted as $\Psi(0) = (J_{\text{opt}} - J_{\text{max}})/\ln P$, where J_{opt} and J_{max} are the minimum and maximum values corresponding to the cost function among all feasible solutions obtained in the first-stage chaotic searching, and the selection range of parameter P is $(0, 1)$.

In this section, the minimum value J_{opt} of the cost function obtained by chaos optimization is used as the optimization weight coefficient of the input layer of the neural network. When chaotic variables are generated, the mapping formula of the chaotic signals generated by logistic mapping is written as:

$$C(K+1) = \rho C(K)[1 - C(K)] \tag{5.4.3}$$

where $\rho \in [3.57, 4]$.

Let the cost function be $J_{\text{opt}} = J(\boldsymbol{f}_{\text{opt}}) = J_{\min}(\boldsymbol{f})$, J and $f \in [d, e]$ represent the cost function and the optimization vector. Suppose that the number of the first-stage iterations of chaos algorithm is N_1 and the number of the second-stage iterations is N_2, then the steps of the weight vector of the input layer for the chaotic optimization neural network equalizer are as follows:

Step 1: In the interval [0,1], n different values are taken as the initial values C (where 0, 0.25, 0.5, 0.75, and 1 could not be taken) and then let the optimal solution be f_{opt} corresponding to J_{opt} and the maximum value of the cost function be $J_{\max}$. Assume $K = 0$ and $t = 0$.

Step 2: n chaotic variables are obtained through iteration of Equation (5.4.3) and each variable has a different trajectory. The following equation is used to change the value range of optimization variables.

$$f(K) = s + rC(K) \tag{5.4.4}$$

where $s = d$ and $r = e - d$.

Step 3: After the chaotic variable is updated, $J(\boldsymbol{f}(K))$ is computed and the value of $J_{\max}$ and J_{opt} is saved.
If $J(\boldsymbol{f}(K)) \leq J_{\text{opt}}$, $J_{\text{opt}} = J(\boldsymbol{f}(K))$ and $\boldsymbol{f}_{\text{opt}} = \boldsymbol{f}(K)$. If $J(\boldsymbol{f}(K)) > J_{\max}$, $J_{\max} = J(\boldsymbol{f}(K)), K = K + 1$. If $K \leq N_1$, then iteration goes back to Step 2.

Step 4: After J_{opt} is obtained in Step 3 and the parameter P is selected, the initial value $\Psi(0)$ of the time-varying parameter is calculated by $\Psi(0) = (J_{\text{opt}} - J_{\max})/\ln P$. After searching in Step 3, the second carrier is calculated by:

$$f(K') = f_{\text{opt}} + \Psi(t)(C(K') - 0.5) \tag{5.4.5}$$

where $\Psi(t+1) = \lambda\Psi(t)$ and $t = t + 1$.

Step 5: Chaos optimization variables are searched after the second carrier is obtained and $J(f(K'))$ is calculated.
If $J(\boldsymbol{f}(K')) \leq J_{\text{opt}}$, $J_{\text{opt}} = J(\boldsymbol{f}(K'))$ and $\boldsymbol{f}_{\text{opt}} = \boldsymbol{f}(K')$, Otherwise, $\boldsymbol{f}(K')$ is given up, and $K' = K' + 1$.

Step 6: If $K' > N_2$, searching is stopped and output J_{opt} and its corresponding global optimal solution f_{opt} are obtained. Otherwise, the searching returns to Step 5.
To sum up, Steps 1–6 are the optimization process of the chaos algorithm. In this section, the chaos optimization algorithm is used to optimize the weight vector between the input layer and the hidden layer of the neural network. The final output $\boldsymbol{f}_{\text{opt}}$ is the initialization value of the weight vector between the input layer and the hidden layer of the multiwavelet neural network. The process is shown in Figure 5.25.

5.4.3 *Multiwavelet double neural network nonlinear blind equalization algorithm based on chaos optimization*

Although the neural network blind equalization algorithm is simple in structure, its nonlinear processing ability is not strong. In this section, the blind equalization algorithm based on double neural network is proposed.

The earliest feedback neural network was put forward by the physics professor Hopfield in 1982. He introduced the concept of "energy function" in the neural network, gave the criterion of network stability, and opened up a new field of the neural network in signal processing. Aiming at the nonlinear channel equalization problem, the neural network is combined with the decision feedback neural network to form the double neural network blind equalization algorithm, as shown in Figure 5.26.

It can be shown from Figure 5.26 that both the prefilter and the feedback filter of the double neural network blind equalization algorithm are a neural network filter. The nonlinear processing ability of the blind equalization algorithm is enhanced by introducing the decision feedback structure and the increase of neurons. $f_{ij}(n)$ is the connection weight between the input layer and the hidden layer of the feedforward neural network, $f_{jp}(n)$ is the connection weight

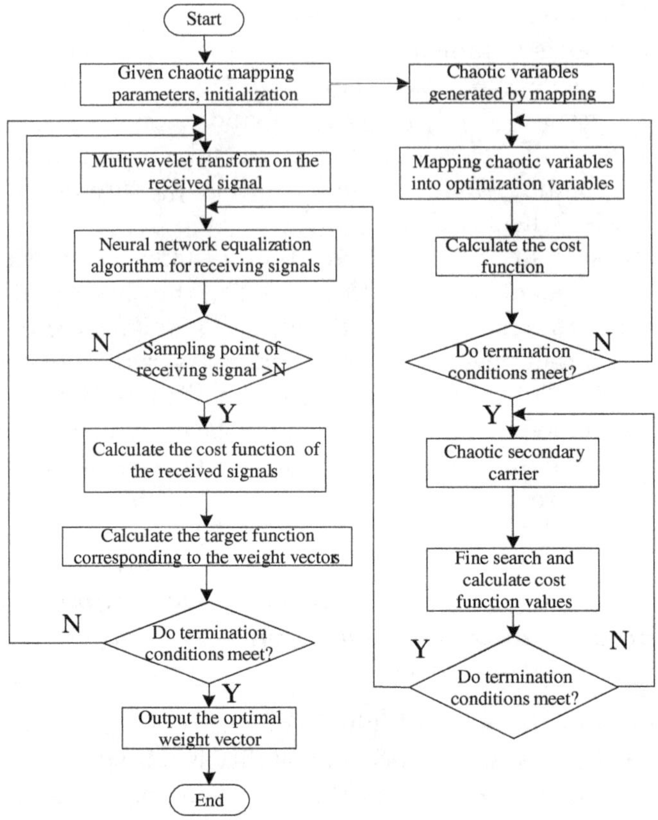

Figure 5.25 Chaotic optimization multi-wavelet neural network blind equalization algorithm.

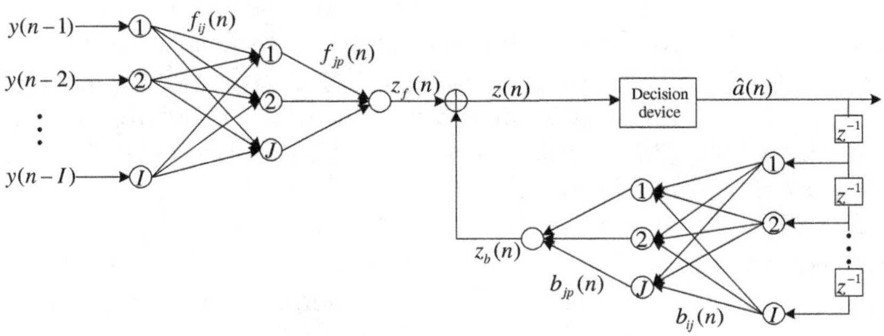

Figure 5.26 Double neural network blind equalization algorithm.

from the hidden layer to the output layer in the feedforward neural network, $z_f(n)$ denotes the output weight of the feedforward neural network, $b_{ij}(n)$ represents the connection weight from the input layer to the hidden layer in the feedback neural network, $b_{jp}(n)$ represents the connection weight of feedforward neural network from hidden layer to output layer, and $z_b(n)$ denotes the output of feedforward neural network.

The output of the prefilter is defined as:

$$z_f(n) = F\left(\sum_{j=1}^{J} f_{jp}(n)F\left(\sum_{i=0}^{I} f_{ij}(n)y(n-i)\right)\right) \tag{5.4.6}$$

The output of the feedback filter is defined as:

$$z_b(n) = F\left(\sum_{j=1}^{J} b_{jp}(n)F\left(\sum_{i=0}^{I} b_{ij}(n)\hat{a}(n-i)\right)\right) \tag{5.4.7}$$

The output of the system is written as:

$$z(n) = z_f(n) - z_b(n) \tag{5.4.8}$$

The iterative equation of the weight vector from the input layer to the hidden layer of the prefilter is obtained by:

$$\boldsymbol{f}_{ij}(n+1) = \boldsymbol{f}_{ij}(n) - 2\mu_{ij}z(n)e(n)F'(u_p^P(n))\boldsymbol{f}_{jp}(n)F'(u_j^J(n))\boldsymbol{y}(n) \tag{5.4.9}$$

The iterative equation of the weight vector from the hidden layer to the output layer is obtained by:

$$\boldsymbol{f}_{jp}(n+1) = \boldsymbol{f}_{jp}(n) - 2\mu_{jp}z(n)e(n)F'(u_p^P(n))\boldsymbol{v}_j^J(n) \tag{5.4.10}$$

where μ_{ij} and μ_{jp} represent the step-size.

The iterative equation of the weight vector from the input layer to the hidden layer in the feedback filter is obtained by:

$$\boldsymbol{b}_{ij}(n+1) = \boldsymbol{b}_{ij}(n) - 2\mu_{bij}z(n)e(n)F'(m_p^P(n))\boldsymbol{b}_{jp}(n)F'(O_j^J(n))\hat{a}(n) \tag{5.4.11}$$

The iterative equation of the weight vector from the hidden layer to the output layer is obtained by:

$$\boldsymbol{b}_{jp}(n+1) = \boldsymbol{b}_{jp}(n) - 2\,\mu_{bjp}\hat{\boldsymbol{a}}(n)e(n)F'(m_p^P(n))O_j^J(n) \qquad (5.4.12)$$

where μ_{bij}, and μ_{bjp} represent the step-size.

$$m_j^J(n) = \sum_{i=0}^{I} \boldsymbol{b}_{ij}(n)\hat{a}(n-i) \qquad (5.4.13)$$

$$O_j^J(n) = F\left(\sum_{i=1}^{I} \boldsymbol{b}_{ij}(n)\hat{a}(n-i)\right) \qquad (5.4.14)$$

$$m_p^P(n) = \sum_{i=1}^{J} \boldsymbol{b}_{jp}(n)O_j^J(n) \qquad (5.4.15)$$

After combining the chaos optimization algorithm with the neural network feedback structure, the principle of the multiwavelet double neural network nonlinear blind equalization algorithm based on chaos optimization is presented in Figure 5.27.

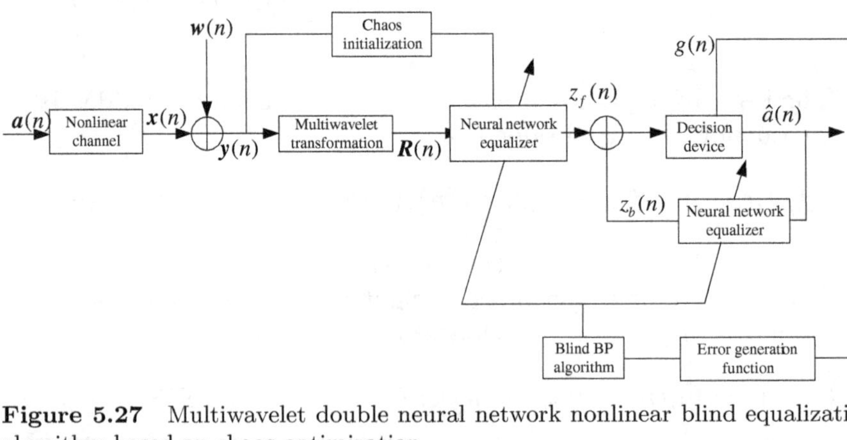

Figure 5.27 Multiwavelet double neural network nonlinear blind equalization algorithm based on chaos optimization.

5.4.4 *Computational complexity*

After introducing the chaos optimization algorithm and the dual neural network structure, the computational complexity will increase to some extent.

Let the length of the input signal be N, the length of the first carrier searching be N_1, that of the secondary carrier searching be N_2, and that of the input layer weight vector of the neural network be L. The complexity of generating mapping signal from Equation (5.4.3) is calculated as L^2+3L. Based on the steps of optimizing the input layer weight vector of the neural network according to the chaos algorithm, its computational load is obtained by $L^2+3L+3N_1^2+6N_1+3N_2^2+6N_2$. According to the fact that the prefilter and the feedback filter are both neural network filters and the Equations (5.4.6)–(5.4.15), the computational complexity of the network filter is equal to $2MN$. So, the computational complexity of the CO-MWT-D-ANN is calculated by $L^2 + 3L + 3N_1^2 + 6N_1 + 3N_2^2 + 6N_2 + 2MN$.

5.4.5 *Algorithm simulation*

In order to verify the effectiveness of the CO-MWT-D-ANN algorithm, the FNN, the MWT-FNN, and the CO-MWT-FNN were compared. The channel model is shown in Figure 5.15. The input was 2PAM signal, the SNR was set to 25 dB, the sampling point was set to 10000, and the length of the weight vector was 16. In the FNN, the step-size was given by $\mu = 0.00001$, and the fourth tap was initialized to 1. In the MWT-FNN, the step-size was set to $\mu = 0.0008$, DB2 multi-wavelet was selected, and the fourth tap was initialized to 1. In the CO-MWT-FNN, $\mu = 0.00002$, the DB2 multi-wavelet was also used, and the fourth tap was initialized to 1. In the CO-MWT-D-ANN, $\mu = \mu_b = 0.00001$, and the fourth tap was initialized to 1500 Monte Carlo simulation results as shown in Figure 5.28.

Figure 5.28(a) shows that the convergence speed of the CO-MWT-D-ANN is faster by about 500 steps, 7500 steps, and 9000 steps than that of the CO-MWT-FNN, the MWT-FNN, and the FNN. The convergence speed of the CO-MWT-FNN has an improvement of about 7000 steps and 8500 steps in comparison with the MWT-FNN and the FNN. As far as the steady-state MSE is considered, the CO-MWT-D-ANN has a drop of about 13 dB and 6 dB in comparison with the FNN and the CO-MWT-FNN, whereas the CO-MWT-FNN has a

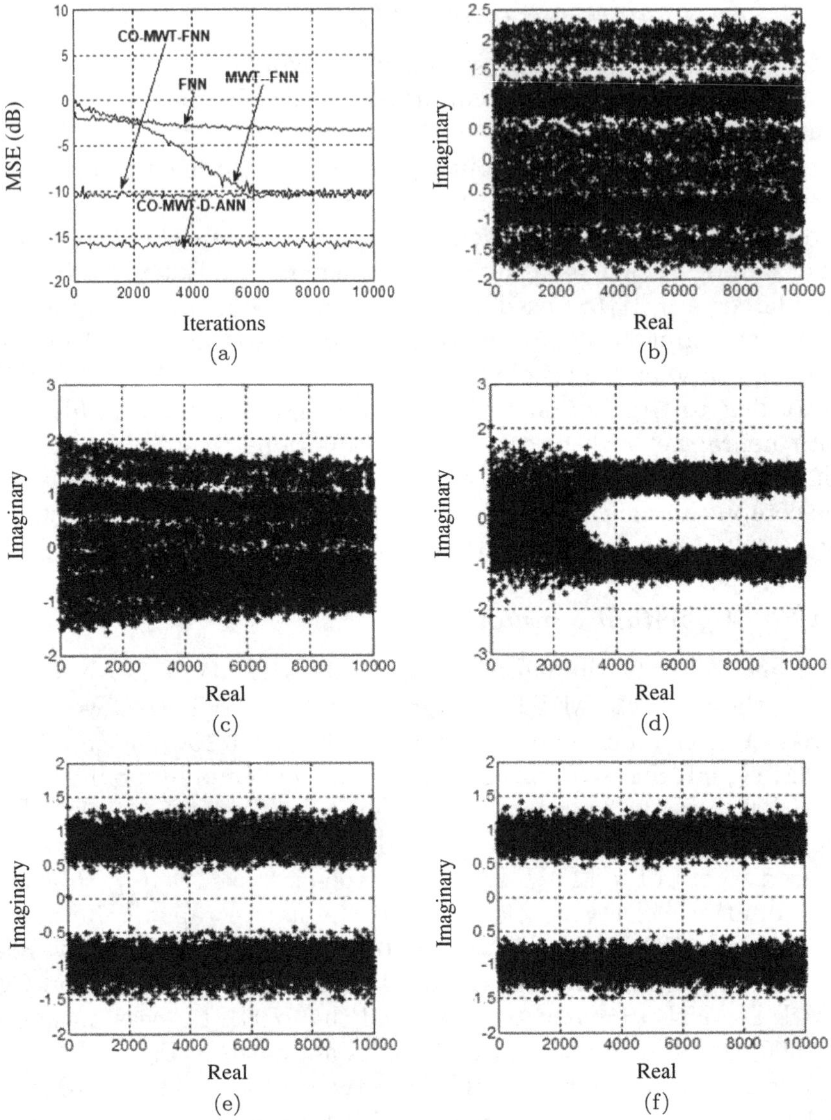

Figure 5.28 Simulation results: (a) MSE curves of all algorithms; (b) input of equalizer; (c) output of the FNN; (d) output of the MWT-FNN; (e) output of CO-MWT-FNN and (f) output of the CO-MWT-D-ANN.

drop of about 7 dB in comparison with the FNN. It can been seen from Figure 5.28(a) to Figure 5.28(f) that the output constellations of the CO-MWT-FNN are clearer and more compact than those of the FNN and the MWT-FNN. The output constellations of the CO-MWT-D-ANN are the clearest and most compact, and the recovery of the signals is more accurate.

5.5 Equalization Algorithm Based on Volterra Filtering Echo State Network and PCA

A key research topic in the field of intelligent signal processing is the study of effective solutions based on dynamic and real-time systems. Due to the increasing complexity of the most relevant practical problems, the development and application of models endowed with feedback loops are virtually a constant demand. Furthermore, it is essential that the proposed models be capable of performing nonlinear mappings and also access the time history of the input and output signals as well as the internal states.

In this context, the recurrent neural network (RNN) allows the feedback connection between different layers of neurons, has good memory, and can be realized by backpropagation through time (BPTT) [45] and real-time recurrent learning (RTRL) [46]. However, all these approaches are associated with well-known drawbacks. For instance, during the training phase, it is difficult to compute the derivatives of the cost function with respect to the parameters. Furthermore, the instability of the network may make it difficult for the network structure function to be effectively played. In these scenarios, the echo state network (ESN), originally proposed by Jaeger [47], represents a robust and promising solution, as they preserve, in a measure, the processing potential of the RNN structure and introduce a significant simplification into the network training process. This is cleverly accomplished by setting the connection weights of the recurrent internal layer, also called the dynamical reservoir, without employing information about the desired signal. Thus, the remaining task is to adapt the weights of the output linear combiner, which can be performed by means of any method that solves a least squares problem.

It is important to stress that the idea of creating the reservoir of dynamics without resorting to an error signal implies a certain reduction in the representation capability of the network, especially if compared with an ideally adjusted similar structure. Therefore, selecting an adequate choice of the reservoir weights is a vital element to the design of the ESN. This observation was the main concern of the work by Ozturk *et al.* [48]. Employing an information-theoretic measure, Ozturk *et al.* analyzed the correlation between the echo states and then proposed a method for selecting the values of the reservoir weights which favors the occurrence of a higher degree of diversity when compared to the original proposal of Jaeger [47]. Nevertheless, both ESNs cannot make a full effective use of the higher-order statistics of the signals coming from the reservoir, since the output layer still corresponds to a linear combiner, for the sake of keeping the training process as simple as possible. Thus, in order to overcome this limitation, moving toward the use of a nonlinear output layer may be an interesting perspective.

In this work, a novel ESN architecture was proposed [49], whose linear combiner is replaced with the structure of a Volterra filter [50], which not only contributes to the better exploitation of the information associated with the internal signals but also preserves the simplicity of the training process, since the problem of adjusting the output weights remains linear with respect to the free parameters. Additionally, aiming to prevent the curse of dimensionality [51], i.e., the occurrence of an excessive growth in the number of weights to be adapted as the number of echo states increases, the principal component of the echo state is extracted by principal component analysis (PCA) [52, 53] to reduce the number of echo states and adjust the length of the weight vector. Since the echo state network can achieve the inherent trade-off between processing power and computational complexity [54, 55], applying an echo state network to the channel equalization problem is a promising and effective solution for both real-time and non-real-time.

5.5.1 *Echo state network*

Consider a discrete-time recurrent neural network with K input units, N internal units, and L output units, as shown in Figure 5.29.

Let $\boldsymbol{u}(n) = [u_1(n), \ldots, u_K(n)]^T$ denote the activation vector of the input unit, which are transmitted to the internal neurons by

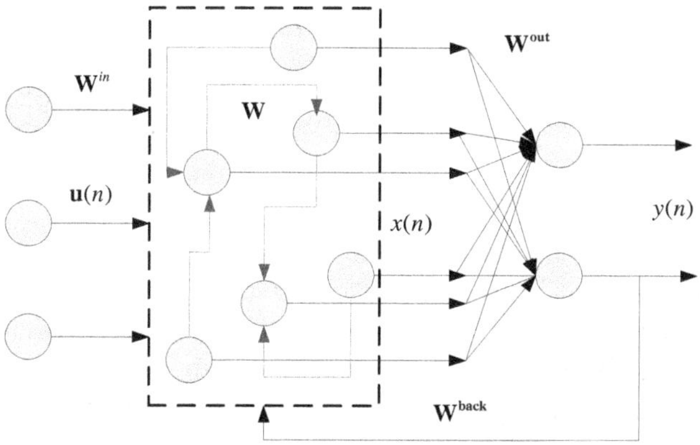

Figure 5.29 The basic recurrent neural network architecture.

means of a linear combination. The weight matrix $\boldsymbol{W}^{\mathrm{in}} \in R^{N\times K}$ specifies the coefficients of such combinations.

The inner layer, also known as the dynamic reservoir, consists of fully connected nonlinear units whose activation, given by $\boldsymbol{x}(n) = [x_1(n), \ldots, x_K(n)]^T$, corresponds to the network state and are updated as:

$$\boldsymbol{x}(n+1) = \boldsymbol{f}(\boldsymbol{W}^{\mathrm{in}}\boldsymbol{u}(n+1) + \boldsymbol{W}\mathbf{x}(n) + \boldsymbol{W}^{\mathrm{back}}\boldsymbol{y}(n)) \qquad (5.5.1)$$

where $\boldsymbol{W} \in R^{N\times N}$ brings the weights of the connections within the reservoir, $\boldsymbol{W}^{\mathrm{back}} \in R^{N\times L}$ contains the weights associated with the connections that project the output samples back to the internal units, and $\boldsymbol{f}(\cdot) = (f_1(\cdot), f_2(\cdot), \ldots, f_N(\cdot))$ represents the activation function of the inner layer unit.

The output vector $\boldsymbol{y}(n) = [y_1(n), y_2(n), \ldots, y_L(n)]^T$ of the network is determined by the following expression:

$$y(n+1) = \boldsymbol{f}^{\mathrm{out}}(\boldsymbol{W}^{\mathrm{out}}\mathbf{x}(n+1)) \qquad (5.5.2)$$

where $\boldsymbol{W}^{\mathrm{out}} \in R^{L\times N}$ represents the output weight matrix and $\boldsymbol{f}^{\mathrm{out}}(\cdot) = \{f_1^{\mathrm{out}}(\cdot), f_2^{\mathrm{out}}(\cdot), \ldots, f_L^{\mathrm{out}}(\cdot)\}$ represents the activation function of the inner unit.

In 2001, Jaeger studied the dynamics of the RNN structure shown in Figure 5.29. Studies have shown that under certain circumstances, the network state $\boldsymbol{x}(n)$ becomes asymptotically independent of the initial conditions. In other words, if the network starts from two different initial states $\boldsymbol{x}(0)$ and $\hat{\boldsymbol{x}}(0)$, and the same sequence of input signals is received, then the obtained sequences of states $\boldsymbol{x}(n)$ and $\hat{\boldsymbol{x}}(n)$ converge close to the values. When this property holds, the effects of the initial state vanish, and the reservoir dynamics depends solely on the input history, so that the network has echo states [47].

In addition, Jaeger demonstrated two sufficient conditions with respect to the echo state: (1) in order to represent the echo state attribute with an RNN, the maximum singular value of the internal weight matrix $\boldsymbol{W}$ must be less than the unit value ($\sigma_{\max}(\mathbf{W}) < 1$) [47,56]; (2) according to the internal weight, the spectral radius of the internal weight matrix $\boldsymbol{W}$ (for the maximum value of the absolute value of the eigenvalue) gives the condition for establishing a network without an echo state and if $||\boldsymbol{W}|| > 1$, then the network has no echo states.

Based on these observations, Jaeger proposed an adaptive process that simplifies the RNN: first, construct a weight matrix $\boldsymbol{W}$ satisfying $||\mathbf{W}|| < 1$; next, arbitrarily define the input weight matrix $\mathbf{W}^{\text{in}}$, since it does not harm the echo state property; finally, by employing a linear combiner as the output layer, the adaptive problem of remaining weights is converted into the problem of minimizing the MSE. Therefore, any linear regression algorithm can be used to perform such tasks, which constitutes the essence of the ESNs [47].

ESNs are a promising and interesting alternative to the traditional RNNs as they are capable of bringing together the ability of accessing the time history of the input and output signals — which they inherit from the RNNs — and a simple training procedure, since the connection weights within the reservoir are predefined. The strategy of maintaining such weights fixed is what provides agility to the training algorithm, in contrast to the standard approach for the adaptation of recurrent neural networks, which involves adjusting the whole set of weights [57–59].

It should be noted that the dynamical reservoir is completely specified by the weight matrix $\boldsymbol{W}$, and it should be as rich as possible in order to allow an adequate approximation of the desired signal. However, in accordance with the spirit of the echo state approach, the reservoir is designed without access to information regarding the particularities of the application, which, in a certain sense, characterizes a reduction in the representation capability of the network when compared to a similar structure with an adjustable weight matrix $\boldsymbol{W}$.

Recently, Ozturk *et al.* used average state entropy (ASE) as a measure of the richness of the dynamic reservoirs [48]. In addition, they observed that the diversity of the echo states reflects the eigenvalue structure of $\boldsymbol{W}$ to some extent. This study led to the proposal of a different method for designing the reservoir, that is, in order to maximize the ASE, the poles of the linearized system derived from the ESN dynamics should be uniformly distributed — given a chosen radius — within the unit circle, which evokes the conceptual framework of Kautz filters. Henceforth, this latter network shall be referred to with the acronym ASE-ESN, whereas ESN will correspond to the original network conceived by Jaeger.

5.5.2 *Average state entropy: echo state network*

The ESN can be explained in terms of a trade-off between performance and simplicity. On the one hand, the learning method in the ESN cannot achieve all the potential from the underlying processing structure, since the recurrent connections within the dynamic reservoir cannot be adaptively adjusted by means of an error signal. On the other hand, the ESN does not adopt an adaptive method, which avoids the inherent difficulties of the classical RNN learning algorithm, which is a significant feature. The core of this trade-off lies in the design of the echo state dynamic reservoir. Since no information regarding an error signal is available, the idea is to produce the highest possible diversity of dynamics, in order to better supply with relevant information the output layer, which is then adapted with the aid of a reference signal. However, in the existent proposals of echo state networks, the output layer corresponds to a linear combiner, which cannot exploit the higher-order statistics of the information coming from the nonlinear dynamics of the neurons of the reservoir.

To avoid this problem, a nonlinear output layer is used here. However, in order to maintain the inherent simplicity of the ESN, it is crucial that the output layer remains linear with respect to the free parameters. After all, in this case, it is possible to find a closed-form solution in the least squares (or Wiener) sense [54]. Taking into account these facts, the proposal in this section is to use a Volterra filter structure for the output layer [49]. Thus, the network outputs shall be composed of linear combinations of polynomial terms, as follows:

$$y_i(n) = h_0 + \sum_{i=1}^{N} h_1(i)x_i(n) + \sum_{i=1}^{N}\sum_{j=1}^{N} h_2(i,j)x_i(n)x_j(n) + \sum_{i=1}^{N}\sum_{j=1}^{N}\sum_{k=1}^{N} h_3(i,j,k)x_i(n)x_j(n)x_k(n) + \cdots \quad (5.5.3)$$

where $x_k(n)$ indicates the kth echo state at the time n and $y_i(n)$ is the output of the ith network. Obviously, the coefficient $h_i(\cdot)$ is linearly related to the output.

However, the Volterra architecture introduces a new concern, that is, as the number of echo states (N) increases, the number of kernels tends to increase dramatically. In fact, it is obvious that if the expression of the number of non-fuzzy kernels N_{ker} is observed.

$$N_{\text{ker}} = 1 + N + \frac{N(N+1)}{2} + \frac{N(N+1)(N+2)}{6} + \cdots \quad (5.5.4)$$

As we can see, depending on the maximum order of the kernels we intend to use, the respective number of coefficients that must be adjusted tends to be severely increased, which threatens the practical application of this structure along with the ESN.

To alleviate this problem, principal component analysis (PCA) is used to reduce the number of effective signals transmitted to the output layer. The principal components are extracted from the active state of the output layer by the PCA. Thus, instead of transmitting the whole set of echo states, only a relatively small number of principal components are used, which reduces the quantity of weights that need to be adjusted without necessarily causing a significant loss of information.

Let $\boldsymbol{R}_{T_s} \in \mathbb{R}^{N\times T_s}$ be the activation matrix of the internal units considering with T_s input samples. The sample covariance matrix of the network state is computed via:

$$\hat{\mathbf{C}} = \frac{\boldsymbol{R}_{T_s}\boldsymbol{R}_{T_s}^T}{T_s} \quad (5.5.5)$$

where $\boldsymbol{V} \in \mathbb{C}^{N \times N_{pc}}$ is a matrix composed of eigenvectors corresponding to the N_{pc} maximum eigenvalues of $\hat{\mathbf{C}}$. Therefore, the principal components are given by [53]:

$$\mathrm{s}_i = \boldsymbol{V}_i^T \boldsymbol{R}_{T_s} \tag{5.5.6}$$

where $i = 1, 2, \ldots, N_{pc}$.

Since there is a non-negligible redundancy among echo states, the PCA can be used to analyze this redundancy. In fact, there was a strong correlation among the echo states [48], which means that it is possible to preserve a few principal components of the echo state $\boldsymbol{x}(n)$. In addition, the PCA can be used to reduce the computational complexity of subsequent operations and the effects of noise.

To summarize, the PCA and Volterra filters are employed in order to achieve compression and a reduction of the number of coefficients to be adapted and also to obtain a more effective exploitation of the information regarding the echo states.

5.5.3 *Principle of channel equalization*

The purpose of channel equalization is to minimize the distortion of the signals as it travels through the channel to the receiver. The channel can be thought of as a filter, its input is the transmitted signal $\boldsymbol{a}(n)$, and its output signal is $\boldsymbol{r}(n)$, as shown in Figure 5.30.

In Figure 5.30, the channel can be either linear or nonlinear, with varying degrees of randomness, and the channel characteristics determine the degree of distortion in which the signal is transmitted. An effective method to reduce the distortion is to use a specially tailored filter or equalizer at the receiver, whose response function is opposed to the channel. Ideally, its output should be the delayed signals of the transmitted signals, i.e.,

$$z(n) = ka(n - d) \tag{5.5.7}$$

Figure 5.30 Model of a communication system.

where $z(n)$ is the output of the equalizer, k is a proportional constant, and d denotes the equalization delay. This equalizer must be forced to zero [54].

The design contexts of the equalizer includes as follows: (1) design a suitable filtering structure; (2) define an effective equalization criterion; (3) give an effective optimization method to adjust the parameters of the filter.

The first choice is crucial. Its design has a direct impact on the equalization performance. The linearity or nonlinearity of the channel determines the memory effect of the channel, which is an inherently random factor. The effects of these factors must be handled by the filtering structure. In some cases, linear channels can be compensated by linear equalizers, and the equalizer with the nonlinear structures can solve some linear problems [55]. Whether it is a linear or nonlinear channel, its memory has a deep influence on the required memory of the equalizer, both for linear and nonlinear models, and in this context the use of recurrent structures arises in a natural way. Finally, the structure of the existing channel noise may require the use of a nonlinear structure – as a consequence, for example, of a pronounced non-Gaussian character or even of a recurrent structure if temporal dependence is a key factor [60].

The second choice amounts essentially to a translation of the aim of the equalization task — with Equation (5.5.7) — into mathematical terms. A classical choice has been to use a MSE criterion, also known as Wiener criterion, which leads to the following cost function:

$$J_{\mathrm{MSE}} = E\{[a(n-d) - z(n)]^2\} \tag{5.5.8}$$

where $E\{\cdot\}$ indicates statistical expectation operator. Under the MSE criterion, if the filter structure is feedforward and linear with respect to its free parameters, then the cost function has a single extreme point, and the Wiener solution based on this cost function can be obtained by direct least squares or iterative LMS and RLS algorithm [54]. On the other hand, if the filter structure is nonlinear or recursive with respect to the free parameters and/or recurrent, the cost function can have multiple local extremum points, that is, the cost function has diversity, which establishes a more complicated search problem.

From the previous analysis, the most ideal equalizer should have a nonlinear recursive structure, because it can meet both memory and mapping in the most general way. From an optimization standpoint, it is difficult to find the optimal MSE solution due to the diversity of the cost functions.

5.5.4 *Algorithm simulation*

Now consider the training and testing methods of the channel equalization algorithm based on the echo state network and the performance of the proposed algorithm in different scenarios. All experiments were carried out in the MATLAB platform.

5.5.4.1 *Method*

In all simulations, the source signal $a(n)$ is selected as an equal probability (BPSK modulation) value $\{+1, -1\}$. The purpose of each echo state network is to estimate the original signal $r(n)$ using the received signal $a(n)$ at the same time instant, meaning that there is no equalization delay and that the number of inputs and outputs was always equal to one ($K = L = 1$). Furthermore, in noisy scenarios, the influence of the SNR over the performance of the ESNs is analyzed. If the presence of noise is not considered, it means that the input $\boldsymbol{u}(n)$ of the ESNs is exactly equal to $\boldsymbol{r}(n)$.

Each network was trained using $T_s = 1100$ samples of the input signal, being the first hundred discarded to eliminate transient effects. The same approach was employed during the testing phase. The MSE was defined as:

$$\text{MSE} = \frac{1}{T_s} \sum_{n=1}^{T_s} [d(n) - z_{\text{ESN}}(n)]^2 \tag{5.5.9}$$

where $d(n)$ denotes the desired signal and $z_{\text{ESN}}(n)$ denotes the network output.

In order to continuously evaluate the performance of the networks, an average MSE (AMSE) for 20 independent runs is regarded as the MSE standard deviation, which reflects the equalization performance of each channel obtained by the echo state networks. The performance of each echo state network is observed under different numbers of internal neurons. Based on the preliminary simulation, the number of internal units remains equal to $N = 40$ in all experiments, whereas the number of principal components varies.

Based on the procedure described in [47], the ESN was constructed as follows: the input weight coefficients were set to -1 or $+1$ with equal probability, and the elements of the reservoir weight matrix $\boldsymbol{W}$ were set to -0.4, 0.4, and 0, respectively, and their corresponding probabilities were 0.025, 0.025 and 0.95.

For both the ASE-ESN and the proposed network, the adjustment of $\boldsymbol{W}^{\text{in}}$ was similar to the procedure in the ESN. In terms of the reservoir weight matrix $\boldsymbol{W}$, the method proposed in [56] was employed for the ASE-ESN [48], using a spectral radius equal to 0.8. Since the proposed network can be easily adapted to different techniques for the reservoir design, its performance was analyzed using the methods offered by [47, 48]. These versions shall be identified as proposed ESN and proposed ASE-ESN, respectively. For the sake of brevity, the presentation of the results for the approach which led to the best performance was restricted. Finally, $\boldsymbol{W}^{\text{back}}$ was set to zero for all networks, as performed in [47, 48].

For all networks, a desired signal $d(n)$ was given, and the output weight matrix $\boldsymbol{W}^{\text{out}}$ was resolved by Wiener method. It is important to emphasize that the output weight coefficients are related to the signals, that is, in the ESN and the ASE-ESN, the echo state $\boldsymbol{x}(n)$ is linearly combined to generate the output. On the other hand, in the proposed network, the PCA and the Volterra filters were used as the basic framework of the output layer. In this case, the output weight coefficients correspond to the coefficients used in the combination of the kernel functions. Another relevant aspect is that in the proposed network, the output was calculated using a third-order Volterra filter, and for the sake of simplicity, the quadratic term was not considered.

5.5.4.2 *First channel*

The transfer function of the first channel is given by:

$$H(z) = 0.5 + z^{-1} \tag{5.5.10}$$

This seems to be a relatively simple channel, but when the equalization delay is zero, the linear filter cannot be used for equalization, since in this case, the corresponding channel states cannot be separated linearly. Table 5.1 gives the performance comparison results for the different network structures. The values presented inside parentheses in Table 5.1 correspond to the standard deviation.

Some interesting remarks can be drawn from Table 5.1. Firstly, when the number of neurons inside the reservoir or the number of principal components increases, the AMSE value decreases and the performance of the ESNs is improved. Secondly, by comparing the AMSE values obtained by the ASE-ESN and the proposed ASE-ESN method, the network performance is significantly improved by using the nonlinear output layer. The benefits acquired with the use of a nonlinear output become evident. In fact, in the ASE-ESN, $N = 60$ and in the proposed ASE-ESN, $N_{pc} = 6$, the number of output weight coefficients to be adjusted in the both networks is roughly equal (60 and 62 weights, respectively), and in this case, the proposed network leads to a significantly better performance. Thirdly, the AMSE values of the ESN and the ASE-ESN are different, which indicates that the dynamic reservoir weight matrix designed using Jaeger's scheme will

Table 5.1 AMSE values obtained with each echo state network considering the first channel.

Channel 1		**AMSE**	
Network	**Parameter**	**Training**	**Testing**
ESN	$N = 10$	$2.66(\pm 3.22)e{-01}$	$2.72(\pm 3.27)e{-01}$
	$N = 40$	$1.87(\pm 2.65)e{-05}$	$2.07(\pm 3.03)e{-05}$
	$N = 60$	$2.37(\pm 2.19)e{-05}$	$2.74(\pm 2.58)e{-05}$
ASE-ESN	$N = 10$	$1.24(\pm 0.60)e{-01}$	$1.28(\pm 0.59)e{-01}$
	$N = 40$	$1.65(\pm 0.97)e{-02}$	$2.54(\pm 3.27)e{-02}$
	$N = 60$	$9.56(\pm 7.08)e{-03}$	$1.72(\pm 2.88)e{-02}$
Proposed ASE-ESN	$N_{pc} = 3$	$1.02(\pm 0.78)e{-02}$	$1.06(\pm 0.78)e{-02}$
	$N_{pc} = 5$	$1.09(\pm 0.81)e{-03}$	$1.58(\pm 1.78)e{-03}$
	$N_{pc} = 6$	$4.09(\pm 4.04)e{-04}$	$8.61(\pm 17.7)e{-04}$
	$N_{pc} = 10$	$4.71(\pm 2.47)e{-06}$	$6.63(\pm 18.2)e{-04}$
Proposed ESN	$N_{pc} = 3$	$4.26(\pm 4.34)e{-03}$	$4.36(\pm 4.36)e{-03}$
	$N_{pc} = 5$	$6.28(\pm 9.72)e{-05}$	$7.38(\pm 11.2)e{-05}$
	$N_{pc} = 6$	$2.56(\pm 3.80)e{-06}$	$3.06(\pm 4.60)e{-06}$
	$N_{pc} = 10$	$3.34(\pm 4.02)e{-10}$	$2.06(\pm 4.86)e{-08}$

achieve a better performance. This observation is corroborated when comparing the AMSE values associated with each version of the proposed network. As shown in Table 5.1, the proposed model achieved a better performance when the reservoir was designed as in the ESN, reaching smaller AMSE values than its counterpart, which, again, indicates the potential of the proposed architecture. One last pertinent comment is related to the variability of the MSE values obtained with each network, that is, it is possible to infer from Table 5.1, with the aid of the standard deviation values, that the proposed model, even in the worst case, offers a significant improvement in the performance, for both training and test sequences, when compared with the ESN and the ASE-ESN.

5.5.4.3 *Second channel*

In 1999, Montalvão *et al.* [61] demonstrated that feedforward structures, either linear or nonlinear, are not suitable tools for the equalization of communication channels that present coincident states, i.e., whose transfer function presents zeros over the unit circle. Therefore, this class of channels demands the use of feedback connections in the equalizer in order to allow a correct separation of the states. For example, for the channel of the transfer function $H(z) = 1 + z^{-1}$, the value $r(n) = 0$ generated by the input sequence of symbols $(a(n) = -1,\ \ a(n-1) = 1)$ and $(a(n) = 1,\ \ a(n-1) = -1)$. Table 5.2 displays the AMSE values obtained for each echo state network.

Table 5.2 ASME values obtained with each ESN for the second channel.

Channel 2		**AMSE**	
Network	**Parameter**	**Network**	**Testing**
ESN	$N = 10$	ESN	$7.58(\pm 4.84)e{-}02$
	$N = 40$		$2.05(\pm 1.78)e{-}03$
	$N = 60$		$7.34(\pm 9.52)e{-}04$
ASE-ESN	$N = 10$	ASE-ESN	$4.48(\pm 2.61)e{-}03$
	$N = 40$		$2.15(\pm 2.74)e{-}04$
	$N = 60$		$2.98(\pm 5.74)e{-}04$
Proposed ASE-ESN	$N_{pc} = 3$	Proposed ASE-ESN	$4.42(\pm 2.36)e{-}02$
	$N_{pc} = 5$		$5.21(\pm 6.07)e{-}03$
	$N_{pc} = 6$		$1.56(\pm 1.83)e{-}03$

Table 5.2 confirms that recurrent connections within the reservoir enable the ESN to preserve the memory of the past received signals, which is decisive to a full recovery of the original information. This means that the fixed dynamic library weights can simplify the training process of the recurrent network and still provide sufficient processing capability for the network. Table 5.2 also shows that the proposed network does not have any improvement in the AMSE and may be related to the essence of this equalization problem, which may be more dependent on the existence of feedback connections rather than on the intrinsic difficulty of the required mapping. However, it is important to emphasize that the proposed network performance is quite satisfactory.

5.5.4.4 *Third channel*

The transfer function of the channel is defined as $H(z) = 0.5 + 0.71z^{-1} + 0.5z^{-2}$. According to Proakis [62], this is a linear channel with three coefficients that introduces the most severe distortions into the transmitted signal.

Firstly, as performed in the previous scenarios, the performance of each echo state network considering the noiseless case shall be analyzed. The obtained results are presented in Table 5.3.

Table 5.3 shows that the nonlinear output layer significantly improves the performance of the ESN and the ASE-ESN. On the other hand, the reservoir weight matrix of the proposed architecture

Table 5.3 ASME values obtained with each ESN for the third channel.

Channel 3		**AMSE**	
Network	**Parameter**	**Training**	**Testing**
ESN	$N = 10$	$3.03(\pm1.07)e{-}01$	$3.08(\pm1.13)e{-}01$
	$N = 40$	$4.73(\pm1.51)e{-}02$	$5.18(\pm1.61)e{-}02$
	$N = 60$	$1.72(\pm0.74)e{-}02$	$2.01(\pm0.68)e{-}02$
ASE-ESN	$N = 10$	$2.55(\pm0.95)e{-}01$	$2.59(\pm0.98)e{-}01$
	$N = 40$	$9.64(\pm4.74)e{-}02$	$1.05(\pm0.43)e{-}01$
	$N = 60$	$7.19(\pm1.96)e{-}02$	$7.53(\pm2.01)e{-}02$
Proposed ESN	$N_{pc} = 3$	$2.30(\pm1.82)e{-}03$	$2.46(\pm1.93)e{-}03$
	$N_{pc} = 5$	$5.56(\pm5.95)e{-}04$	$6.60(\pm7.20)e{-}04$
	$N_{pc} = 6$	$9.00(\pm8.40)e{-}07$	$1.01(\pm1.88)e{-}05$

designed by the method in [48] also provides good performance, but the improvement of the performance is not very significant.

Second, in the case of Gaussian additive noise, the input of the echo state network is $u(n) = r(n) + \sigma N(0,1)$, where σ denotes the standard deviation of the noise and $N(0,1)$ corresponds to a zero mean and unity variance Gaussian random variable.

Assuming $N = 60$ for both the ESN and the ASE-ESN and $N_{pc} = 6$ for the proposed network. For each of the five different SNRs for each network, 20 independent experiments were performed. The obtained AMSE values are given in Table 5.4. The dynamic reservoir layer is designed by the method given in [47], because this choice achieves a better performance in the absence of the noise.

Table 5.4 shows that as the noise becomes more obvious, the error increases significantly. However, when SNR value is fixed, the difference between AMSE values of each network is relatively small, but the performance of the proposed network is the best.

Table 5.4 AMSE value associated with each echo state network versus the SNR.

Channel 3		**AMSE**	
SNR	**Network**	**Training**	**Testing**
0 dB	ESN	8.0714e−01	9.3295e−01
	ASE-ESN	8.0659e−01	1.0009e−00
	Proposed ESN	8.0567e−01	9.1407e−01
5 dB	ESN	6.6104e−01	7.5504e−01
	ASE-ESN	6.6828e−01	7.7295e−01
	Proposed ESN	6.5706e−01	7.4966e−01
10 dB	ESN	3.8648e−01	4.4366e−01
	ASE-ESN	4.5590e−01	5.3292e−01
	Proposed ESN	3.7231e−01	4.2315e−01
15 dB	ESN	1.6712e−01	1.9371e−01
	ASE-ESN	2.5561e−01	2.9896e−01
	Proposed ESN	1.4976e−01	1.7238e−01
20 dB	ESN	7.6860e−02	9.0315e−02
	ESN	1.3896e−01	1.6526e−01
	ASE-ESN	5.8613e−02	6.8043e−02

5.5.4.5 *Fourth channel*

The transfer function was given as $H(Z) = 0.38 + 0.6z^{-1} + 0.6z^{-2} + 0.38z^{-3}$, which causes severe distortion of the transmitted signal. In the simulation, the number of training and test samples each network was set to 5500. To avoid transient effects, the first 500 samples were excluded. The obtained results are shown in Table 5.5.

As it can been seen for Table 5.5, the proposed network has a better equalization effect than the ESN and the ASE-ESN, which means that, in this case, not only the recursive structure is necessary, but also the nonlinear output layer is essential. When the distortions of the transmitted signals become more pronounced, the proposed network can indeed achieve more efficient channel equalization.

5.5.4.6 *Fifth channel*

Nonlinear channel is defined as

$$y_{\text{channel}}(n) = y_{(1)}(n) + 0.25y_{(1)}^2(n) \tag{5.5.11}$$

where $y_{(1)}(n)$ is the output of the first linear channel defined by Equation (5.5.10).

The nonlinear nature of the channel poses additional difficulties to the equalization task, so that the ability of performing nonlinear

Table 5.5 ASME value obtained with each ESN for the fourth channel.

Channel 4		AMSE	
Network	**Parameter**	**Training**	**Testing**
ESN	$N = 10$	$5.24(\pm 0.89)e{-}01$	$5.25(\pm 0.90)e{-}01$
	$N = 40$	$1.74(\pm 0.16)e{-}01$	$1.73(\pm 0.19)e{-}01$
	$N = 60$	$1.37(\pm 0.11)e{-}01$	$1.39(\pm 0.11)e{-}01$
ASE-ESN	$N = 10$	$4.32(\pm 0.44)e{-}01$	$4.31(\pm 0.43)e{-}01$
	$N = 40$	$2.95(\pm 0.30)e{-}01$	$3.00(\pm 0.30)e{-}01$
	$N = 60$	$2.62(\pm 0.22)e{-}01$	$2.70(\pm 0.24)e{-}01$
Proposed ESN	$N_{pc} = 3$	$3.59(\pm 0.57)e{-}01$	$3.56(\pm 0.56)e{-}01$
	$N_{pc} = 5$	$7.84(\pm 2.36)e{-}02$	$7.82(\pm 2.37)e{-}02$
	$N_{pc} = 6$	$3.78(\pm 1.33)e{-}02$	$3.85(\pm 1.35)e{-}02$

mappings represents a decisive element for enabling an adequate recovery of the information. Therefore, this scenario is very useful for the analysis of the processing capability of each ESN. Ultimately, the importance of this scenario is also emphasized by the evidence that, in certain contexts, like satellite communications, nonlinear channels may be good representations of naturally occurring channels. In Table 5.6, the performance obtained with the ESNs is presented.

The results shown in Table 5.6 clearly reveal the advantages of the proposed network in this case, that is, indeed, the achieved AMSE values are many orders of magnitude smaller than those associated with the ESN and the ASE-ESN. Therefore, based on these evidences, it is possible to affirm that the use of a more flexible structure as the output layer of the echo state network, like the Volterra filter, was certainly a key factor for the good performance achieved by the proposed ESN architecture, especially in the cases in which the distortions became more severe. Furthermore, with the aid of the PCA technique, the verified improvement in the quality of the equalization was reached without a significant increase in the complexity of the network training process, which is another attractive feature offered by the proposal.

Table 5.6 Performance of ESNs for nonlinear channels.

Channel 5		AMSE	
Network	**Parameter**	**Training**	**Testing**
ESN	$N = 10$	$5.66(\pm1.84)e{-}01$	$5.70(\pm1.85)e{-}01$
	$N = 40$	$9.97(\pm3.94)e{-}02$	$1.04(\pm0.38)e{-}01$
	$N = 60$	$4.38(\pm2.26)e{-}02$	$5.34(\pm2.77)e{-}02$
ASE-ESN	$N = 10$	$5.05(\pm0.43)e{-}01$	$5.10(\pm0.66)e{-}01$
	$N = 40$	$4.05(\pm0.23)e{-}01$	$4.52(\pm0.39)e{-}01$
	$N = 60$	$3.79(\pm0.24)e{-}01$	$4.64(\pm0.46)e{-}01$
Proposed ESN	$N_{pc} = 3$	$4.65(\pm4.26)e{-}02$	$4.65(\pm4.28)e{-}02$
	$N_{pc} = 5$	$2.94(\pm3.44)e{-}04$	$3.10(\pm3.55)e{-}04$
	$N_{pc} = 6$	$2.19(\pm2.20)e{-}05$	$2.86(\pm3.25)e{-}05$
	$N_{pc} = 10$	$1.22(\pm1.34)e{-}08$	$3.21(\pm5.19)e{-}07$

References

[1] Meloni A. and Murroni M. On the genetic optimization of APSK constellations for satellite broadcasting. *Processing of the IEEE International Symposium on Broadband Multimedia Systems and Broadcasting (BMSB)*, IEEE Press, Beijing, China, 2014, pp. 1–6.

[2] Haiquan Z. and Jiashu Z. Adaptive combinatorial neural network equalization of nonlinear channels in chaotic communication systems. *Acta Physica Sinica*, 2008, 57(7): 3996–4006 (in Chinese).

[3] Deleu T., Dervin M., Kasai K., *et al.* Iterative predistortion of the non-linear satellite channel. *IEEE Transactions on Communications*, 2014, 62(8): 2916–2926.

[4] Yecai G. and Ran X. Improved equalizer of nonlinear satellite channel. *Journal of Computer Application*, 2013, 33(11): 3039–3041 (in Chinese).

[5] Singerl P. and Koeppl H. A low-rate identification method for digital predistorters based on Volterra kernel interpolation. *Proceeding of The 48th Midwest Symposium on Circuits and Systems*, Ohio, 2005, pp. 1533–1536.

[6] Bolstad A., Miller B. A., Goodman J., *et al.* Identification and compensation of Wiener-Hammerstein systems with feedback. *Processing of the IEEE International Conference on Speech and Signal Processing(ICASSP)*, Acoustics, 2011, pp. 22–27.

[7] Ogunfunmi T. and Drullinger T. Equalization of non-linear channels using a Volterra-based non-linear. *Processing of the IEEE 54th International Midwest Symposium on Circuits and Systems (MWS CAS)*, IEEE Press, Seoul, Korea (South), 2011, pp. 1–4.

[8] Benammar B., Thomas N., Poulliat C., *et al.* On linear MMSE based turbo-equalization of non-linear Volterra channels. *Processing of the IEEE International Conference on Acoustics, Speech and Signal Processing (ICASSP)*, Vancouver, BC, 2013, pp. 4703–4707.

[9] Zerdoumi Z., Chikouche D. and Benatia D. Adaptive decision feedback equalizer based neural network for non-linear channels. *Processing of the 3rd International Conference on Systems and Control(ICSC)*, Algiers, 2013, pp. 850–855.

[10] Shunlan, L. and Shunan, J. Research on LMS newton equalization algorithm for nonlinear channel. *Journal of Circuits and Systems*, 2011, 16(1): 108–115 (in Chinese).

[11] Yecai, G. and Zhenxing, L. Blind equalization algorithms based on balance orthogonal multiwavelet transform. *Acta Armamentar II*, 2010, 31(3): 279–284 (in Chinese).

[12] Giacoumidis E., Matin A., Wei J., *et al.* Blind Nonlinearity Equalization by machine learning based clustering for single- and multi-channel coherent optical OFDM. *Journal of Lightwave Technology*, 2018, (99): 1–1.
[13] Abdulkader H., Benammar B., Poulliat C., *et al.* Neural networks-based turbo equalization of a satellite communication channel. *IEEE, International Workshop on Signal Processing Advances in Wireless Communications*, IEEE Press, Toronto, ON, Canada, 2014, pp. 494–498.
[14] Vapnik V. *Statistical Learning Theory*. New York: John Wiley & Sons, 1998.
[15] Cortes C. and Vapnik V. Support vector networks. *Machine Learning*, 1995, 20(3): 273–297.
[16] Bing L. and Weisun J. Chaos optimization method and its application. *Control Theory and Applications*, 1997, 14(4): 613–615 (in Chinese).
[17] Grigoras V. and Rigoras C. Chaos parameter modulation equalization. *International Conference on Communications(COMM)*, Bucharest, 2010, 33–36.
[18] Zhang Y., Zuo F., Zhai Z., *et al.* A new image encryption algorithm based on multiple chaos system. *International Symposium on Electronic Commerce and Security, IEEE Computer Society*, IEEE Press, Guangzhou, China, 2008, pp. 347–350.
[19] Jaeger H. The "echo state" approach to analysing and training recurrent neural networks-with an erratum note. *Bonn, Germany: German National Research Center for Information Technology GMD Technical Report*, 2001, 148(34): 13.
[20] Boccato L., Lopes A., Attux R., *et al.* An echo state network architecture based on Volterra filtering and PCA with application to the channel equalization problem. *Neural Networks (IJCNN), The 2011 International Joint Conference on IEEE*, IEEE Press, San Jose, CA, USA, 2011, pp. 580–587.
[21] Liu D. N. and Fitz M. P. Iterative equalization in non-linear satellite channels. *Turbo Codes and Iterative Information Processing (ISTC), 2012 7th International Symposium on IEEE*, IEEE Press, Bologna, Italy, 2012, pp. 220–224.
[22] Penghui C., Yuhong Y., Yijun Z., *et al.* Associated processing algorithm for 16APSK over nonlinear satellite channel. *Journal of Data Acquisition and Processing*, 2013, 28(6): 823–827 (in Chinese).
[23] Xiaoqin F. and Yan Z. Impact of amplifier nonlinearities on QAM transmission system performance. *Communications Technology*, 2014, 47(5): 504–507 (in Chinese).

[24] Guo Y. C. and Ji J. Variable momentum factor decision feedback blind equalization algorithm based on constant parameter error function. *Information and Automation*. Berlin Heidelberg: Springer, 2011, pp. 282–288.
[25] Benedetto S., Biglieri E. and Daffara R. Modeling and performance evaluation of nonlinear satellite links–a Volterra series approach. *IEEE Transactions on Aerospace & Electronic Systems*, 2007, AES-15(4): 494–507.
[26] Yecai G., Sainan F. and Hui W. Nonlinear satellite channel blind equalization algorithm based on multiwavelet double transformation. *Acta Electronica Sinica*, 2016, 44(10): 2384– 2391 (in Chinese).
[27] Cheng Z. X. and Zhang L. L. Multiwavelet analysis and application. *Journal of Engineering Mathematics*, 2001, 18(1): 99–107 (in Chinese).
[28] Guo Y. C. *Blind Equalization Theory, Algorithm and Implementation of Fuzzy Wavelet Neural Network*. Beijing: Science Press, 2011 (in Chinese).
[29] Sainan F. *A Nonlinear Channel Blind Equalization Algorithm Based on Multi-Wavelet Neural Network*. Nanjing: Nanjing University of Information Science and Technology, 2015 (in Chinese).
[30] Li C., Jing C., Xintao G., *et al.* Research on classification algorithm based on support vector machine and inverse K nearest neighbor. *Computer Engineering and Applications*, 2010, 46(24): 135–138 (in Chinese).
[31] Yecai G., Wencai X. and Fang X. Chaos and support vector machines optimization based wavelet transform weighted multi-modulus blind equalization algorithm. *Journal of System Simulation*, 2013, 25(3): 451–455 (in Chinese).
[32] Yecai G., Jing S. and Wencai X. Orthogonal wavelet multi-mode blind equalization method based on chaos optimization. Chinese Patent: CN201110208286.9, 2011-11-30 (in Chinese).
[33] Yecai G., Wencai X. and Fang X. Orthogonal wavelet constant modulus blind equalization method optimized by chaos and steepest descent method. Chinese Patent: ZL20111 0208373.4, 2014 March 26 (in Chinese).
[34] Yecai G. and Wencai X. Wavelet weighted multi-mode blind equalization based on chaotic support vector machine optimization. Chinese Patent: CN201110159986.3, 2012 February 22 (in Chinese).
[35] Xuejie D. *Orthogonal Wavelet Blind Equalization Algorithm Based on Diversity Technique*. Nanjing: Nanjing University of Information Science and Technology, 2015 (in Chinese).
[36] Jie Z. *Research on Blind Equalization Algorithm and Simulation Based on Diversity Technology*. Huainan: Anhui University of Science and Technology, 2008 (in Chinese).

[37] Yecai G., Lingling H. and Rui D. An orthogonal wavelet weighted multi-mode blind equalization algorithm based on quantum particle swarm optimization. *Acta Physica Sinica*, 2012, 61(5): 54304-054304 (in Chinese).
[38] Yecai G. and Zhenxing L. Blind equalization method of fuzzy neural network based on balanced orthogonal multiwavelet transform. Chinese Patent: ZL201010267953.6, 2013 October 30 (in Chinese).
[39] Yecai G. and Lihua W. Blind equalization method based on hybrid wavelet neural network controlled by fuzzy neural network. Chinese Patent: ZL201010267951.7, 2013 July 03 (in Chinese).
[40] Yecai G. and Lihua W. Dynamic wavelet neural network feedback blind equalization method based on fuzzy control. Chinese Patent: ZL201010216296.2, 2013 October 30 (in Chinese).
[41] Yali P. *Research on Multi-user Detection Technology Based on Chaotic Neural Network*. Chongqing: Chongqing University, 2006 (in Chinese).
[42] Jing S. *Blind Equalization Algorithm of Orthogonal Wavelet Transform Based on Chaos Theory*. Nanjing: Nanjing University of Information Science and Technology, 2012 (in Chinese).
[43] Yecai G. and Jing S. Blind equalization algorithm of orthogonal wavelet transform based on chaotic system. *Control Engineering of China*, 2012, 19(3): 443–447 (in Chinese).
[44] Min G. and Yecai G. Wavelet multi-mode blind equalization algorithm based on chaotic firefly optimization. *Computer Engineering*, 2014, 40(1): 213–218 (in Chinese).
[45] Werbos P. J. Backpropagation through time: What it does and how to do it. *Proceedings of the IEEE*, 1990, 78(10): 1550–1560.
[46] Williams R. J. and Zipser D. A learning algorithm for continually running fully recurrent neural networks. *Neural Computation*, 1989, 1(2): 270–280.
[47] Jaeger H. The echo state approach to analyzing and training recurrent neural networks. Technical Report GMD Report 148. Bremem: German National Research Center for Information Technology, 2001.
[48] Ozturk M. C., Xu D. and Príncipe J. C. Analysis and design of echo state networks. *Neural Computation*, 2007, 19(1): 111–138.
[49] Boccato L., Lopes A., Attux R., *et al.* An echo state network architecture based on volterra filtering and PCA with application to the channel equalization problem. *Proceedings of International Joint Conference on Neural Networks*, San Jose, California, USA, July 31–August 5, 2011, pp. 580–588.
[50] Mathews V. J. Adaptive polynomial filters. *IEEE Signal Processing Magazine*, 1991, 8(3): 10–26.
[51] Bellman R. *Dynamic Programming*. Courier Corporation, Princeton, New Jersey, USA, 2013.

[52] Wold S., Esbensen K. and Geladi P. Principal component analysis. *Chemometrics and Intelligent Laboratory Systems*, 1987, 2(1–3): 37–52.

[53] Hyvärinen A., Hurri J. and Hoyer P. O. Independent component analysis. Natural Image Statistics. London: Springer, 2009, pp. 151–175.

[54] Haykin S. S. *Adaptive Filter Theory*. India: Pearson Education, 2008.

[55] Adali T. Why a nonlinear solution for a linear problem? *Neural Networks for Signal Processing IX, 1999. Proceedings of the 1999 IEEE Signal Processing Society Workshop*, IEEE Press, Madison, WI, USA, 1999, pp. 157–165.

[56] Jaeger H. Short term memory in echo state networks. GMD-Forschungszentrum Informationstechnik, 2001.

[57] Jaeger H. Adaptive nonlinear system identification with echo state networks. *Advances in Neural Information Processing Systems*, 2003, 609–616.

[58] Jaeger H. and Haas H. Harnessing nonlinearity: Predicting chaotic systems and saving energy in wireless communication. *Science*, 2004, 304(5667): 78–80.

[59] Salmen M. and Ploger P. G. Echo state networks used for motor control. *Proceedings of the 2005 IEEE, International Conference on IEEE*, IEEE Press, Barcelona, Spain, 2005, pp. 1953–1958.

[60] Erdogmus D. and Principe J. C. From linear adaptive filtering to nonlinear information processing-The design and analysis of information processing systems. *IEEE Signal Processing Magazine*, 2006, 23(6): 14–33.

[61] Montalvao J., Dorizzi B. and Mota J. C. M. Some theoretical limits of efficiency of linear and nonlinear equalizers. *Journal of the Brazilian Telecommunications Society*, 1999, 14(2): 85–92.

[62] Proakis J. G. *Digital Communications*. McGraw-Hill, New York, 1995.

Chapter 6

Nonlinear Volterra Channel Blind Equalization Algorithm

Summary

The nonlinear Volterra channel is researched. By transforming the third-order nonlinear memory term of the truncated Volterra equalizer into a second-order memory term, the nonlinear channel equalization algorithm based on the improved structure of Volterra equalizer is proposed. Based on the analysis of minimum mean square error (MMSE)-based precise equalization algorithm and time-invariant MMSE coefficients, a linear MMSE-based turbo equalization and iterative equalization algorithm is analyzed. From the perspective of information theory, the nonlinear channel equalization algorithm based on steady-state analysis of maximum correlation entropy Volterra filter is discussed. By designing fuzzy neural network controller and combining with complex neural network polynomial, the complex neural network polynomial Volterra channel blind equalization algorithm based on fuzzy neural network control is studied.

Satellite communication is one of the most important communication technologies currently. This technology can overcome the problem of long distance communication and is not sensitive to the interference of various radio waves on the ground. However, there are various distortions and fading in the satellite channel. Among them, the nonlinearity caused by both the satellite internal traveling wave tube amplifier (TWTA) and the group delay of the satellite transmission signals is a typical distortion in the satellite channel [1], which results in inter-symbol interference (ISI), phase rotation, and amplitude distortion of the signals, and seriously affects the quality of the communication. To eliminate the distortion and fading of the signals, it is necessary to study the signal distortion compensation method based on the nonlinear satellite channel.

6.1 Nonlinear Channel Adaptive Equalization Algorithm

6.1.1 *Nonlinear channel adaptive equalization model*

The goal of channel equalization is to recover the original transmitted signals from the received distortion signal. The equivalent baseband model of the adaptive channel equalizer is shown in Figure 6.1 [2–4].

In Figure 6.1, χ represents the delay order. The adaptive equalizer has the ability to dynamically adjust the filter parameters at the current moment based on the filter parameters at the previous moment so that the filter continually approximates the inverse characteristics of the channel. This self-adjusting and tracking ability of the adaptive equalizer makes it suitable for real-time signal processing.

On the one hand, the channel can be divided into the linear channel and the nonlinear channel in an actual equalization method. It is feasible for linear equalizer to be used to equalize the linear channel, whereas it often cannot equalize the nonlinear channel well. In order to equalize the nonlinear channel, the nonlinear equalizer and the inverse characteristic of the nonlinear channel must perfectly match.

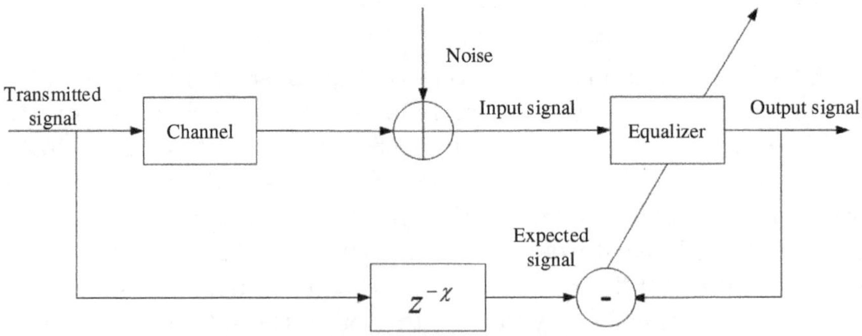

Figure 6.1 Equivalent baseband model of adaptive channel equalizer.

On the other hand, the basic form of the linear channel can be roughly determined, that is, the proportional relationship between the outputs and the inputs manifests is a fixed constant, while the nonlinear form of the channels varies greatly, and the causes of nonlinearity are also different, and the analytical expressions of nonlinear are completely different. Therefore, for different nonlinear channels, the different structures of nonlinear equalizers are needed for equalization, but it depends on the specific nonlinearity.

Studies have shown that the Volterra series can describe the nonlinear characteristics of a variety of systems. Benedetto *et al.* proposed a nonlinear satellite channel Volterra series model based on MPSK signals according to Salah's amplitude-phase conversion model and Volterra series and gave the corresponding Volterra equalizer structure [5]. When the input is MPSK signals, its complex envelope can be defined as:

$$x(t) = \sum_{n=-\infty}^{+\infty} a(n)h(t - nT) \tag{6.1.1}$$

where $a(n)$ is the transmitted symbol sequence and $h(t)$ denotes the impulse response of the shaping filter. Let $y(n)$ be the output of the discrete channel. By substituting Equation (6.1.1) into the Volterra series, the Volterra series model of nonlinear satellite channel is obtained, i.e.,

$$\begin{aligned} y(n) &= \sum_{i=0}^{N-1} h(i)a(n-i) \\ &\quad + \sum_{i=0}^{N-1}\sum_{j=0}^{N-1}\sum_{k=0}^{N-1} h(i,j,k)a(n-i)a(n-j)a^*(n-k) + \cdots + w(n) \end{aligned} \tag{6.1.2}$$

where $a(n) = e^{j\varphi n}$, $h(i)$ is the first-order linear kernel coefficient of the nonlinear satellite channel, $h(i, j, k)$ denotes the third-order nonlinear kernel coefficient of the nonlinear satellite channel, and $w(n)$ is additive noise. When the amplifier works in the linear region, there is only linear portions and additive noise in Equation (6.1.2). If the amplifier operates in or near the saturation region, one or more high-order terms will be included in Equation (6.1.2), whose order depends on the nonlinearity of the satellite channel. Equation (6.1.2) contains only odd-order terms, and even-order terms are eliminated during the derivation process. This is because the satellite channel is regarded as a bandpass filter and the even-order input signals are out-of-band, which is filtered out during transmission. In addition, i, j, and k represent the memory of the channel, as the memory of the channel enhances and the nonlinear term coefficients of the channel will increase geometrically, which will greatly aggravated the nonlinear distortion of the signals. So, the distortional channel must be compensated.

Benedetto pointed out that a nonlinear Volterra series satellite channel can be equalized by a nonlinear filter with the nonlinear tap delay structure, a series of complex multipliers, and an accumulating unit. Taking the third-order truncated Volterra equalizer as an example, its structure is shown in Figure 6.2.

In Figure 6.2, $f(0), f(1), \ldots, f(N-1)$ denote linear weight coefficients of the third-order truncated Volterra equalizer, and $f(0,0,0), f(0,0,1), \ldots, f(N-1, N-1, N-1)$ stand for the nonlinear weight coefficients of the third-order truncated Volterra equalizer. $\boldsymbol{y}(n) = [y(n), y(n-1), \ldots, y(n-N+1),\ y^2(n)y^*(n), \ldots, y^2y^*(n-N+1), \ldots, y(n-N+1)]$ is the input vector of the equalizer. From

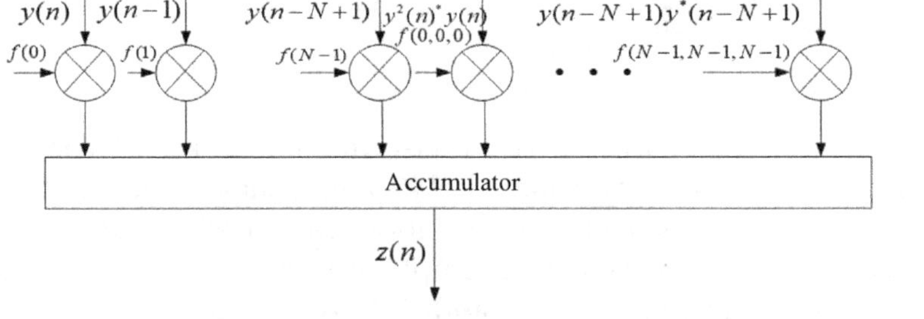

Figure 6.2 Structure of third-order truncated Volterra equalizer.

Figure 6.2, the following output of equalizer can be obtained.

$$z(n) = \sum_{i=0}^{N-1} f(i)y(n-i) + \sum_{i=0}^{N-1}\sum_{j=0}^{N-1}\sum_{k=0}^{N-1} f(i,j,k)y(n-i)y(n-j)y^*(n-k). \quad (6.1.3)$$

The third-order truncated Volterra equalizer can be used to compensate for the general weak nonlinear distortion. When the nonlinearity of the channel is enhanced, it is necessary for a higher-order Volterra equalizer to be used to better compensate for the nonlinear distortion channel. However, the experiments show that as the nonlinearity of the channel enhances, the equalization effect of the Volterra equalizer is not obvious, and as the order of the equalizer increases, the computation complexity of the Volterra equalizer increases geometrically. Therefore, in practical applications, a third-order or fifth-order Volterra equalizer is generally used.

Table 6.1 shows the simplified coefficients of the nonlinear channel Volterra series model, which are obtained from statistical experiments based on the PSK signals. Since the PSK signals are not very sensitive to the nonlinearity of satellite channel, this set of data for this coefficient is often used to describe a weak nonlinear satellite

Table 6.1 Simplified coefficients of Volterra equalizer.

Orders	**Volterra Kernel Coefficients**	**Value of Kernel Coefficients**
First-order linear part	$h(0)$	$1.22 + 0.646j$
	$h(1)$	$0.063 - 0.001j$
	$h(2)$	$-0.024 - 0.014j$
	$h(3)$	$0.036 - 0.031j$
Third-order nonlinear part	$h(0,0,2)$	$0.039 - 0.022j$
	$h(3,3,0)$	$0.018 - 0.018j$
	$h(0,0,1)$	$0.035 - 0.035j$
	$h(0,0,3)$	$-0.040 - 0.009j$
	$h(1,1,0)$	$-0.01 - 0.017j$
Fifth-order nonlinear part	$h(0,0,0,1,1)$	$0.039 - 0.022j$

channel. Researchers usually use this set of data as the kernel coefficient of the nonlinear satellite channel Volterra series model and the third-order Volterra equalizer to design the nonlinear satellite channel equalization algorithm.

6.1.2 *Nonlinear channel adaptive equalization algorithm*

The least mean square (LMS) algorithm for a nonlinear satellite channel is the same as that for the linear channel and its cost function is defined as:

$$J = E\{|z(n) - a(n)|^2\} \tag{6.1.4}$$

where $z(n)$ is the output signal of the equalizer and $a(n)$ is the desired signal. After minimizing the cost function in Equation (6.1.4), the iterative equation of the equalizer weight vector may be given by:

$$\boldsymbol{f}(n+1) = \boldsymbol{f}(n) + \mu \boldsymbol{y}(n) e^*(n) \tag{6.1.5}$$

where $e(n)$ is the error function, and

$$e(n) = a(n) - z(n) \tag{6.1.6}$$

The step-size factor μ in Equation (6.1.5) should satisfy:

$$0 < \mu < \frac{2}{\lambda_{\max}} \tag{6.1.7}$$

where $\lambda_{\max}$ is the maximum value of the eigenvalue of the autocorrelation matrix of the input signals [6–8]. We call Equations (6.1.5)–(6.1.7) as a Volterra equalizer algorithm based on the LMS. Since it has the smallest computation loads and the least storage space, it is easy to implement it in hardware. However, Figure 6.2 shows that the input components of a Volterra equalizer are much more complex than those of the linear equalizers. The weight vector length of the third-order Volterra equalizer is computed by:

$$N = \sum_{n=1}^{2} M^{2n-1} \tag{6.1.8}$$

It can be seen from Equation (6.1.8) that the length of the weight vector increases exponentially with the increase of the channel memory length or depth, which inevitably increases the dimension and the autocorrelation of the input signals and further makes the eigenvalue expansion of the autocorrelation matrix quite serious (i.e., the eigenvalues of the autocorrelation matrix of the input signal have multiple different values). The expansion of the eigenvalue of the autocorrelation matrix is represented by the increase of $\lambda_{\max}$ or the decrease of the step-size μ, which results in the slow convergence of the equalizer. For many years, researchers have been trying to find some equalization algorithms with a fast convergence rate and good equalization effect of nonlinear satellite channel such as Volterra series structure-based equalization algorithm and other structure-based equalization algorithm.

6.1.2.1 *Frequency domain Volterra series equalization algorithm*

Linear adaptive algorithms have proved that orthogonal transforms can reduce the autocorrelation of input signals. Converting the time domain signal into a frequency domain signal is one of the orthogonal transform methods. Compared with the time domain signal, the equalizer processes the frequency domain signal quickly and the variable step-size in the equalization algorithm can further improve the convergence speed. In [9–11], these two methods were combined to obtain the frequency domain variable step-size adaptive Volterra equalizer. This frequency domain Volterra equalizer structure is shown in Figure 6.3.

In Figure 6.3, the input signals are transformed by DFT, and the frequency domain signal is obtained as follows:

$$\boldsymbol{Y} = \mathrm{DFT}(\boldsymbol{y}) \tag{6.1.9}$$

At this time, Equation (6.1.5) becomes

$$\boldsymbol{F}(n+1) = \boldsymbol{F}(n) + \mu(n)\,\boldsymbol{Y}(n)\mathrm{DFT}[e^{*}(n)] \tag{6.1.10}$$

where $\boldsymbol{F}$ is the transform domain coefficient vector of $\boldsymbol{f} \cdot \mu(n)$ denotes the modified variable step-size and is defined as:

$$\mu(n) = \alpha/[\beta + \sigma_s^2(n)] \tag{6.1.11}$$

$$\sigma_Y^2(n) = \lambda\sigma_Y^2(n-1) + (1-\lambda)\mathrm{DFT}[e^2(n)] \tag{6.1.12}$$

where $\sigma_Y^2(n) = E(Y^2)$, $0 < \alpha < 2$, $0 < \lambda < 1$.

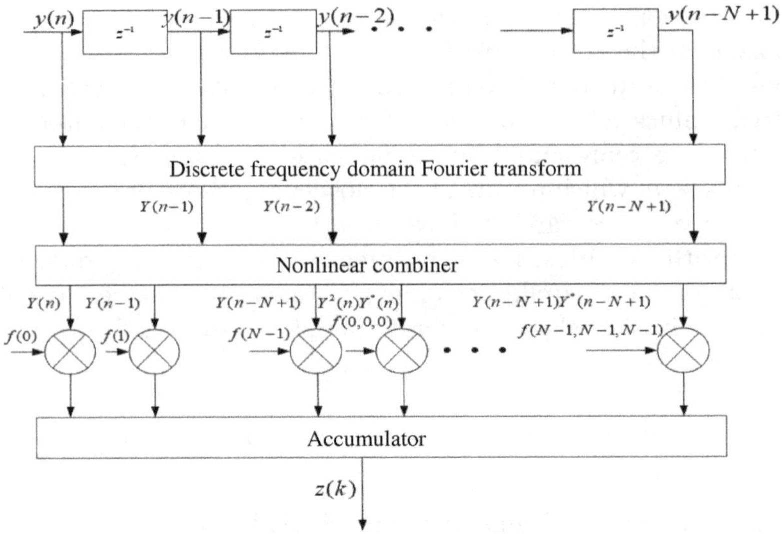

Figure 6.3 Structure of frequency domain Volterra equalizer.

6.1.2.2 *Equalization algorithm based on compression mapping*

The basic idea of the equalization algorithm based on compression mapping is to use both the compression mapping method and iterative method to calculate the original satellite transmitted sequence according to the output of the satellite channel Volterra series model at each time. For Equation (6.1.2), the process of solving the transmitted signals from receiving the signals can be regarded as the solution of the following equation:

$$T(a(n)) = a(n) + \beta[y(n) - V(a(n))] \tag{6.1.13}$$

where $V(a(n))$ is the satellite channel Volterra series. When $a(n)$ in Equation (6.1.13) is the fixed point, $T(a(n)) = a(n)$, the function $T(a(n))$ is converted to $y(n) = V(a(n))$, that is, it is the solution of Equation (6.1.2).

According to the compression mapping fixed-point iterative method, we get the following formula:

$$z_{k+1}(n) = z_k(n) + \beta(y(n) - V(z_k(n))) \tag{6.1.14}$$

where k represents the number of iterations, $z_{k+1}(n)$ denotes the output signal of the equalizer after k iterations, $z_k(n)$ denotes the output signal of the first $k-1$ iteration equalizer, and $y(n)$ is the output signal of the satellite channel. All transmitted signal points can be obtained by multiple iterations.

The equalization effect of the compression mapping method is shown in Figure 6.4.

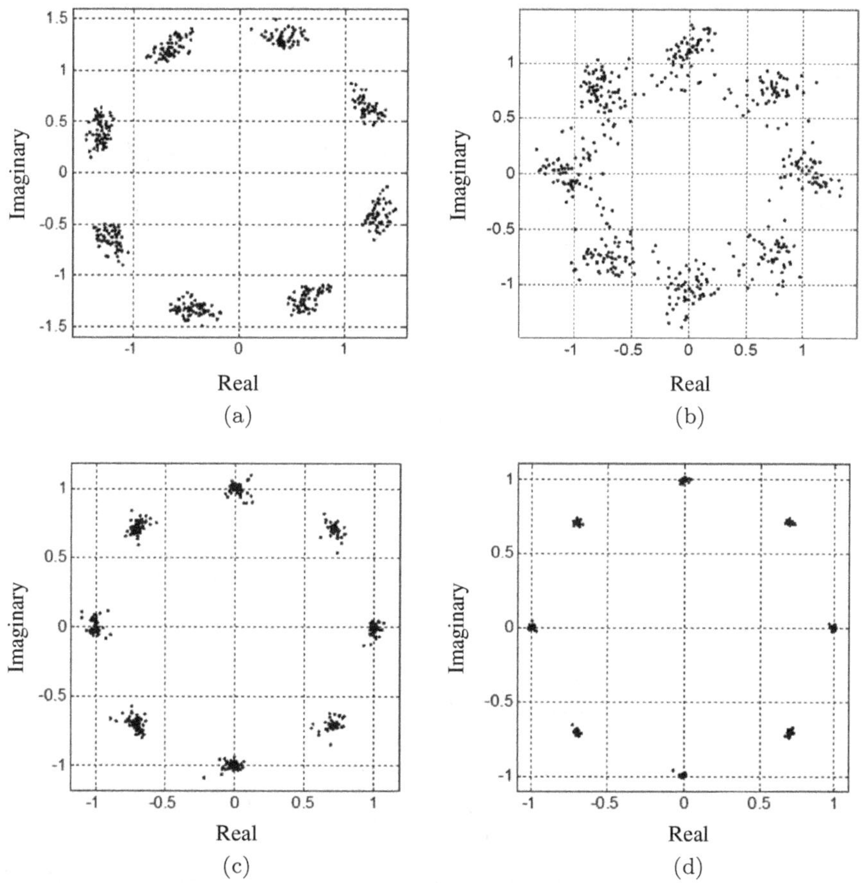

Figure 6.4 Simulation results of compression mapping equalization algorithm: (a) constellations of receive signals; (b) constellation of first iteration output signals; (c) two iteration constellations of output signals and (d) third iteration constellations of output signals.

6.2 Improved Volterra Equalizer for Nonlinear Channel

A third-order Volterra equalizer can be used to compensate for general weak nonlinear distortion. When the nonlinearity of the channel is enhanced, it is necessary for a higher-order Volterra equalizer to be used to compensate for its effect. Experiments have shown that when the nonlinearity of the channel is enhanced, the equalization effect of the higher than third-order Volterra equalizer is not obvious, and the increase of the order of the equalizer will sharply improve the computational complexity. Therefore, in practical applications, a third-order or fifth-order Volterra equalizer is generally used.

6.2.1 *Improved nonlinear channel Volterra equalizer*

The output of the most frequently used Volterra equalizer based on Volterra series model for the satellite channel can be written as:

$$\begin{aligned} y(n) &= \sum_{i=0}^{N-1} h(i)a(n-i) \\ &\quad + \sum_{i=0}^{N-1}\sum_{j=0}^{N-1}\sum_{k=0}^{N-1} h(i,j,k)a(n-i)a(n-j)a^*(n-k) \\ &= V(a(n), a(n-1), \ldots, a(n-N+1)) \end{aligned} \tag{6.2.1}$$

The output of the corresponding third-order truncated Volterra equalizer can be written as:

$$\begin{aligned} z(n) &= \sum_{m_1=0}^{m_{1\max}} f(m_1)y(n-m_1) \\ &\quad + \sum_{m_1=0}^{m_{1\max}}\sum_{m_2=0}^{m_{2\max}}\sum_{m_3=0}^{m_{3\max}} f(m_1,m_2,m_3) \\ &\quad \times y(n-m_1)y(n-m_2)y^*(n-m_3) \end{aligned} \tag{6.2.2}$$

It has been seen from Equation (6.2.1) that the satellite channel distortion mainly comes from its nonlinearity and memory, which introduces the ISI in the receiving signals, whereas the ISI will ultimately affect the correct decision output of the decision device, so the ISI must be compensated. The values of $m_{1\,\max}$, $m_{2\,\max}$, and $m_{3\,\max}$ depend on the nonlinear satellite channel characteristics. Ideally, $m_{1\,\max}-1$, $m_{2\,\max}-1$, and $m_{3\,\max}-1$ should all take the same value as the channel memory depth, but the memory depth of the channel cannot be determined in advance. In order to better compensate for the nonlinearity, $m_{1\,\max}-1$, $m_{2\,\max}-1$, and $m_{3\,\max}-1$ of the satellite channel should be as large as possible for the memory depth of the channel, that is, $m_{1\,\max}$, $m_{2\,\max}$, and $m_{3\,\max}$ should be taken as large as possible. However, such a value would increase the computational complexity of the equalizer. Therefore, feasible measures must be taken to reduce the computational complexity.

It is known from Equation (6.2.2) that the computational complexity of the truncated Volterra equalizer mainly comes from the product of the third-order nonlinear term at the time of m_1, m_2, and m_3 $(m_1 = 0, 1, 2, \ldots, m_{1\,\max}-1, m_2 = 0, 1, 2, \ldots, m_{2\,\max}-1, m_{3\,\max}-1)$. In the ideal state, $m_{1\,\max}-1$, $m_{2\,\max}-1$, and $m_{3\,\max}-1$ should be equal to the maximum value of the memory depth of each part of the third-order nonlinear term corresponding to the nonlinear satellite channel and the truncated Volterra equalizer. For third-order nonlinear terms, if $m_{1\,\max}-1$, $m_{2\,\max}-1$, and $m_{3\,\max}-1$ are all N, the multiplication of the memory terms of the third-order nonlinear term requires a total of $2N^3$ complex multiplications, and the number of complex multiplications required for each item $f(m_1, m_2, m_3)$ to be multiplied by the respective equalizer coefficients is included. Therefore, without considering the number of complex multiplication of iterative equations of the equalizer, in order to obtain each output signal point, $3N^3$ complex multiplication are required when input signals pass through the truncated third-order nonlinear Volterra equalizer.

In order to reduce this computational complex, it is necessary to improve the Volterra equalizer. The idea of the improvement is to transform the third-order nonlinear term of the truncated Volterra equalizer into the second-order nonlinear term by a certain transformation. After improving the structure of the Volterra equalizer, the number of complex multiplications required for the signal to pass through the equalizer will be reduced from $O(N^3)$ to $O(N^2)$.

According to this idea, the relationship in structure between the first-order linear term and the third-order nonlinear term of the

Volterra equalizer is noted, and the output of the improved Volterra equalizer can be written as:

$$\begin{aligned}
z(n) &= \sum_{m_1=0}^{N-1} f(m_1)y(n-m_1) \\
&\quad + \sum_{m_1=0}^{N-1}\sum_{m_2=0}^{N-1}\sum_{m_3=0}^{N-1} f(m_1,m_2,m_3)y(n-m_1)y(n-m_2)y^*(n-m_3) \\
&= \sum_{m_1=0}^{N-1} f(m_1)y(n-m_1) \\
&\quad + \sum_{m_1=0}^{N-1}\sum_{m_2=0}^{N-1}\sum_{m_3=0}^{N-1} \frac{f(m_1,m_2,m_3)}{f(m_1)} y(n-m_1)y(n-m_2)y^*(n-m_3)
\end{aligned} \tag{6.2.3}$$

Let $\frac{f(m_1,m_2,m_3)}{f(m_1)} = f(m_2,m_3)$, Equation (6.2.3) can be written as:

$$\begin{aligned}
z(n) &= \sum_{m_1=0}^{N-1} f(m_1)(y(n-m_1) \\
&\quad + \sum_{m_2=0}^{N-1}\sum_{m_3=0}^{N-1} f(m_2,m_3)y(n-m_1)y(n-m_2)y^*(n-m_3)) \\
&= \sum_{m_1=0}^{N-1} f(m_1)y(n-m_1) \\
&\quad \times \left(1 + \sum_{m_2=0}^{N-1}\sum_{m_3=0}^{N-1} f(m_2,m_3)y(n-m_2)y^*(n-m_3)\right) \\
&= \sum_{m_1=0}^{N-1} f(m_1)y(n-m_1)b(n)
\end{aligned} \tag{6.2.4}$$

where $f(m_2, m_3)$ represents the weight coefficient of the nonlinear part of the equalizer. Equation (6.2.4) is an improved structure of the nonlinear channel equalizer, as shown in Figure 6.5(a), and it can be seen from Figure 6.5(a) that this improved structure is a cascade of a nonlinear equalizer and a linear equalizer. Figure 6.5(b)

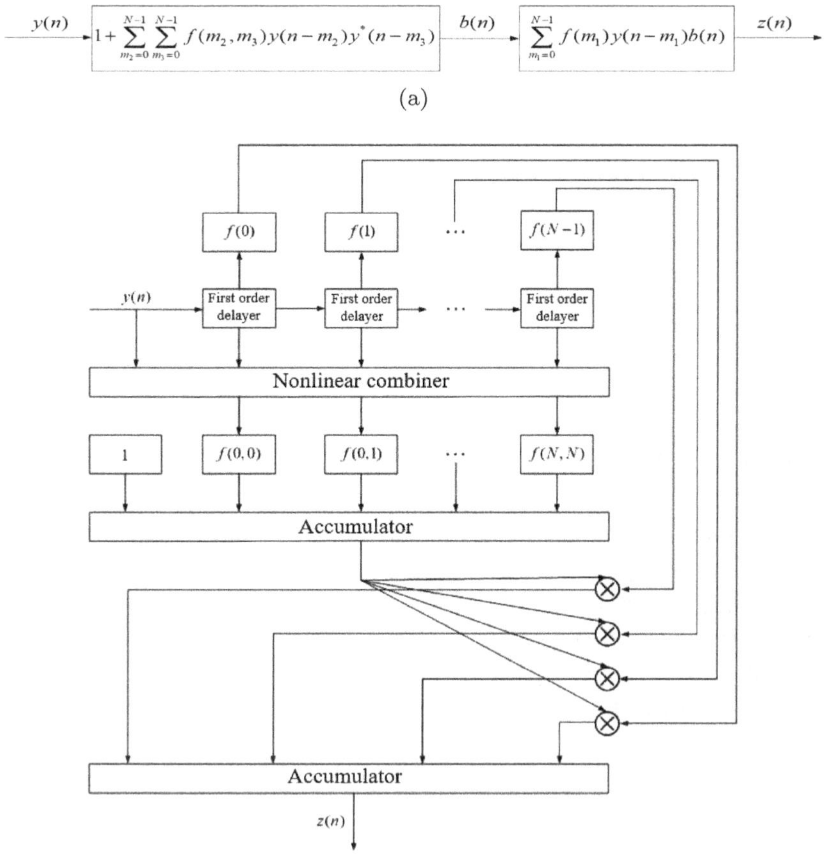

(a)

(b)

Figure 6.5 Improved structure of nonlinear satellite channel equalizer: (a) improved cascade of equalizers; (b) improved internal structure of the equalizer.

shows the internal structure of the improved equalizer, where $y(n)$ is the input signal of the equalizer, $b(n)$ is the output signal of the nonlinear part of the equalizer, and $z(n)$ is the output signal of the linear part of the equalizer. The structure of the improved equalizer shows that the essence of the improvement is to transform the third-order nonlinear memory multiplication of the truncated Volterra equalizer into the second-order memory multiplication. Compared with the conventional truncated Volterra equalizer, the number of complex multiplications required by the improved equalizer is reduced by $3N^3 - 2N^2 - N$ times without considering the number of complex multiplications in computing the weight vector of the equalizer. Let

$$C_{\text{load}} = 3N^3 - 2N^2 - N \qquad (6.2.5)$$

$$C_{\text{load}-1}(N) = \frac{C_{\text{load}-1}(N)}{N} = 3N^2 - 2N^1 - 1 \qquad (6.2.6)$$

$$C_{\text{load}-2}(N) = N \qquad (6.2.7)$$

When $N_1 > N_2$, there is the following relationship:

$$C_{\text{load}-1}(N_1) - C_{\text{load}-1}(N_2) = [3(N_1 + N_2) - 2](N_1 - N_2) \qquad (6.2.8)$$

Since $N_1 - N_2$ is a positive integer, and N_1 is greater than or equal to 1, and N_2 is greater than or equal to zero, $C_{\text{load}-1}(N_1)$ monotonically increases and is always greater than zero, and $C_{\text{load}}(N)$ is also a function of a single increase and greater than or equal to zero. When $N > 1$ $C_{\text{load}}(N)$ is always a positive integer. It can be seen that the computational load of the improved equalizer is greatly reduced and confirmed by the subsequent simulation experiments.

The cost function is now defined as:

$$J(n) = E\{|e(n)|^2\} \qquad (6.2.9)$$

where $e(n) = a(n) - z(n)$ is the error function. Let the weight vector of nonlinear part of the equalizer be $\boldsymbol{F}_{\text{non}}(n)$ and that of the linear part be $\boldsymbol{F}_{\text{lin}}(n)$. The instantaneous mean square error(MSE) is used instead of the statistical MSE [12], and the vectors of the equalizer tap coefficients are updated by the stochastic gradient descent method as follows:

$$\boldsymbol{F}_{\text{non}}(n+1) = \boldsymbol{F}_{\text{non}}(n) + \lambda_1 e(n) \frac{\lambda e^*(n)}{\partial \boldsymbol{F}_{\text{non}}(n)} \qquad (6.2.10)$$

$$\boldsymbol{F}_{\text{lin}}(n+1) = \boldsymbol{F}_{\text{lin}}(n) + \lambda_2 e(n) \frac{\lambda e^*(n)}{\partial \boldsymbol{F}_{\text{lin}}(n)} \qquad (6.2.11)$$

where $\nabla_1(n)$ and $\nabla_2(n)$ are the gradient parameters of the nonlinear part and the linear part tap coefficient of equalizer, respectively.

The derivation process of $\nabla_1(n)$ is as follows:

$$\nabla_1(n) = \frac{\partial z(n)}{\partial \boldsymbol{F}_{\text{non}}} = \frac{\partial z(n)}{\partial b(n)} \frac{\partial b(n)}{\partial \boldsymbol{F}_{\text{non}}} \tag{6.2.12}$$

where

$$\frac{\partial z(n)}{\partial b(n)} = \sum_{m_1=0}^{N-1} f(i) y(n-m_i) \tag{6.2.13}$$

$$\frac{\partial z(n)}{\partial \boldsymbol{F}_{\text{non}}(n)} = \{y(n)y^*(n), y(n)y^*(n-1), \ldots, y(n-N+1)y^*(n-N+1)\} = \boldsymbol{Y}(n) \tag{6.2.14}$$

The derivation process of $\nabla_2(n)$ is as follows:

$$\nabla_2(n) = \frac{\partial z(n)}{\partial \boldsymbol{F}_{\text{non}}(n)} \tag{6.2.15}$$

$$\frac{\partial z(n)}{\partial \boldsymbol{F}_{\text{non}}(n)} = \{y(n), y(n-1), \ldots, y(n-N+1)\} b(n) = \boldsymbol{B}(n) \tag{6.2.16}$$

The step-size of the nonlinear and linear parts satisfies the following conditions, respectively.

$$0 < \mu < 2/\mu_{\max} \tag{6.2.17}$$

$$0 < \lambda < 2/\lambda_{\max} \tag{6.2.18}$$

Let $\boldsymbol{Y} = \{y(n)y^*(n), y(n)y^*(n-1), \ldots, y(n-N+1)y^*(n-N+1)\}$ be the input signal vector of the nonlinear equalizer and $\boldsymbol{B} = b(n)\{y(n), y(n-1), \ldots, y(n-N+1)\}$ be the input signal vector of the nonlinear equalizer. $\mu_{\max}$ and $\lambda_{\max}$ are the maximum eigenvalues of the autocorrelation matrix of $\boldsymbol{Y}$ and $\boldsymbol{B}$, respectively. It can be seen by analyzing $\boldsymbol{Y}$ and $\boldsymbol{B}$ that the nonlinear part of the equalizer converges slower than the linear part because the input signal length of the improved nonlinear part is much larger than that of the linear part. In the latter simulation, it is found that if the step-sizes of the linear equalizer and the nonlinear equalizer are different, the performance of the equalizer is unstable, that is to say, the equalization effect of the equalizer varies greatly each time. In order to speed up the convergence and keep the stability performance of the equalizer, the step-size of the nonlinear equalizer should take a larger value, whereas the step-size of the linear equalizer takes the same value as the step-size of the nonlinear equalizer.

6.2.2 *Algorithm simulations*

In simulation tests, the transmission signal was 32APSK, and its length was 7000. The channel memory depth was set to 3, the iteration step-size of the weight vector for the Volterra equalizer was set to 0.005, and the iteration step-size of the weight vector for the improved nonlinear and linear of equalizer was selected to be 0.008. The first tap coefficient of the Volterra equalizer was initialized to 1, and the other taps were initialized to 0. The center tap of the improved linear and nonlinear portions of equalizer was set to 1. The results of 50 Monte Carlo simulations were shown in Figure 6.6. The MSE in Figure 6.6 is defined as:

$$\mathrm{MSE} = 20\log(\mathrm{MSE}(n)) \tag{6.2.19}$$

It is can be known from Figure 6.6(a) that the MSE curves of the Volterra equalizer were very jittery. The MSE of the improved equalizer is smaller than that of the conventional Volterra equalizer, and the robustness of the curve is better, but its convergence speed is slower.

Figure 6.6(b) shows that the phase rotation of the inner circle constellations of the 32APSK signal output from the conventional Volterra equalizer still exists, while that of the 32APSK signal constellations of the improved equalizer are compact and symmetrical, and the phase rotation of constellations has been completely eliminated.

6.2.3 *Computational complexity*

The computational complexity of the above two equalizers is now analyzed. Since the computation loads of performing a complex multiplication are much larger than that of performing a complex addition, the number of complex multiplications required to produce

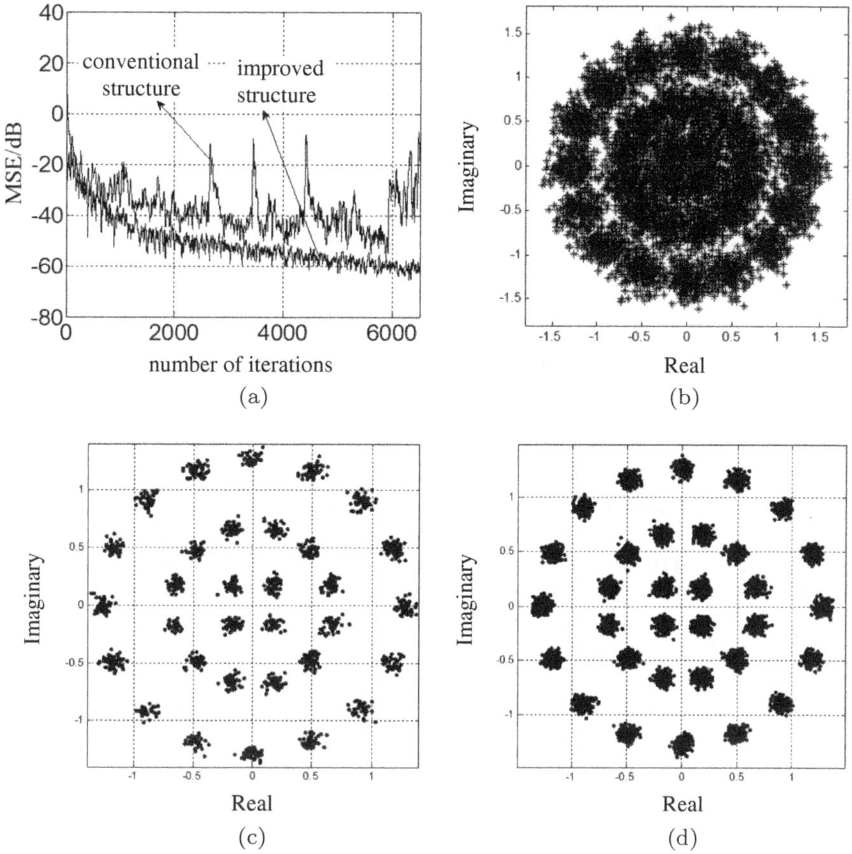

Figure 6.6 Comparison of equalization effects: (a) MSE curves; (b) input constellations of equalizer; (c) output of conventional Volterra equalizer and (d) output of improved Volterra equalizer.

each output signal point is used as a measure of the computational complexity.

For a conventional Volterra equalizer, $3N^3 + N$ complex multiplications are required for each output signal point, and $N^3 + N + 1$ complex multiplications are required for the weight vector. For the improved equalizer, $2N^2 + 2N$ and $2N^2 + 3N + 2$ complex multiplications are required for the two processes, respectively. Figure 6.7 shows that as the depth of channel memory increases, the number of complex multiplications of the conventional Volterra equalizer increases

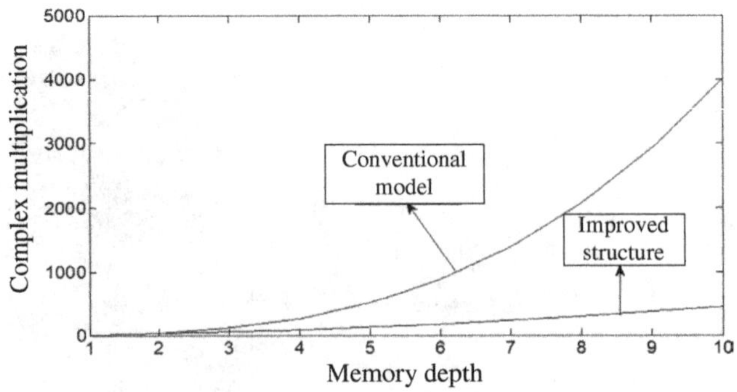

Figure 6.7 Computation complexity.

Table 6.2 Comparison of running time.

Equalizer Model	**Channel Memory Depth**	**Operation Time(/s)**
Conventional	3	42.726
Volterra equalizer	5	154.626
Improved	3	29.790
Volterra equalizer	5	78.348

exponentially, while that of the improved Volterra equalizer grows slowly. When the channel memory depth is large, the complex multiplication of the improved equalizer is much smaller than that of the conventional Volterra equalizer. For example, when the channel memory depth is set to 10, the complex multiplication of the improved Volterra equalizer is about 1/9 of that of the conventional Volterra equalizer.

The running time of the equalizer before and after improvement under different channel memory depths is shown in Table 6.2 when the length of the transmitting signals is given as 1000. Table 6.2 shows that the computational speed of the improved Volterra equalizer is faster than that of the conventional Volterra equalizer when the length of the transmitting signals is shorter and the channel memory is not particularly serious.

6.3 Nonlinear Channel Turbo Blind Equalization Algorithm Based on Linear MMSE

The error correction ability of turbo code is close to the limit [13] in the Gaussian white noise channel, which is the best performance of all the coding techniques so far. Satellite channels are time-varying, frequency-varying, amplitude-varying, and phase-varying. For coherent detection, the receiver must estimate the channel to obtain optimal performance and can obtain channel estimation information based on pilot symbols. However, the pilot symbols can reduce spectral efficiency. Blind equalization technology is one of the effective methods to improve spectral efficiency. In this section, turbo code blind equalization technology is used to study the nonlinear satellite channel turbo blind equalization algorithm based on linear MMSE criterion [14–16].

6.3.1 *System specification*

The turbo equalization principle regards the ISI channel (equivalent to a convolutional code), interleaver, and channel coding as a serial cascaded convolutional code, which is used to perform the iterative equalization and decoding at the receiving end. After one of the two parallel soft-input soft-output decoders is slightly modified, that is, its original decoding function is modified into the equalization function of soft-input soft-output, another soft-input soft-output decoder is still used for decoding [14]. In this way, both the improved equalizer and decoder can process the input prior information and the output posterior information, and the output posterior information can be interleaved or de-interleaved as the other party's prior information to complete the iterative transmission of the external information between the channel decoder and the equalizer and further obtain the approximation optimal performance.

A channel coding communication system with Volterra channel and additive noise is taken as an example and shown in Figure 6.8. The binary sequence $\{d(n)\}$ is encoded by an error-correcting code to produce an encoded bit stream $\{b_n(n)\}$, which is sent to the bit interleaver, and the interleaved bit stream $\{c(n)\}$ is input into the linear modulator (each L bit forms a transmission symbol) to obtain a complex data symbol $\{x(n)\}$. Finally, each data symbol passes through

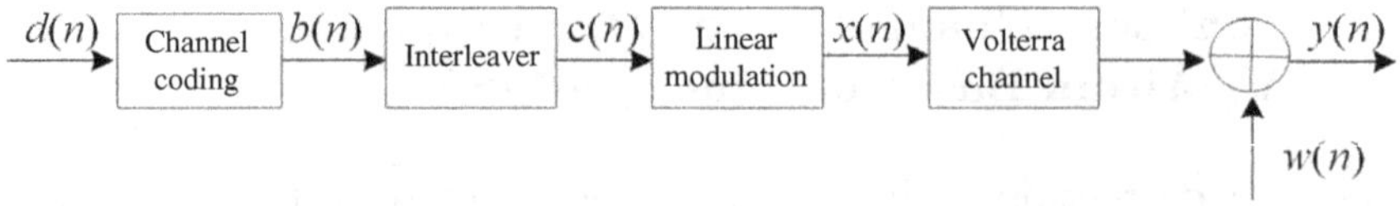

Figure 6.8 Coded communication system model through Volterra channel and additive noise.

the Volterra channel at a time. Suppose that each data block contains M symbols, the relationship between discrete input and output of the Volterra channel can be described by a finite linear impulse response model, i.e.,

$$\boldsymbol{y}(n) = \boldsymbol{H}\boldsymbol{x}(n) + \sum_i \sum_j \sum_k \boldsymbol{H}_{ijk}\boldsymbol{x}^{ijk}(n) + \boldsymbol{w}(n) \tag{6.3.1}$$

where $\boldsymbol{y}(n) = [y(n-N),\ldots,y(n+N_2)]^{\mathrm{T}}$, $\boldsymbol{x}(n) = [x(n-N_1-L_{01}),\ldots,x(n-N_2-L_{02})]^{\mathrm{T}}$, $\boldsymbol{w}(n) = [w(n-N_1),\ldots,w(n+N_2)]^{\mathrm{T}}$ are represented as a received signal at the nth sampling time, the input signal of the Volterra channel, and Gaussian white noise, respectively.

Assume that the complex symbols $\boldsymbol{x}(n)$ are totally independent of each other and selected from a finite character set A $= \{a_0, a_1, \ldots, a_{|A|}\}$, the prior probability of the symbols $\boldsymbol{x}(n)$ is given by:

$$p_{ni} = P(x(n) = a(n)), n = 0, 1, \ldots, M-1; i = 0, 1, \ldots, |A| \tag{6.3.2}$$

When *a priori* information is available, $p_{ni} = 1/|A|$, i.e., all symbols are equal probability.

Further assume that the additive noise sequence $w(n)$ is a complex independent and identically distributed Gaussian random variable with zero mean and is independent of the symbol sequence $\boldsymbol{x}(n)$. The variance of each noise sample value is σ^2.

Let $\boldsymbol{y} = \{y(0), y(1), \ldots, y(M-1)\}$, in the absence of the channel information $\boldsymbol{h}$ and noise parameters, the posterior probability $p(x(n) = a(i)|\boldsymbol{y})$ of the transmitted symbols based on the received signal $\boldsymbol{y}$ and the prior information $p_{n,i}$ may be estimated.

The estimated posterior probability is used to compute the output information bits $d(n)$ of the decoder.

The job of the receiver is to estimate the transmitted data using channel information and encoded redundancy information. Although

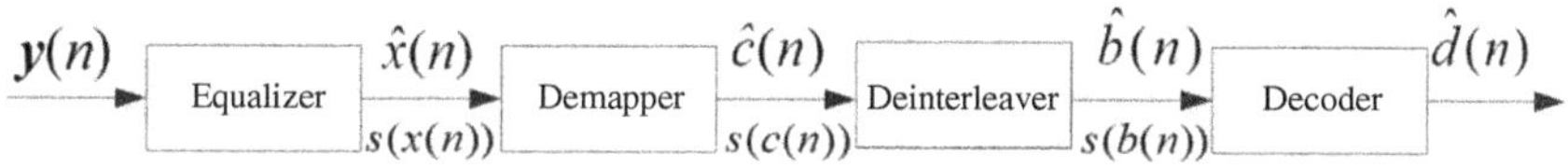

Figure 6.9 Receiver structure.

the receiver is generally designed under additive noise, it also has a good inhibitory effect on the symbol correlation caused by ISI. Generally, the process of reducing the influence of the ISI between transmission data is referred to as "Equalization" and "Detection", while the process of recovering data from the equalized symbols is referred to as "Decoding". The receiver structure is shown in Figure 6.9.

In the receiver, $\hat{x}(n)$, $\hat{c}(n)$, and $\hat{b}(n)$ are sequentially obtained from the equalizer to the decoder as estimates of $x(n)$, $c(n)$, and $b(n)$, respectively, or the probability information of $x(n)$, $c(n)$, and $b(n)$ is obtained by calculating the reliability value or the soft information $s(x(n))$, $s(c(n))$, and $s(b(n))$.

6.3.2 *Nonlinear channel Volterra–Turbo equalization algorithm based on linear MMSE*

According to the minimum mean square error (MMSE) criterion, let

$$\boldsymbol{y}(n) = [y(n - N_1), \ldots, y(b + N_2)]^T \tag{6.3.3}$$

$$\boldsymbol{x}(n) = [x(n - N_1 - L_{01}), \ldots, x(n + N_2 + L_{02})]^T \tag{6.3.4}$$

$$\boldsymbol{w}(n) = [w(n - N_1), \ldots, w(n + N_2)]^T \tag{6.3.5}$$

where $\boldsymbol{x}(n)$ is the transmitted signal vector obtained by channel coding and the interleaving mapping of data bit $a(n)$, $\boldsymbol{y}(n)$ is the equalizer receiving signal vector, and $\boldsymbol{w}(n)$ is the additive white Gaussian noise (AWGN) vector. The relationship between these variables is given as:

$$\boldsymbol{y}(n) = \boldsymbol{H}\boldsymbol{x}(n) + \sum_i \sum_j \sum_k \boldsymbol{H}_{ijk}\boldsymbol{x}^{ijk}(n) + \boldsymbol{w}(n) \tag{6.3.6}$$

where $\boldsymbol{H} = \text{Toeplitz}([h(L_{01}), \ldots, h(0), \ldots, h(-L_{02}), \boldsymbol{0}_{1\times N_1+N_2}])$ is a linear convolution matrix with $[h(L_{01}), \ldots, h(0), \ldots, h(-L_{02}),$

$\boldsymbol{0}_{1\times N_1+N_2}]$ as the first row, and $\boldsymbol{0}_{1\times N_1+N_2}$ is a zero vector. $\boldsymbol{H}_{ijk} = h_{i,j,k}\boldsymbol{I}_N$ is a nonlinear ISI matrix, where $N = N_1 + N_2 + 1$, $\boldsymbol{I}_N$ is $N \times N$ unit matrix; $\boldsymbol{x}^{ijk}(n)$ is a third-order ISI term, and

$$\boldsymbol{x}^{ijk}(n) = \begin{bmatrix} x(n-N_1-i)x(n-N_1-j)x^*(n-N_1-k) \\ \vdots \\ x(n+N_2-i)x(n+N_2-j)x^*(n+N_2-k) \end{bmatrix} \tag{6.3.7}$$

6.3.2.1 *Exact MMSE-based equalization algorithm*

The output signal of the MMSE equalizer is written as:

$$\hat{x}(n) = \boldsymbol{a}^H(n)\boldsymbol{y}(n) + b(n) \tag{6.3.8}$$

The MMSE $E[|\hat{x}(n) - x(n)|^2]$ of the signal $x(n)$ can be calculated from Equation (6.3.8). The time-varying MMSE coefficient is computed as follows:

$$\boldsymbol{a}(n) = \mathrm{Cov}(\boldsymbol{y}(n), \boldsymbol{y}(n))^{-1}\mathrm{Cov}(\boldsymbol{y}(n), \boldsymbol{x}(n)) \tag{6.3.9}$$

$$b(n) = E[x(n)] - \boldsymbol{a}^H(n)E[\boldsymbol{y}(n)] \tag{6.3.10}$$

According to $\mathrm{Cov}(\boldsymbol{x}, \boldsymbol{y}) \triangleq E[\boldsymbol{x}\mathbf{y}^H] - E[\boldsymbol{x}]E[\boldsymbol{y}^H]$, get

$$\hat{x}(n) = \boldsymbol{a}^H(n)(\boldsymbol{y}(n) - E[\boldsymbol{y}(n)]) + E[x(n)] \tag{6.3.11}$$

where

$$E[\boldsymbol{y}(n)] = \boldsymbol{H}E\left[\boldsymbol{x}(n) + \sum_i \sum_j \sum_k \boldsymbol{H}_{ijk}E[\boldsymbol{x}^{ijk}(n)]\right] \tag{6.3.12}$$

The structure of the soft interference canceler is analyzed according to Equation (6.3.11) and Figure 6.10, whether linear and nonlinear ISI terms are removed in $E[\boldsymbol{y}(n)]$ is considered. In order not to cancel the current symbol to be estimated, the *a priori* log-likelihood ratio (LLR) (L_a) for the nth symbol should not be considered in terms of linear ISI, only for a linear ISI term, $E[x(n)] = 0$ and $\mathrm{Var}(x(n)) \triangleq v_n = 1$.

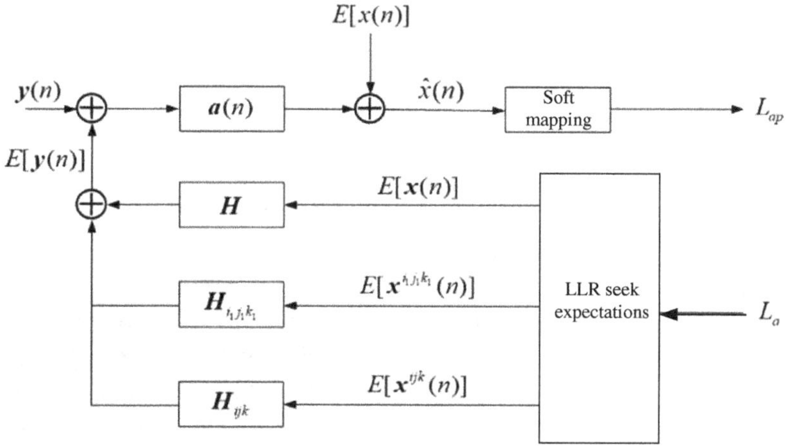

Figure 6.10 MMSE equalizer model.

For the MMSE equalizer, the terms in Equations (6.3.10) and (6.3.11) are:

$$\begin{aligned}\mathrm{Cov}(\boldsymbol{y}(n),\boldsymbol{y}(n)) &= \sigma_w^2\boldsymbol{I}_N + \boldsymbol{H}\mathrm{Cov}(\boldsymbol{x}(n),\boldsymbol{x}(n))\boldsymbol{H}^H \\ &\quad + (1-v(n))\boldsymbol{h}(n)\boldsymbol{h}^H(n) \\ &\quad + \sum_{i,j,k}\boldsymbol{H}_{ijk}\mathrm{Cov}(\boldsymbol{x}^{ijk}(n),\boldsymbol{x}(n))\boldsymbol{H}^H \\ &\quad + \sum_{i.j.k}\boldsymbol{H}\mathrm{Cov}(\boldsymbol{x}(n),\boldsymbol{x}^{ijk}(n))\boldsymbol{H}_{ijk}^H \\ &\quad + \sum_{i,j,k}\sum_{i_1,j_1,k_1}\boldsymbol{H}_{ijk}\mathrm{Cov}(\boldsymbol{x}^{ijk}(n),\boldsymbol{x}^{i_1j_1k_1}(n))\boldsymbol{H}_{i_1j_1k_1}^H\end{aligned} \tag{6.3.13}$$

$$\begin{aligned}\mathrm{Cov}(\boldsymbol{y}(n),x(n)) &= \boldsymbol{H}\mathrm{Cov}(\boldsymbol{x}(n),x(n)) + (1-v(n))\boldsymbol{h}(n) \\ &\quad + \sum_{i,j,k}\boldsymbol{H}_{ijk}\mathrm{Cov}(\boldsymbol{x}^{ijk}(n),x(n))\end{aligned} \tag{6.3.14}$$

$$E[\boldsymbol{y}(n)] = \boldsymbol{H}E[\boldsymbol{x}(n)] - \boldsymbol{h}(n)E[\boldsymbol{x}(n)] + \sum_{i,j,k}\boldsymbol{H}_{ijk}E[\boldsymbol{x}^{ijk}(n)] \tag{6.3.15}$$

where $\boldsymbol{h}(n) = \boldsymbol{H}\times[\boldsymbol{0}_{1\times N_1+L_{01}}, \boldsymbol{1}, \boldsymbol{0}_{1\times N_2+L_{02}}]^T$.

The following equations can be obtained by Equations (6.3.13) and (6.3.14).

$$\hat{x}(n) = \boldsymbol{a}^H(n)(\boldsymbol{y}(n) - E[\boldsymbol{y}(n)]) + E[\boldsymbol{x}(n)] \tag{6.3.16}$$

$$E[\boldsymbol{y}(n)] = \boldsymbol{H}E[\boldsymbol{x}(n)] + \sum_i \sum_j \sum_k \boldsymbol{H}_{ijk}E[\boldsymbol{x}^{ijk}(n)] \tag{6.3.17}$$

According to the above analyses, the linear and nonlinear ISI terms in $E[\boldsymbol{y}(n)]$ are eliminated.

Since there is an interleaver between the decoder and the equalizer, whose main function is to reset the bit order of the input signal sequence so that the sequence correlation before and after the interleaving is zero, so that the input characters are independent of each other. Therefore, the mean value of the conjugate product between p characters and q–p characters is obtained as follows:

$$\begin{aligned} &E[x(n-i_1)x(n-i_2)\cdots x(n-i_p)x^*(n-i_p+1)\cdots x^*(n-i_p+q)] \\ &= \prod_j E[x^{v(j)}(n-i_j)x^{*v^*(j)}(n-i_j)] \\ &= \prod_j \sum_{m=1}^{M} s^{v(j)}(m)s^{*v^*(j)}(m)P[x(n-i_j)=s(m)] \end{aligned} \tag{6.3.18}$$

where $v(i)$(or $v^*(i)$) indicates the number of occurrences of the character $x(n-i)$(or $x^*(n-i)$). $s(m)$ is the mth constellation symbol ($m \in [1,\ldots,M]$) with a probability

$$P(x(n) = x(m)) = \prod_{i=1}^{\log_2(M)} P(c_i(n) = s_i(m)) \tag{6.3.19}$$

where $c_i(n)$ denotes the ith coded bit of the character $x(n)$, and $s_i(m)$ denotes the ith coded bit of the constellation symbol $s(m)$.

The coded bit probability calculated by the input LLRsL_a from the decoder is computed as follows:

$$L_a(c_i(n)) = \ln\frac{p(c_i(n)=0)}{p(c_i(n)=1)} = \ln\frac{p(c_i(n)=0)}{1-p(c_i(n)=0)} \tag{6.3.20}$$

Equation (6.3.21) is used at each step of the equalization in the calculation process, and matrix transposition also contains an amount of computational complexity. However, since there is a common submatrix in the continuous instantaneous covariance matrix in Equation (6.3.21), the computation complexity can be significantly reduced by the fast recursion method.

6.3.2.2 *Time-invariant MMSE coefficient*

The calculation of the exact MMSE equalizer can be approximated by the computation load of computing the minimum RMS coefficient, or the computation load of the time-invariant equalizer can be determined without prior information [17]. This section proposes two time-invariant approximation algorithms. The main difference between them is reflected in the MMSE coefficients.

6.3.2.2.1 MMSE approximation algorithm without prior information

At the beginning of equalizing, it is assumed that there is no *a priori* information to calculate the MMSE coefficients. In this case, for any n, $E[x(n)] = 0, v(n) = 1$. However, since the nonlinear ISI term is related to the modulation order, the channel coefficient, the modulation method (PSK, QAM, APSK, ...), etc., there is no the expression of the simplified MMSE coefficient.

When the symbols are extracted from the constant modulus modulation, the third-order kernel (i, j, k) $(i \neq k, j \neq k)$ of the channel is present, otherwise it contributes to the linear ISI portion. Therefore, this can be reflected in the modulated signals of the second order or more. According to this conclusion, for $\forall n, \forall(i, j, k)$, there are:

$$\forall n, \forall(i,j,k) \quad E[\boldsymbol{x}^{ijk}(n)] = \boldsymbol{O}_{N\times 1} \tag{6.3.21}$$

$$\forall n, \forall(i,j,k) \quad \mathrm{Cov}(\boldsymbol{x}(n), \boldsymbol{x}^{ijk}(n)) = \boldsymbol{O}_{N+L_{01}+L_{02}\times N} \tag{6.3.22}$$

$$\forall n, \forall(i,j,k) \quad \mathrm{Cov}(\boldsymbol{x}^{ijk}(n), \boldsymbol{x}(n)) = \boldsymbol{O}_{N\times N+L_{01}+L_{02}} \tag{6.3.23}$$

$$\forall n, \forall(i,j,k) \neq (i',j',k') \quad \mathrm{Cov}(\boldsymbol{x}^{ijk}(n), \boldsymbol{x}^{i'j'k'}(n)) = \boldsymbol{O}_N \tag{6.3.24}$$

$$\forall n, \forall(i,j,k) \quad \mathrm{Cov}(\boldsymbol{x}^{ijk}(n), \boldsymbol{x}^{ijk}(n)) = \boldsymbol{I}_N \tag{6.3.25}$$

These simplify Equation (6.3.10). Now, the NA-MMSE coefficient can be written as:

$$\boldsymbol{a}_{NA} = (\sigma_w^2 \boldsymbol{I}_N + \boldsymbol{H}\boldsymbol{H}^{\mathrm{H}} + \sum_{ijk} |h_{ijk}|^2 \boldsymbol{I}_N)^{-1} \boldsymbol{h}(n) \tag{6.3.26}$$

6.3.2.2.2 MMSE approximation algorithm for Low Complexity

On the basis of no *a priori* information, the idea of the time-invariant MMSE approximation with the LC-MMSE is to compute the MMSE filter with the minimum $\frac{1}{L}\sum_{n=0}^{L-1} E[|\hat{x}(n)-x(n)|^2]$.

According to this idea, the optimal solution is given by:

$$\boldsymbol{a}_{LC} = \left(\frac{1}{L}\sum_{n=0}^{L-1}\mathrm{Cov}(y(n), y(n))\right)^{-1}\left(\frac{1}{L}\sum_{n=0}^{L-1}\mathrm{Cov}(y(n), \boldsymbol{x}(n))\right) \tag{6.3.27}$$

where $\frac{1}{L}\sum_{n=0}^{L-1}\mathrm{Cov}(y(n), y(n))$ is a diagonal matrix and can be further simplified by assuming that the secondary effect is negligible, which results in $\frac{1}{L}\sum_{n=0}^{L-1}\mathrm{Cov}(y(n), y(n)) = vI_N$, $v = \frac{1}{L}\sum_{n=0}^{L-1}\mathrm{Var}(x(n))$.

As far as the third-order covariance matrix is concerned, it can only show a Toeplitz structure [18]. Since only one matrix inversion operation under every L equalized symbol is required, the computational complexity is slightly reduced. Therefore, compared with the NA-MMSE algorithm, the computation loads of the LC-MMSE are only $1/L$ of that of the NA-MMSE algorithm.

6.3.2.2.3 Soft demapper

To map the output of the equalizer to code LLRs, the residual output error of the equalizer is defined as $e(n) = \hat{x}(n) - x(n)$. Assume that the residual error $e(n)$ is a Gaussian type with zero mean, its variance is written as:

$$\mathrm{Var}(e(n)) = 1 + \mathrm{Cov}(\hat{x}(n), \hat{x}(n)) - 2\mathrm{Re}(\mathrm{Cov}(\hat{x}(n), x(n)))$$

For the exact MMSE algorithm, its variance can be written as:

$$\mathrm{Var}(e(n)) = 1 - \mathrm{Cov}(x(n), y(n))a(n)$$

For the NA-MMSE algorithm, the computational complexity of its error variance is greater than that of the exact MMSE algorithm, and it may be called as NA-MMSE-A. In literature [19], the role of the third-order Volterra term is ignored, and a simplified NA-MMSE error variance method, called as NA-MMSE-B, is given. The computational complexity of the NA-MMSE-B is lower than that of

NA-MMSE-A, but the nonlinear term is not eliminated. When the SNR is very high, the NA-MMSE-B will no longer be accurate.

The equalizer output $\hat{x}(n) \sim N(x(n), \mathrm{Var}(e(n)))$ and the output *a posteriori* LLR L_{ap} can be written as:

$$\begin{aligned} &L_{ap}(c_i(n)|\hat{x}(n)) \\ &= \ln \frac{P(c_i(n)=0|\hat{x}(n))}{P(c_i(n)=1|\hat{x}(n))} \\ &= \ln \frac{\sum_{s(j):s_i(j)=0}\left(\exp\left(\frac{-|\hat{x}(n)-s(j)|^2}{\mathrm{Var}(e(n))}\right)\right)\prod_k P(c_k(n)=s_k(j))}{\sum_{s(j):s_i(j)=1}\left(\exp\left(\frac{-|\hat{x}(n)-s(j)|^2}{\mathrm{Var}(e(n))}\right)\right)\prod_k P(c_k(n)=s_j(j))} \end{aligned} \tag{6.3.28}$$

The *a posteriori* LLRs are made extrinsic L_e by subtracting the prior probability fed by the decoder.

$$L_e(c_i(n)) = L_{ap}(c_i(n)|\hat{x}(n)) - L(c_i(n)) \tag{6.3.29}$$

6.3.2.3 *Algorithm simulation*

In the simulation, the convolutional code of the generated polynomial is used, the generated polynomial is [5, 7], code rate is 1/2, the modulation signal is 8PSK, and the coding bits are randomly interleaved and mapped. At the receiver, the MMSE equalizer parameters are $N_1 = 5$ and $N_2 = 3$. The output of the soft demapper is de-interleaved and then forwarded to the maximum *a posteriori* probability (MAP) decoder. The output of the decoder is interleaved again and transmitted to the equalizer. The channel is a typical nonlinear Volterra satellite channel, and its parameter expression is shown in Equation (6.3.31), and the simulation results are shown in Figure 6.11.

$$\begin{aligned} y(n) = w(n) &+ \sum_{i=o}^{L_0} h(i)x(n{-}\mathrm{i}) + h(0,0,2)x^2(n)x^*(n-2) \\ &+ h(3,3,0)x^2(n-3)x^*(n) + h(0,0,1)x^2(n)x^*(n-1) \\ &+ h(0,0,3)x^2(n)x^*(n-3) + h(1,1,0)x^2(n-1)x^*(n) \end{aligned} \tag{6.3.30}$$

The channel parameters are shown in Table 6.3.

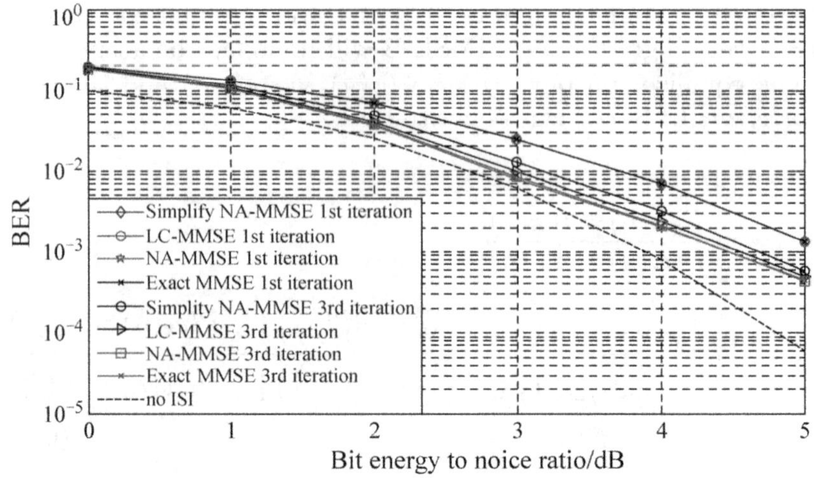

Figure 6.11 Simulation curves of bit error rate.

Table 6.3 Channel Volterra kernel.

First-order Kernel	Third-order Kernel
$h(0) = 0.8529 + j0.4502$	$h(0,0,1) = 0.0979 - j0.0979$
$h(1) = 0.0881 - j0.0014$	$h(0,0,2) = 0.1091 - j0.0615$
$h(2) = -0.0336 - j0.0196$	$h(0,0,3) = -0.1119 - j0.0252$
$h(3) = 0.0503 + j0.0433$	$h(1,1,0) = -0.028 - j0.0475$
	$h(3,3,0) = 0.0503 - j0.0503$

Figure 6.11 shows the bit error rate (BER) curves of the four equalizers include accurate MMSE linear equalizer (exact MMSE), MMSE equalizer with no *a priori* information for accurate error variance (NA-MMSE-A), MMSE equalizer with simplified error variance without *a priori* information (NA-MMSE-B), and low complexity MMSE equalizer (LC-MMSE). It can be seen from Figure 6.11 that when iterating once, there is no obvious difference in the performance of the four equalizers, but when the number of iterations increases to three times, their BERs (bit error rates) are significantly reduced, which is more and more close to the performance without ISI channel, and among them, the BER of the accurate MMSE is the best. The BER performance of the approximation algorithms

such as NA-MMSE and LC-MMSE is very close to that of the accurate MMSE, whereas the BER performance of the accurate MMSE, NA-MMSE, and LC-MMSE is significantly better than that of the simplified NA-MMSE. If the iteration continues, their performance can be further improved, but in terms of the time required by the iteration and performance improvement, it is unnecessary to continue to iterate.

6.3.3 *Iterative blind equalization algorithm based on linear MMSE*

This section mainly discusses a blind iterative equalizer, which consists of a soft feedback equalizer with linear complexity and a MMSE channel estimator with linear complexity [20]. The inherent block processing structure of turbo receiver and its careful selection of filter parameters are used to overcome the shortcomings of low bandwidth utilization and numerical instability for blind equalization algorithm based on linear MMSE.

6.3.3.1 *Iterative blind equalization system model*

The structure of the iterative blind equalization receiver is shown in Figure 6.12. A binary data with length K is denoted as $\boldsymbol{a} = [a(1), a(2), \ldots, a(K)]$. It is encoded by a convolutional code to obtain

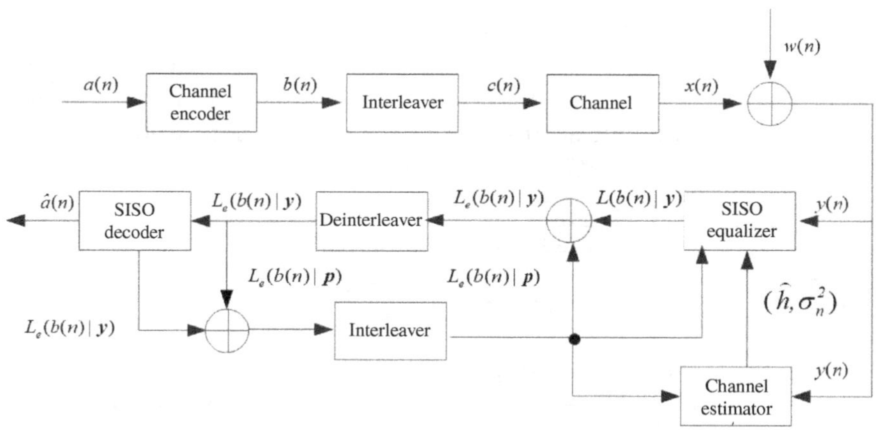

Figure 6.12 Structure of blind iterative equalization receiver.

$\boldsymbol{b} = [b(1), b(2), \ldots, b(N)]$, and the code rate is $R = K/N$. In order to improve performance, $\boldsymbol{c} = [c(1), c(2), \ldots, c(N)]$ is obtained by using a random interleaver.

The channel model can be composed of a transmit filter, a channel, and a receive filter. The output of the equivalent discrete-time channel model can be expressed as:

$$\boldsymbol{y}(n) = \sum_{l=0}^{L} h(l)x(n-l) + \boldsymbol{w}(n) = \boldsymbol{h}^T \boldsymbol{x}(n) + \boldsymbol{w}(n) \quad 0 \le k \le N \tag{6.3.31}$$

where L is the channel memory length, $\boldsymbol{h} = [h(1), h(2), \ldots, h(N)]^T$ is a channel coefficient vector, $\boldsymbol{x}(n) = [x(n), x(n-1), \ldots, x(n-L)]^T$ represents the state transition of the kth mesh segmentation [19], and $\boldsymbol{w}(n)$ is the noise sample vector.

6.3.3.2 *SISO equalizer*

The tapped delay line model of the ISI channel and the lattice response of the channel are shown in Figures 6.13 and 6.14, respectively. Assume that the length of the impulse response is L, the model contains $L-1$ delay units. Thus, for a given binary input character set $\{+1, -1\}$, the channel has 2^{L-1} states $s(i)(i = 0, 1, \ldots, 2^{L-1}-1)$, and a collection of possible states may be represented as $\boldsymbol{S} = \{s(0), s(1), \ldots, s(2^{L-1}-1)\}$. At every moment $n = 0, 1, 2, \ldots, N-1$, the channel state is a random variable $s(n) \in S$. Given a state $s(n), s(n+1)$, according to one of only two possible states of input $+1$ and -1, if there are only eight branches in this section, the channel can be any of the four states, so the state transition probability can

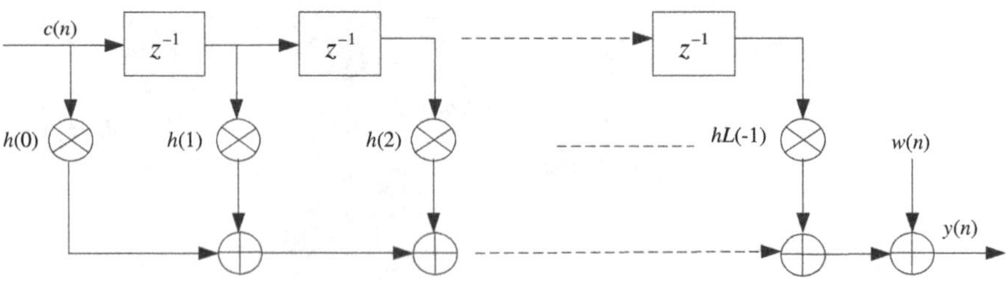

Figure 6.13 ISI channel tap delay line model.

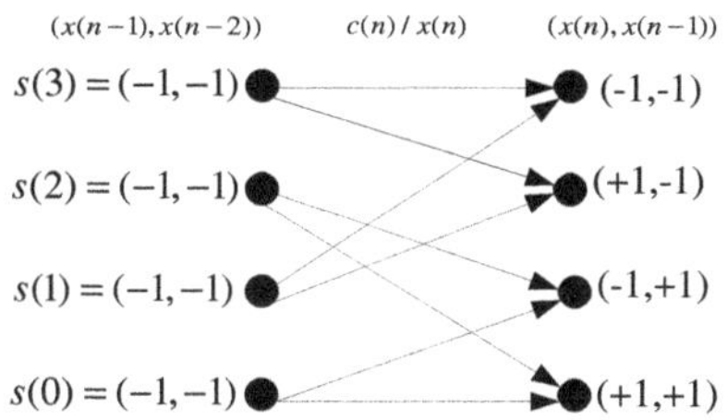

Figure 6.14 Lattice response map of the channel.

be written as:

$$p(s(n-1), s(n)) = P_r(c(n) = c)P(y(n)|x(n)), \quad c \in [0, 1] \quad (6.3.32)$$

Based on (6.3.30) and the noise distribution, we have

$$f(y(n)|x(n)) = \frac{1}{\sqrt{2\pi\sigma_w^2}} e^{\frac{(y(n)-x(n))^2}{2\sigma_w^2}} \quad (6.3.33)$$

The maximum LLR is defined as:

$$L(c(n)) = \ln(Pr(c(n) = 0)/Pr(c(n) = 1))$$

Thus,

$$Pr(c(n) = c) = \frac{e^{-cL(c(n))}}{1 + e^{-L(c(n))}}, \quad c \in \{0, 1\} \quad (6.3.34)$$

Therefore, Equation (6.3.32) can be written as:

$$p(s(n-1), s(n)) = e^{-c \cdot L(c(n))} \cdot e^{\frac{(y(n)-x(n))^2}{2\alpha_w^2}} \quad (6.3.35)$$

The forward feedback term $\alpha(s)$ and the backward feedback term $\beta(s)$ can be recursively calculated, i.e.,

$$\alpha(s(n)) = \sum_{\forall s(n) \in s} p(n)(s(n-1), s(n))\alpha(s(n-1)) \quad (6.3.36)$$

$$\beta(s(n)) = \sum_{\forall s(n) \in s} p(s(n), s(n+1))\beta(s(n+1)) \quad (6.3.37)$$

For all $s(n) \in S$, the initial value $\alpha(0) = p(s_0 = s)$ represents the distribution of the state at time $k = 0$, the initial value $\beta(N) = 1/2^{L-1}$.

Based on mapping operations, the condition LLR of code bits $c(n)$ can be written as:

$$L(c(n)|y) = \ln \frac{\Pr(c(n)=0|y)}{\Pr(c(n)=1|y)}$$
$$= \ln \frac{\sum_{\forall s(n)\in s, c(n)=0} \alpha(s(n-1))p(s(n-1),s(n))\beta(s(n))}{\sum_{\forall s(n)\in s, c(n)=1} \alpha(s(n-1))p(s(n-1),s(n))\beta(s(n))} \tag{6.3.38}$$

6.3.3.3 *SISO decoder*

The grid representation of the convolutional code used in the ISI decoder process is shown in Figure 6.15. State transition probability is denoted as:

$$p(s(n-1),s(n)) = P_r(a(n)=a)\cdot P_r(b_2(n-1)|y)\cdot P_r(b_2(n)|y) \tag{6.3.39}$$

where $a \in \{0,1\}$. Assume that the data bits are subject to independent and identical distribution, then $P_r(a(n)=a) = 1/2$. According to Equation (6.3.31), we have

$$P_r(b(n)=b) = \frac{e^{-bL(b(n))}}{1+e^{-L(b(n))}}, \quad b\in\{0,1\} \tag{6.3.40}$$

The conditional LLRs can be calculated according to the following equations.

$$L(a(n)|\boldsymbol{p}) = \ln \frac{\sum_{\forall s(n)\in s, a(n)=0} \alpha(s(n-1))p(s(n-1),s(n))\beta(s(n))}{\sum_{\forall s(n)\in s, a(n)=1} \alpha(s(n-1))p(s(n-1),s(n))\beta(s(n))} \tag{6.3.41}$$

$$L(b_2(n-1)|\boldsymbol{p}) = \ln \frac{\sum_{\forall s(n)\in s, b_2(n-1)=0} \alpha(s(n-1))p(s(n-1),s(n))\beta(s(n))}{\sum_{\forall s(n)\in s, b_2(n)=1} \alpha(s(n-1))p(s(n-1),s(n))\beta(s(n))} \tag{6.3.42}$$

$$L(b_2(n)|p) = \ln \frac{\sum_{\forall s(n)\in s, b_2(n)=0} \alpha(s(n-1))p(s(n-1),s(n))\beta(s(n))}{\sum_{\forall s(n)\in s, b_2(n)=1} \alpha(s(n-1))p(s(n-1),s(n))\beta(s(n))} \tag{6.3.43}$$

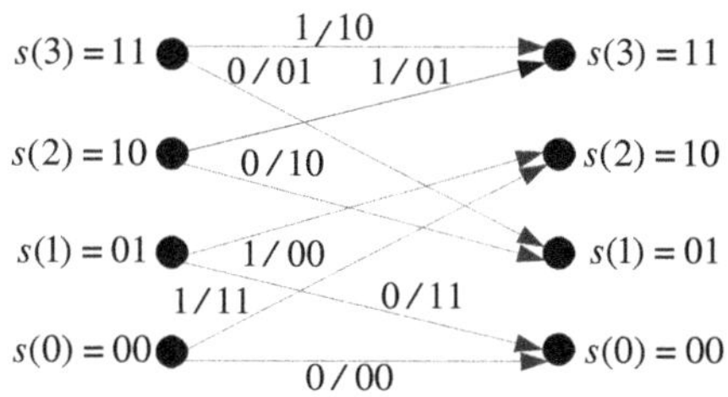

Figure 6.15 Half rate four state transition diagram.

where $\boldsymbol{p} = [P_r(b_1|\boldsymbol{y}), P_r(b_2|\boldsymbol{y}), \ldots, P_r(b_N|\boldsymbol{y})]$ is the probability input vector of the decoder. The formula for estimating $\hat{a}$(n) from the original data is:

$$\hat{a}(n) = \begin{cases} 0, & L(a(n)|\boldsymbol{p}) = \ln \dfrac{P_r(a(n) = 0|p)}{P_r(a(n) = 1|\boldsymbol{p})} \geq 0 \\ 1, & L(a(n)|\boldsymbol{p}) = \ln \dfrac{P_r(a(n) = 0|p)}{P_r(a(n) = 1|\boldsymbol{p})} < 0 \end{cases} \tag{6.3.44}$$

6.3.3.4 *Algorithm simulation*

In simulation tests, the impulse response of the Proakis' B channel $h(n) = 0.408\sigma(n{+}1){+}0.815\sigma(n){+}0.408\sigma(n{-}1)$ was selected [21]. The coding method adopts the R = 1/2 system recursive convolutional code, the constraint length was set to 3, the generated vector was $\boldsymbol{g} = [111, 110] = [7, 5]$, random interleaving was adopted, the size of information bit block was 128bit, and BPSK modulation ($M = 256$) was used, the SNR was defined as $\mathrm{SNR} = 10\log(\mathrm{E}/\mathrm{N_0R}) = 10\log(1/2\sigma^2\mathrm{R})$dB, and the number of iterations was set to 3. The simulation results are shown in Figure 6.16.

Figure 6.16 shows that as the number of iterations increases, the performance of iterative blind equalization algorithm based on linear MMSE gets better and better. The initial iterative coding gain is higher, but as the number of iterations increases, the gain growth will be relatively slow, and the BER of the system will change slightly on an order of magnitude and get closer and closer to the performance of the AWGN.

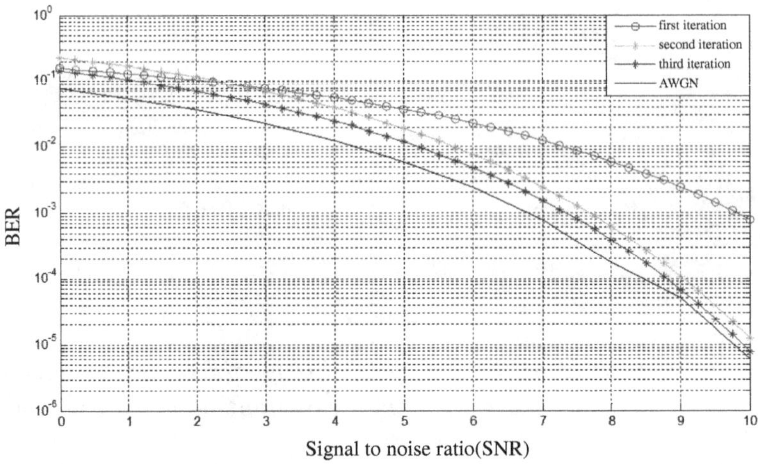

Figure 6.16 BER curves of iterative blind equalization algorithm based on linear MMSE.

6.4 Linear Frequency Domain Turbo Equalization Algorithm Based on Nonlinear Volterra Channel

The increasing demand for high data rates urges new communication systems to provide both high spectral and power efficiency. In recent satellite communications standards, this trade-off can be achieved using multi-level amplitude constellations such as amplitude and phase-shift keying (APSK). However, such modulations increase the signal fluctuations and thus give rise to nonlinear distortions when being passed through a high power amplifier (HPA) operating near saturation. These distortions can be either mitigated at the transmitter using signal or symbol pre-distortion or at the receiver using equalization. To do so, analytical expressions for the nonlinear channel have been widely investigated [22–24]. [22, 23] use a triple-effect representation of the nonlinear channel, consisting of additive noise, a scalar multiplication, and warping. This approximation, although practical, is not accurate for higher-order modulations since the additive noise cannot be considered circular for outer ring symbols. Another category of channel models consists of a decomposition of the nonlinear channel by means of Taylor series leading to the so-called Volterra representation. The Volterra series representation has been applied to different transmission channels such as the nonlinear magnetic channel and differential impulse radio UWB systems in [25, 26].

As far as satellite channels are concerned, a Volterra series expansion or a decomposition depends on the chain filters, the HPA model, the modulation type, and the input back-off (IBO). Based on this Volterra series expansion, the satellite channel can be represented as a finite state machine, which allows for optimal symbol and sequence detection, maximum *a posteriori* probability (MAP) and maximum likelihood sequence detection (MLSD). However, as the complexity of these optimal equalizers is exponential in the channel memory, the low complexity equalizers have been investigated among which we will focus on iterative solutions. Indeed, new equalization trends are based on iterative receivers for their excellent Shannon-bound approaching abilities [27–29]. In [29], an iterative time domain MMSE-based equalizer was presented to mitigate the channel linear and nonlinear interference, and different low complexity approximations were investigated. The investigated performances show that a No *a priori* approximation has minor performance degradation but far less complexity than the exact MMSE solution. Non-iterative frequency domain equalization algorithm for nonlinear channels [30] based on an adaptive block LMS equalizer can reduce the computational complexity further. Obviously, a complexity comparison shows the advantage of using frequency rather than time domain equalization. In this section, we use the frequency domain Volterra channel model to derive an iterative linear MMSE FDE. As argued for the time domain solution in [29], time domain linear equalization is able to cope with both linear and nonlinear interference. Similarly, the proposed iterative FDE is able to cancel both linear and nonlinear soft symbols which asymptotically approaches the AWGN ISI-free performance. The remainder of this section is organized as follows: first, the frequency domain Volterra channel model is proposed; second, the iterative linear MMSE-FDE based on the No A priori MMSE approximation is derived; finally, the performance of the MMSE equalizer implementations is investigated.

6.4.1 *Available symbols in the loop model*

Let us introduce some useful notations and results. First, II_m is used to represent an identity matrix of size m and $\boldsymbol{1}_{1\times m}$ is an all-in-one vector of $1 \times m$. $(\cdot)_m$ refers to the modulo-m operator. Let $\boldsymbol{F}$ be the normalized one-dimensional discrete Fourier transform (1D – DFT) matrix of size N. The matrix element $F_{i,j}$ is given as:

$$F_{i,j} = \frac{1}{\sqrt{N}} W_N^{-ij} \tag{6.4.1}$$

where $W_N^k = e^{\frac{j2\pi k}{N}}$ and $\boldsymbol{F}^H = \boldsymbol{F}^{-1}$. Let $\boldsymbol{H}$ and $\boldsymbol{G}$ be two circulant matrix of size $N \times N$. A circular matrix can be diagonalized using the DFT matrix $\boldsymbol{F}$, more precisely:

$$\boldsymbol{H} = \boldsymbol{F}^H \boldsymbol{H}_d \boldsymbol{F}, \quad \boldsymbol{G} = \boldsymbol{F}^H \boldsymbol{G}_d \boldsymbol{F} \tag{6.4.2}$$

where $\boldsymbol{H}_d$ and $\boldsymbol{G}_d$ represent diagonal matrix containing the eigenvalues of $\boldsymbol{H}$ and $\boldsymbol{G}$, respectively. $\boldsymbol{H}$ and $\boldsymbol{G}$ satisfy the characteristics as follows:

(1) $\boldsymbol{HG}$ is a circular matrix and $\boldsymbol{HG} = \boldsymbol{F}^H \boldsymbol{H}_d \boldsymbol{G}_d \boldsymbol{F}$
(2) $\boldsymbol{H}^H$ is a circular matrix and $\boldsymbol{H}^H = \boldsymbol{F}^H \boldsymbol{H}_d^H \boldsymbol{F}$
(3) $\boldsymbol{H}^{-1}$ is a circular matrix and $\boldsymbol{H}^{-1} = \boldsymbol{F}^H \boldsymbol{H}_d^{-1} \boldsymbol{F}$
(4) $\forall (n, m) \in \{0, \ldots, N-1\}^2, \boldsymbol{u}_m^T \boldsymbol{H} \boldsymbol{u}_m = \boldsymbol{u}_n^T \boldsymbol{H} \boldsymbol{u}_n$

where $\boldsymbol{u}_m = [0_m 1\ 0\ 0_{N-m-1}]^T$.

For $m, n, l \in \{0, \ldots, N-1\}$, the $N-3$D normalized DFT of the three-dimensional symbols $y^{(3)}(m, n, l)$ is written as:

$$Y^{(3)}(p, q, r) \triangleq \frac{1}{\sqrt{N}^3} \sum_{m=0}^{N-1} \sum_{n=0}^{N-1} \sum_{l=0}^{N-1} y^{(3)}(p, q, r) W_N^{-mp} W_N^{-np} W_N^{-lr} \tag{6.4.3}$$

Similarly, the N-3D normalized DFT of 3D symbols $Y_{p,q,r}^{(3)}$ for $m, n, l \in \{0, \ldots, N-1\}$ can be expressed as:

$$y^{(3)}(p, q, r) \triangleq \frac{1}{\sqrt{N}^3} \sum_{p=0}^{N-1} \sum_{q=0}^{N-1} \sum_{r=0}^{N-1} Y^{(3)}(p, q, r) W_N^{mp} W_N^{np} W_N^{lr} \tag{6.4.4}$$

6.4.2 *Frequency domain nonlinear Volterra channel model*

A satellite high power amplifier (HPA) is a memoryless device and can be described by two frequency independent functions

$A(\cdot)(AM/AM)$ and $\Phi(\cdot)(AM/PM)$ [32]. The relationship between both the power and phase of the output signal $\boldsymbol{y}$ and the amplitude of the input signal $\boldsymbol{x}$ is as follows:

$$y = c(|x|)\exp(j\varphi(x)) = A(|x|)\exp(j(\Phi(|x|) + \varphi(x))) \qquad (6.4.5)$$

The memory function is added to the receiving response transmitting filter. In the whole satellite transponder response, the symbols received in the receiver are decomposed into Volterra odd series. In this way, the decomposition expression of the received symbol $y(n)$ is represented by the Volterra odd series as:

$$\begin{aligned}\tilde{y}(n) = \sum_{m=0}^{v} \sum_{n_1=-\infty}^{\infty} \cdots \sum_{n_{2m+1}}^{\infty} & \tilde{x}(n-n_1)\cdots \\ & \tilde{x}(n-n_{m+1})\tilde{x}^*(n-n_{m+2})\cdots \\ & \tilde{x}^*(n-n_{2m+1})h(n_1,\ldots,n_{2m+1}) + w(n) \qquad (6.4.6)\end{aligned}$$

where v defines the decomposition order of the Volterra series. $h(n_1,\ldots,n_{2m+1})$ is called the Volterra kernel and $w(n)$ is the filtered sample additive noise with variance σ_w^2. The overall system can be described by Figure 6.17. The Volterra decomposition of satellite transponders with Butterworth filters is given in [27], and orders higher than three are ignored. Therefore, this has been selected as

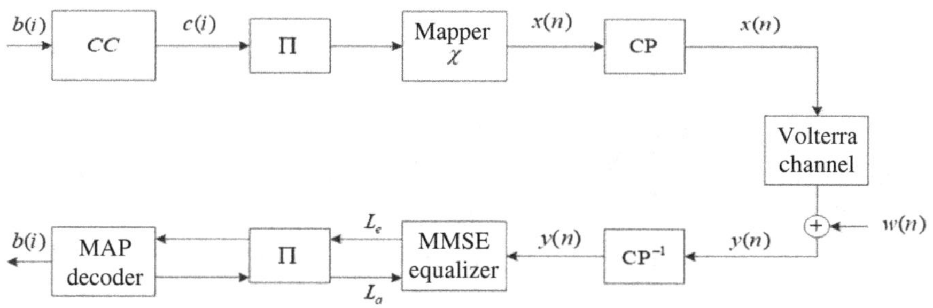

Figure 6.17 System model overview.

the order of Volterra series decomposition in this section.

$$\tilde{y}(n) = \sum_{i=0}^{M-1} h(i)\tilde{x}(n-i) + \sum_{i=0}^{M-1}\sum_{j=0}^{M-1}\sum_{k=0}^{M-1} h(i,j,k)\tilde{x}(n-i)\tilde{x}(n-j)\tilde{x}^*(n-k) + w(n) \tag{6.4.7}$$

where M is the memory length. After removing the cyclic prefix, the received samples in Equation (6.4.7) is written as:

$$\begin{aligned} y(n) &= \sum_{i=0}^{M-1} h(i)x((n-i)M) \\ &\quad + \sum_{i=0}^{M-1}\sum_{j=0}^{M-1}\sum_{k=0}^{M-1} h(i,j,k)x((n-i)M) \\ &\quad \times x((n-j)M)x^*((n-k)M) + w(n) \\ &= y^{(1)}(n) + y^{(3)}(n,n,n) + w(n) \end{aligned} \tag{6.4.8}$$

The right-hand side (RHS) of Equation (6.4.8) consists of a sum of both a circular convolution $y^{(1)}(n)$ and a third-order circular convolution $y^{(3)}(n,n,n)$ can be expressed as follows:

$$y^{(1)}(n) = \sum_{i=0}^{M-1} h(i)x((n-i)M) \tag{6.4.9}$$

$$\begin{aligned} y^{(3)}(m,n,l) &= \sum_{i=0}^{M-1}\sum_{j=0}^{M-1}\sum_{k=0}^{M-1} h(i,j,k) \\ &\quad \times x((m-i)M)x((n-j(M)x^*((l-k)M) \end{aligned} \tag{6.4.10}$$

In the frequency domain, the 1D-DFT of the cyclic convolution in Equation (6.4.9) is converted to an element-wise multiplication of the 1D-DFT linear filter coefficients, and the 1D-DFT of the symbols

$x(n)$ is denoted as follows:

$$\boldsymbol{Y}^{(1)}(m) = \boldsymbol{H}_d(m)\boldsymbol{X}(m) \tag{6.4.11}$$

where $0 \leq m \leq N-1$. For a general value of (m, n, r), it can be proved that the 3D-DFT of $y(m, n, l)$ is:

$$y^{(3)}(p, q, r) = \sqrt{N}^3 H^{(3)}(p, q, r)X(p)X(q)\overline{X}(r) \tag{6.4.12}$$

where $\overline{X}$ is the N-1D-DFT of conjugate symbols $x^*(n)$ and $H^{(3)}(p, q, r)$ is the N-3D-DFT of the third-order Volterra kernels $h(i, j, k)$. Equivalently for $m = n = l$, the time domain symbols $y^{(3)}(n, n, n)$ can be expressed as:

$$\begin{aligned} y^{(3)}(n, n, n) &= \frac{1}{\sqrt{N}^3} \sum_{p=0}^{N-1}\sum_{q=0}^{N-1}\sum_{r=0}^{N-1} Y^{(3)}(n, n, n)W_N^{n(p+q+r)} \\ &= \sum_{\alpha=0}^{N-1} W_N^{n\alpha} \sum_{p=0}^{N-1}\sum_{q=0}^{N-1}\sum_{r=0}^{N-1} H^{(3)}(p, q, r)X(p)X(q) \\ &\quad \times \overline{X}(r)\delta_N(p + q + r - \alpha) \end{aligned}$$

where the delta function module N is defined as:

$$\delta_N(m) = 1 \qquad \text{if}(m)_N = 0 \tag{6.4.13}$$

Equation (6.4.13) shows that $y^{(3)}(n, n, n)$ is a mono-dimensional $N-1\text{D} - \text{IDFT}$ of a combination of $N-3\text{D} - \text{DFT}$ Volterra kernels. Therefore, the nonlinear interference can be expressed as a $1\text{D} - \text{DFT}$ instead of a $3\text{D} - \text{DFT}$. The mth N-1D-DFT output of the received symbols can thus be written as follows:

$$\begin{aligned} Y(m) = H_d(m)X(m) + \sqrt{N} \sum_{p=0}^{N-1}\sum_{q=0}^{N-1}\sum_{r=0}^{N-1} H^{(3)}(p, q, r)X(p)X(q)\overline{X}(r) \\ \delta_N(p + q + r - m) \end{aligned} \tag{6.4.14}$$

It should be noted that triplets (p, q, r) satisfying $\delta_N(p + q + r - m) = 1$ are disjoint for different frequency indexes m.

6.4.3 *Linear frequency domain Volterra–MMSE equalizer*

In this section, the frequency domain MMSE equalizer will be derived. Let us define the following notations:

$$\boldsymbol{y} \triangleq [y(0), \ldots, y(N-1)]^T, \quad \boldsymbol{x} \triangleq [x(0), \ldots, x(N-1)]^T$$
$$\boldsymbol{w} \triangleq [w(0), \ldots, w(N-1)]^T$$

It follows that:

$$\boldsymbol{y} = \boldsymbol{H}\boldsymbol{x} + \sum_i \sum_j \sum_k \boldsymbol{H}_{ijk}\boldsymbol{x}^{ijk} + \boldsymbol{w}$$

where $\boldsymbol{H} \triangleq \text{Circ}([h_0 \boldsymbol{0}_{1\times N-M} h_{M-1} \ldots h_1])$ is the circular convolution matrix having $[h_0 \boldsymbol{0}_{1\times N-M} h_{M-1} \ldots h_1]$ in its first row; $H_{ijk} = h(i,j,k)\text{II}_N$ are the nonlinear ISI matrices and x^{ijk} are the third-order ISI terms:

$$\boldsymbol{x}^{ijk} \triangleq \begin{bmatrix} x((0-i)M)x((0-j)M)x^*((0-k)M) \\ \cdots \\ x((N-1-i)M)x((N-1-j)M)x^*((N-1-k)M) \end{bmatrix}$$

6.4.3.1 *Turbo MMSE FDE*

The linear MMSE equalizer consists of an affine estimation of the received symbols [33].

$$\hat{\boldsymbol{x}}(n) = \boldsymbol{a}^H(n)\boldsymbol{y} + \boldsymbol{b}(n) \tag{6.4.15}$$

which computes estimates that minimize the MSE with the transmitted symbols $E[|\hat{x}(n) - x(n)|^2]$, so the time-varying MMSE coefficients are given as follows:

$$\begin{cases} \boldsymbol{a}(n) = \text{Cov}(\ \boldsymbol{y},\boldsymbol{y})^{-1}\text{Cov}(\boldsymbol{y}, \boldsymbol{x}(n)) \\ b(n) \ = E[x(n)] - \boldsymbol{a}^H(n)E[\boldsymbol{y}] \end{cases} \tag{6.4.16}$$

where $\text{Cov}(\boldsymbol{x}, \boldsymbol{y})E[\boldsymbol{x}\boldsymbol{y}^H] - E[\boldsymbol{x}]E[\boldsymbol{y}^H]$, leading to:

$$\hat{x}(n) = \boldsymbol{a}^H(n)(\boldsymbol{y} - E[\boldsymbol{y}]) + E[x(n)] \tag{6.4.17}$$

In order to eliminate all linear and nonlinear interference, the symbol $x(n)$ can only remove linear terms. This allows for a perfect reconstruction of the nonlinear soft interference and thus its full cancellation at perfect priors. The estimated symbols can thus be expressed as:

$$\hat{x}(n) = \boldsymbol{a}^H(n)(\boldsymbol{y} - E[\boldsymbol{y}]) + \boldsymbol{a}^H(n)\boldsymbol{H}\boldsymbol{u}(n)E[x(n)] \tag{6.4.18}$$

As discussed in [31], the computational complexity of time-varying exact MMSE is high. This has resulted into investigating some low-complexity MMSEs, among which we select the No *A priori* implementation. The MMSE approximation solution with No *A priori* information is written as follows:

$$C_{YY} \triangleq \mathrm{Cov}(\boldsymbol{y},\boldsymbol{y}) = \sigma_w^2 \mathrm{II}_N + \boldsymbol{H}\boldsymbol{H}^H + \sum_{(i,j,k)} |h(i,j,k)|^2 \mathrm{II}_N$$

$$C_{Yx(n)} \triangleq \mathrm{Cov}(\boldsymbol{y}, x(n)) = \boldsymbol{H}\boldsymbol{u}(n)$$

Nonlinear interference can be described by Gaussian additive white noise with a variance of $\sigma_i^2 = \sum_{(i,j,k)} |h(i,j,k)|^2$. Therefore, by defining the overall noise variance, $\sigma_{\tilde{w}}^2 = \sigma_i^2 + \sigma_w^2$,

$$C_{YY} = \sigma_{\tilde{w}}^2 \mathrm{II}_N + \boldsymbol{H}\boldsymbol{H}^H \tag{6.4.19}$$

The NA – MMSE solution can thus be written as:

$$\hat{x}(n) = \boldsymbol{u}^T(n)\boldsymbol{H}^H C_{YY}^{-1}(\boldsymbol{y} - E[\boldsymbol{y}]) + \boldsymbol{u}^T(n)\boldsymbol{H}^H C_{YY}^{-1}\boldsymbol{H}\boldsymbol{u}(n)E[x(n)] \tag{6.4.20}$$

where $C = \boldsymbol{u}^T(n)\boldsymbol{H}^H C_{YY}^{-1}\boldsymbol{H}\boldsymbol{u}(n)$ is a constant $\forall n \in \{0, \ldots, N-1\}$. Therefore, the succinct form of the estimated symbols can be written as:

$$\hat{\boldsymbol{x}} = \boldsymbol{H}^H C_{YY}^{-1}(\boldsymbol{y} - E[\mathbf{y}]) + CE[\boldsymbol{x}] \tag{6.4.21}$$

where

$$\boldsymbol{C}_{YY}^{-1} = (\boldsymbol{F}^H(\sigma_{\tilde{w}}^2 \mathrm{II}_N + \boldsymbol{H}_d\boldsymbol{H}_d^H)\boldsymbol{F})^{-1} = \boldsymbol{F}^H\boldsymbol{C}_{YY,d}^{-1}\boldsymbol{F} \tag{6.4.22}$$

where $\boldsymbol{C}_{YY,d} = \sigma_{\tilde{w}}^2 \mathrm{II}_N + \boldsymbol{H}_d\boldsymbol{H}_d^H$. Its frequency domain estimated symbols are:

$$\hat{\boldsymbol{X}} \triangleq \boldsymbol{F}\hat{\boldsymbol{x}} = \boldsymbol{H}_d^H\boldsymbol{C}_{YY,d}^{-1}(\boldsymbol{Y} - E[\boldsymbol{Y}]) + CE[\boldsymbol{X}] \tag{6.4.23}$$

where $E[\boldsymbol{Y}] = F[E[\boldsymbol{y}]]$, $E[\boldsymbol{X}] = F[E[\boldsymbol{x}]]$. The first soft frequency domain symbol $E[Y_i]$ is expressed as:

$$E[Y_i] = H_d(i)E[X_i] + \sqrt{N}\sum_{p=0}^{N-1}\sum_{q=0}^{N-1}\sum_{r=0}^{N-1} H^{(3)}(p,q,r) \cdot E[X(p)X(q)\bar{X}(r)]\delta_N(p+q+r-i) \tag{6.4.24}$$

The constant C can be computed using $\boldsymbol{u}(0)$:

$$C = \boldsymbol{u}^T(0)\boldsymbol{H}^H\boldsymbol{C}_{YY}^{-1}\boldsymbol{H}\boldsymbol{u}(0) = \boldsymbol{u}^T(0)\boldsymbol{F}^{\boldsymbol{H}}\boldsymbol{H}_d^H\boldsymbol{C}_{YY,d}^{-1}\boldsymbol{H}_d\boldsymbol{F}\boldsymbol{u}(0) = \frac{1}{N}\sum_{i=0}^{N-1}\frac{|H_d(i)|^2}{\sigma_{\tilde{w}}^2 + |H_d(i)|^2} \tag{6.4.25}$$

where the factor $\frac{1}{N}$ comes from $\boldsymbol{u}^T(0)F^H = \frac{1}{\sqrt{N}}\boldsymbol{1}_{1\times N}$. Thus, the output of the frequency domain equalizer is estimated as follows:

$$\hat{X}_i = \frac{H_d^*(i)}{\sigma_{\tilde{w}}^2 + |H_d(i)|^2}Y_i + \left(C - \frac{|H_d(i)|^2}{\sigma_{\tilde{w}}^2 + |H_d(i)|^2}\right)E[X_i] - \frac{H_d^*(i)}{\sigma_{\tilde{w}}^2 + |H_d(i)|^2}\sum_{p=0}^{N-1}\sum_{q=0}^{N-1}\sum_{r=0}^{N-1} H_{p,q,r}^{(3)}E[X(p)X(q)\bar{X}(r)] \cdot \sqrt{N}\delta_N(p+q+r-i) \tag{6.4.26}$$

Equation (6.4.26) and Figure 6.18 show that a soft ISI canceller can eliminate both linear ISI $E[X_i]$ and nonlinear ISI $E[X(p)X(q)\bar{X}(r)]$ in $E[Y_i]$. In order to avoid the complexity

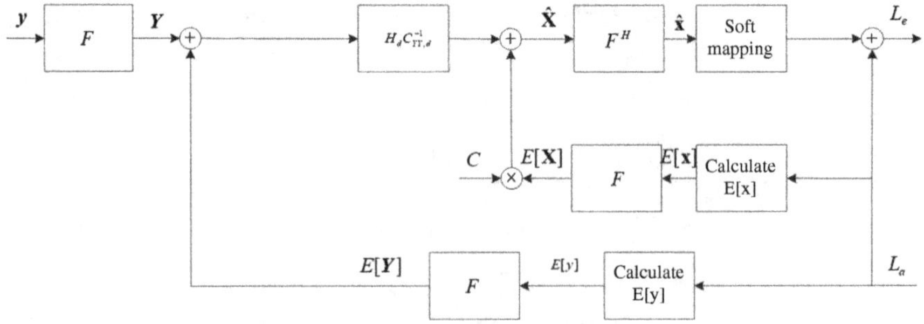

Figure 6.18 Soft interference canceller MMSE enhanced equalizer.

of computing third-order nonlinear interference associated with $E[X(p)X(q)\bar{X}(r)]$, first the nonlinear interference can be computed in the time domain, then the nonlinear equivalent frequency domain interference can be computed. In other words, let:

$$\begin{aligned} E[Y^{(3)}] &= \sqrt{N}\sum_{p=0}^{N-1}\sum_{q=0}^{N-1}\sum_{r=0}^{N-1} H^{(3)}(p,q,r)E[X(p)X(q)\bar{X}(r)] \\ &\quad\times \Delta_N(p+q+r) \\ &= F\sum_i\sum_j\sum_k H_{ijk}E[\boldsymbol{x}^{ijk}] \end{aligned} \tag{6.4.27}$$

where we define $\Delta_N(p+q+r) = [\delta_N(p+q+r-0),\ldots,\delta_N(p+q+r-N-1)]^T$. Computing soft nonlinear symbols is then less computationally complex.

In order to compute the MMSE coefficients, one needs to compute expectations of $E[x^{ijk}]$ and $E[x]$, i.e., expectation of products of three symbols and symbol conjugates at different time instants. Due to the presence of an interleaver between the decoder and the equalizer, symbols can be considered mutually independent. Therefore, the expectation of a product of p symbols and $q-p$ symbol conjugates is computed by:

$$\begin{aligned} &E[x(n-i_1)x(n-i_2)\ldots x(n-i_p)x^*(n-i_{p+1})\ldots x^*(n-i_q)] \\ &= \prod_j\sum_{m=1}^{M} s^{v(j)}(m)s^{*v^*(j)}(m)P[x(n-i_j)=s(m)] \end{aligned} \tag{6.4.28}$$

where $v(i)$ (or $v^*(i)$) represents the number of occurrences of symbol $x(n-i_j)$(or $x^*(n-i_j)$) in the product average, and $s(m), m\in[1\ldots M]$ is the mth constellation symbol with probability computed as:

$$P(x(n)=s(m)) = \prod_{i=1}^{\log_2(M)} P(c_j(n)=s_i(m))$$

where $c_i(n)$ is the ith coded bit of a symbol $x(n)$, and $s_i(m)$ is the ith of the constellation symbol $s(m)$. The encoding bit probabilities can be computed from the input LLRs L_a from the decoder as:

$$L_a(c_i(n)) = \ln\frac{P(c_i(n)=0)}{P(c_i(n)=1)}$$

6.4.3.2 *Soft demapper*

To map the output of the equalizer to an encoded LLR, the distribution of the estimation error $e(n) = \hat{x}(n) - k(n)x(n)$ should be computed instead of the distribution of $\hat{x}(n)$ given $x(n)$ turns out to be a practical choice, since one does not need to track the occurrences of $x(n)$ in third-order covariances. For practical consideration, this error is assumed to be Gaussian. More precisely:

$$k(n) = \text{Cov}(\hat{x}(n), x(n)) = \boldsymbol{u}^T(n)\boldsymbol{F}^H\boldsymbol{H}_d^H\boldsymbol{C}_{YY,d}^{-1}\boldsymbol{H}_d\boldsymbol{F}\boldsymbol{u}(n) = C \tag{6.4.29}$$

It can be seen that the expectation of the residual error is $E[e(n)] = 0$. The variance of the residual error is written as follows:

$$\begin{aligned} v_e(n) &\triangleq \text{Cov}(e(n), e(n)) = \text{Cov}(\hat{x}(n), \hat{x}(n)) - |\boldsymbol{C}|^2 \\ &= \boldsymbol{u}^T(n)\boldsymbol{F}^H\boldsymbol{H}_d^H\boldsymbol{C}_{YY,d}^{-1}(\boldsymbol{H}_d\boldsymbol{F}(\boldsymbol{V} + (1 - v(n))\boldsymbol{u}(n)\boldsymbol{u}^T(n)) \\ &\quad \times \boldsymbol{F}^H\boldsymbol{H}_d^H + \sigma_{\tilde{w}}^2\text{II}_N)\boldsymbol{C}_{YY,d}^{-1}\mathbf{H}_d\mathbf{F}\mathbf{u}(n) - |C|^2 \end{aligned}$$

where $\boldsymbol{V} = \text{diag}(v(0), \ldots, v(N-1))$ and $v(i)$ is the covariance of the ith symbol. The term $(1 - v(n))\boldsymbol{u_n}\boldsymbol{u}_n^T$ is added due to the subtraction of the contribution of the symbol $x(n)$ which translates to $v(n) = 1$. In the computation of the Cov($\mathbf{Y},\mathbf{Y}$), the covariance of the nonlinear ISI terms was approximated by σ_i^2. The equalizer output $\hat{x}(n) \sim \text{N}(Cx(n), v_e(n))$ and the output extrinsic *LLR* L_e can be expressed as:

$$\begin{aligned} &L_e(c_i(n)|\hat{x}(n)) \\ &\triangleq \ln\frac{P(c_i(n) = 0|\hat{x}(n))}{P(c_i(n) = 1|\hat{x}(n))} - L_a(c(n), i) \\ &= \ln\frac{\sum_{s(j):s_i=0(j)}\left(\exp\left(\frac{-|\hat{x}(n) - Cs(j)|^2}{v_e(n)}\right)\right)\prod_{k\neq i}P(c_k(n) = s_k(j))}{\sum_{s(j):s_j=1(j)}\left(\exp\left(\frac{-|\hat{x}(n) - Cs(j)|^2}{v_e(n)}\right)\right)\prod_{k\neq i}P(c_k(n) = s_k(j))} \end{aligned}$$

The *a posteriori* LLRs are made extrinsic(L_e) by subtracting the *a priori* probability fed by the decoder:

$$L_e(c_i(n)|\hat{x}(n)) = L_{ap}(c_i(n)|\hat{x}(n)) - La(c_i(n))$$

6.4.4 *Simulation verification*

The independent and identically distributed bits are encoded using a 1/2 rate (7,5) non-recursive convolutional encoder. The code words are then interleaved and mapped into 8PSK symbols using a gray mapping. A cyclic prefix of length 3 symbols is appended to each block of $N = 512$ symbols. The overall baud rate equivalent channel is represented by the Volterra kernels in Table 6.4 with a stronger nonlinear ISI. Figure 6.19 plots the BER performance for four turbo iterations for both the time and frequency domain MMSE equalizers,

Table 6.4 Volterra kernel channel.

First-order Kernel	Third-order Kernel
$h(0) = 0.8529 + j0.4502$	$h(0,0,1) = 0.0979 - j0.0979$
$h(1) = 0.0881 - j0.0014$	$h(0,0,2) = 0.1091 - j0.0615$
$h(2) = -0.0336 - j0.0196$	$h(0,0,3) = -0.1119 - j0.0252$
$h(3) = 0.0503 + j0.0433$	$h(1,1,0) = -0.028 - j0.0475$
	$h(3,3,0) = 0.0503 - j0.0503$

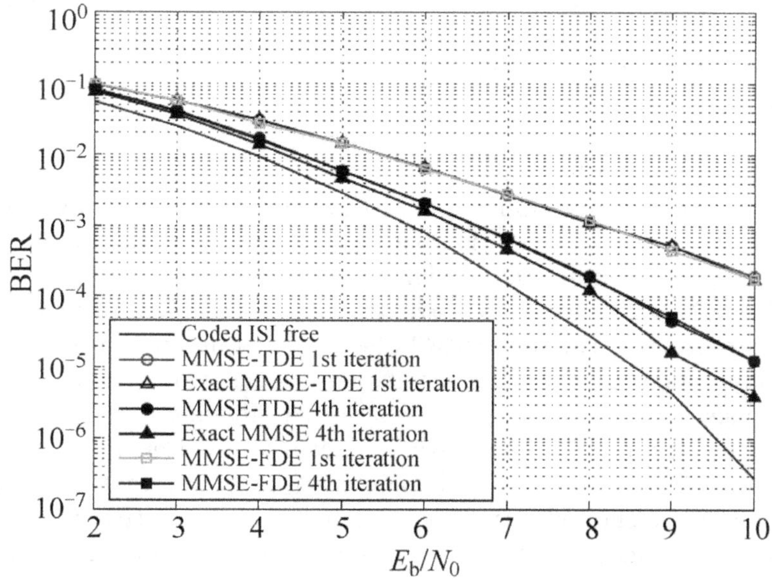

Figure 6.19 Comparison of BERs for different enhanced equalizers.

namely, the exact MMSE linear equalizer, the time domain MMSE equalizer (MMSE-TDE), and the frequency domain MMSE equalizer (MMSE-FDE). The exact MMSE refers to the full complexity time domain equalizer, whereas the MMSE-TDE refers to the time domain NA-MMSE-B. All time domain equalizers were simulated without CP, and hence the MMSE-FDE is shifted by 0.0254 dB to account for the loss of spectral efficiency. The performance of the frequency domain equalizer is comparable to that of the time domain equalizer.

6.5 Nonlinear Channel Equalization Steady-State Algorithm Based on Maximum Correlation Entropy Volterra Filter

Over the past decade, several improved algorithms based on l_2-norm have been developed to improve the performance of recursive least squares(RLS) algorithm [34, 35]. However, if the second-order moments of additive noise are infinite, the ℓ_2 norm-based algorithm may fail to work [36,37]. If the additive noises in underwater acoustic channel and VHF radio can be well-modeled by a stable α distribution model [37], the performance of the RLS algorithm based on the stable distributed noise background is degraded because it has no finite second-order or higher-order moments. For the minimum kernel based on Wirtingers calculus, the LMS algorithm and the enhanced LMS kernel algorithm based on linear estimation have been used to solve the nonlinear channel equalization problem. For non-Gaussian signals, the MSE criterion is a suboptimal criterion because it does not consider the influence of higher moments.

Recently, the maximum correntropy criterion (MCC) provides a measure of the local similarity between two random variables. Due to its simplicity and low sensitivity to singular value noise, MCC is often used as a basic criterion for adaptive filtering.

In this section, a complex-valued recursive Volterra filter, called Volterra–CMCC, is described. The filter is based on the complex maximum correlation entropy criterion and still effectively equalizes nonlinear channels in the presence of a large number of singular values or under α-stable noise.

6.5.1 *Algorithm theory*

Figure 6.20 shows a discrete-time model of a communication system, where transmission occurs over a nonlinear channel and an adaptive equalizer is employed in front of the receiver. The error signal of the adaptive equalizer can be defined as:

$$e(n) = d(n) - y(n) = a(n - \chi) - y(n) \tag{6.5.1}$$

where $a(n)$ is the transmitted signal (i.e., known training sequence), $d(n) = a(n - \chi)$ is the desired signal or reference signal, χ is the propagation delay associated with the communication channel, and $y(n)$ is the output of the equalizer.

The output of the nonlinear channel is defined as:

$$x(n) = y\{a(n)\} \tag{6.5.2}$$

The input to the equalizer (disturbed signal) is written as:

$$y(n) = x(n) + w(n) \tag{6.5.3}$$

where $w(n)$ is the additive noise. Considering the computation complexity of the Volterra filter, a second-order Volterra (SOV) expansion [38] is applied here. The expanded input vector $\boldsymbol{x}(n)$ of the SOV filter is expressed as:

$$\begin{aligned}\boldsymbol{x}(n) = &\underbrace{[x(n), x(n-1), \ldots, x(n-1+M)]}_{\text{linear kernels}},\\ &\times \underbrace{[x^2(n), x(n)x(n-1), \ldots, x^2(n-1+M)]^T}_{\text{quadratic kernels}}\end{aligned} \tag{6.5.4}$$

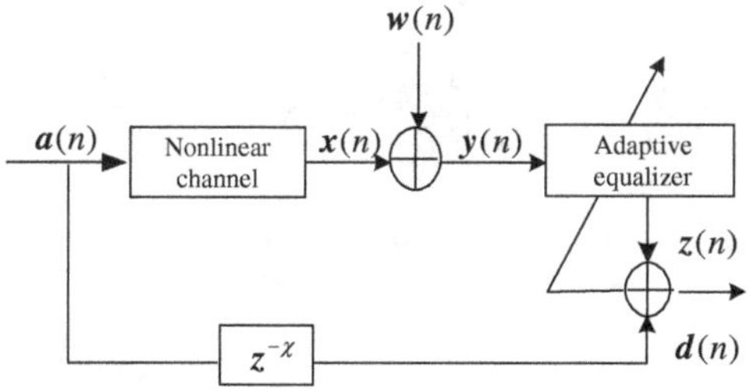

Figure 6.20 Equalizer block diagram in the digital transmission system.

where M represents the length of the linear kernel. The length L of the second-order Volterra filter can be computed by $L = M(M+3)/2$. Therefore, the input signal $x(n)$ can be equalized by the equalizer to reduce the influence of the channel and the transmitted signal can be recovered according to the received signal samples. An α-stabilized noise distribution model describing the impulse noise can be denoted by a probability density function (PDF) without a closed solution. The characteristic function of a symmetric α-stable distribution, which is the Fourier transform of the PDF, can be described by:

$$\Phi(t) = \exp\{-\gamma|t|^{\alpha}\} \tag{6.5.5}$$

where $0 < \alpha < 2$ represents the characteristic index and γ denotes the *dispersion* parameter whose behavior is similar to the variance of the Gaussian distribution. A small α indicates the highly impulse noise. In particular, if $\alpha = 2$, Equation (6.5.5) becomes the characteristic function of the Gaussian distribution.

6.5.2 *Volterra–CMCC algorithm*

Let D and Y be two random variables with the same dimensions, so the correlation entropy between D and Y is defined as [39]:

$$V(D,Y) = E\{\kappa(D,Y)\} = \int \kappa(d,y)d\Re_{D,Y}(d,y) \tag{6.5.6}$$

where $\kappa(\cdot,\cdot)$ is *a* shift-invariant Mercer kernel and $\Re_{D,Y}(d,y)$ stands for the joint distribution function of (d,y). The correlation entropy kernel function in this section is selected as the Gaussian kernel, i.e.,

$$\kappa(D,Y) = \frac{1}{\sqrt{2\pi}\sigma}\exp\left\{-\frac{|D-Y|^2}{2\sigma^2}\right\} \tag{6.5.7}$$

where σ represents the kernel size. The cost function of the MCC-based algorithm is defined as follows [39]:

$$J(n) = E\left\{\exp\left(-\frac{|e(n)|^2}{2\sigma^2}\right)\right\} \tag{6.5.8}$$

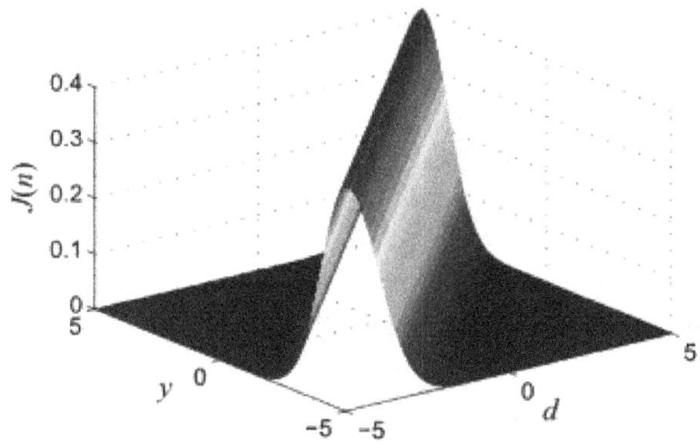

Figure 6.21 Cost function of MCC in joint space.

The MCC is used to measure the similarity between two random variables, whose value is mainly determined by the kernel function. Figure 6.21 shows the cost function $\zeta(d, y)$ of the MCC in the joint space of d and y. For a Gaussian kernel, the MCC divides the joint space into three regions. When the two points are close, the MCC behaves like the l_2-norm distance in Euclidean space. When the two points are separated, the MCC behaves like the l_1-norm. When the two points are gradually separated, they eventually become similar to the l_0-norm [39]. This property shows the robust performance of the correntropy for outliers. In order to achieve the improved performance of the algorithm, the RLS-based algorithm is usually preferred. Therefore, the cost function of the Volterra−CMCC is defined as follows:

$$\begin{aligned} J_{\text{Volterra-CMCC}}(n) &= \sum_{i=1}^{n} \lambda^{n-i} \kappa(d(i), y(i,n)) \\ &= \frac{1}{\sqrt{2\pi}\sigma} \sum_{i=1}^{n} \lambda^{n-i} \exp\left\{ -\frac{e(i,n)e^*(i,n)}{2\sigma^2} \right\} \end{aligned} \tag{6.5.9}$$

where $0 \leq \lambda < 1$ represents the forgetting coefficient. The output signal $y(i, n)$ and the error signal $e(i, n)$ are expressed as:

$$y(i,n) = \boldsymbol{f}^H(n)\boldsymbol{x}(i) \tag{6.5.10}$$

$$e(i,n) = d(i) - \boldsymbol{f}^H(n)\boldsymbol{x}(i) \tag{6.5.11}$$

where H represents a transpose conjugate. Taking the gradient of Equation (6.5.9) with respect to the weight vector $\boldsymbol{f}(n)$, we obtain

$$\frac{\partial J(n)}{\partial \boldsymbol{f}^*(n)} = -\frac{1}{2\sigma^2}\frac{1}{\sqrt{2\pi}\sigma}\sum_{i=1}^{n}\lambda^{n-i}\exp\left\{-\frac{e(i,n)e^*(i,n)}{2\sigma^2}\right\}\frac{\partial e(i,n)e^*(i,n)}{\partial \boldsymbol{f}^*(n)} \tag{6.5.12}$$

Let Equation (6.5.12) be zero, one gets

$$\sum_{i=1}^{n}\lambda^{n-i}\psi(i,n)\boldsymbol{x}(i)\boldsymbol{x}^H(i)\boldsymbol{f}(n) = \sum_{i=1}^{n}\lambda^{n-i}\psi(i,n)\boldsymbol{x}(i)d^*(i) \tag{6.5.13}$$

where $\psi(i,n)$ is the weighting factor and can be expressed as:

$$\psi(i,n) = \exp\left\{-\frac{e(i,n)e^*(i,n)}{2\sigma^2}\right\} \tag{6.5.14}$$

Then, the expression of $\boldsymbol{f}(n)$ is obtained as follows:

$$\boldsymbol{f}(n) = \boldsymbol{F}_{\boldsymbol{f}}(n)\boldsymbol{\pi}(n) = \boldsymbol{R}^{-1}(n)\boldsymbol{\pi}(n) \tag{6.5.15}$$

where

$$\boldsymbol{F}(n) = \mathbf{R}^{-1}(n), \mathbf{R}(n) = \sum_{i=1}^{n}\lambda^{n-i}\psi(i,n)\mathbf{x}(i)\mathbf{x}^H(i)$$

and

$$\boldsymbol{\pi}(n) = \sum_{i=0}^{n}\lambda^{n-i}\psi(i,n)d^*(i)\mathbf{x}(i)$$

For $\psi(i,n) = 1$, this algorithm becomes the RLS algorithm. When $\psi(i,n) \neq 1$, $\mathbf{R}(n)$ and $\boldsymbol{\pi}(n)$ are weighted autocorrelation matrix and the weighted cross-correlation vector of the optimal weights via $\psi(i,n)$. At each iteration, Equation (6.5.15) needs to be recalculated. In order to avoid this inconvenience, a sliding window method is proposed in [40]. However, the main disadvantage of this sliding window strategy algorithm is that all previous samples within a window need to be saved in memory. To overcome this disadvantage, $\mathbf{R}(n)$ and $\boldsymbol{\pi}(n)$ are updated by a recursive method as follows:

$$\mathbf{R}(n) \approx \sum_{i=0}^{n} \lambda^{n-i}\psi(i,i)\mathbf{x}(i)\mathbf{x}^H(i) = \lambda\mathbf{R}(n-1) + \psi(n,n)\mathbf{x}(n)\mathbf{x}^H(n) \tag{6.5.16}$$

$$\boldsymbol{\pi}(n) \approx \sum_{i=0}^{n} \lambda^{n-i}\psi(i,i)d^*(i)\mathbf{x}(i) = \lambda\boldsymbol{\pi}(n-1) + \psi(n,n)\mathbf{x}(n)d^*(n) \tag{6.5.17}$$

It should be emphasized that the adaptive results of Equations (6.5.15) and (6.5.17) are similar to the Wiener solution and it is necessary to carry out the matrix inverse operations in order to obtain the equalization results. So, the computation complexity of Equation (6.5.15) is very high and is seldom used in practice. By using the matrix inversion lemma [41], $\mathbf{F}(n)$ can be expressed as:

$$\boldsymbol{F}(n) = \lambda^{-1}\boldsymbol{F}(n-1) - \lambda^{-1}\boldsymbol{\Phi}(n)\boldsymbol{x}^H(n)\boldsymbol{F}(n-1) \tag{6.5.18}$$

when $\mathbf{F}(0) = \rho^{-1}\boldsymbol{I}$, ρ is a small positive number. The gain factor is defined by:

$$\boldsymbol{\Phi}(n) = \frac{\psi(n,n)\mathbf{F}(n-1)\mathbf{x}(n)}{\lambda + \psi(n,n)\mathbf{x}^H(n)\mathbf{F}(n-1)\mathbf{x}(n)}$$

Therefore, $\boldsymbol{f}(n)$ can be updated as:

$$\boldsymbol{f}(n) = \boldsymbol{f}(n-1) + \boldsymbol{\Phi}(n)[d^*(n) - \boldsymbol{x}^H(n)\boldsymbol{f}(n-1)] \tag{6.5.19}$$

***Note 1*:** Note that Equations (6.5.14) and (6.5.19) indicate an implicit relationship between $\boldsymbol{f}(n)$ and $\psi(n,n)$ that cannot be solved in one step. Therefore, the algorithm requires an iterative approximation to the solution, where $\psi(n,n)$ is calculated by using $\boldsymbol{f}(n-1)$, and the new value for $\boldsymbol{f}(n)$ is obtained via the value of $\psi(n,n)$.

***Note 2*:** Since no *a priori* information about the noise characteristics is required, the proposed algorithm is nearly blind and can be implemented using only σ and λ.

6.5.3 *Steady-state performance*

The steady-state performance of the Volterra–CMCC algorithm is now analyzed. Since the output of the Volterra depends on the linearity of the coefficients of the filter itself, the steady-state excess MSE (EMSE) of the Volterra–CMCC algorithm is calculated by the method in [40, 41]. Considering the expected response arising from the model,

$$d(n) = \boldsymbol{f}_{\text{opt}}^{H}\boldsymbol{x}(n) + w(n) \tag{6.5.20}$$

where $\boldsymbol{f}_{\text{opt}}$ is a vector containing the best coefficient values.

Define the weight deviation vector as:

$$\boldsymbol{\Omega}(n) = \boldsymbol{f}_0 - \boldsymbol{f}(n) \tag{6.5.21}$$

Now, some assumptions are given as:

(1) The input signal $\mathbf{x}(n)$ is a zero-mean independent and identically distributed (*i.i.d.*) and is approximately independent of the *a priori* excess error.
(2) The noise signal $\boldsymbol{w}(n)$ is *i.i.d.* with zero mean and variance σ_{w}^2.
(3) $\boldsymbol{w}(n)$ and $\mathbf{x}(n)$ are mutually independent.

The iterative equation of the weight vector deviation vector of the Volterra–CMCC algorithm can be given by:

$$\begin{aligned}\boldsymbol{\Omega}(n) = \boldsymbol{\Omega}(n-1) - &\frac{\psi(n,n)\boldsymbol{F}(n-1)\boldsymbol{x}(n)}{\lambda + \psi(n,n)\boldsymbol{x}^H(n)\boldsymbol{F}(n-1)\boldsymbol{x}(n)} \\ &\cdot [d^*(n) - \boldsymbol{x}^H(n)\boldsymbol{f}(n-1)]\end{aligned} \tag{6.5.22}$$

According to the literature [42], applying the matrix inversion formula, we have

$$\boldsymbol{F}^{-1}(n) = \lambda^{n+1}\rho\boldsymbol{I} + \sum_{i=0}^{n}\lambda^{n-i}\psi(i,i)\boldsymbol{x}^H(i)\boldsymbol{x}(i) \tag{6.5.23}$$

Therefore, Equation (6.5.22) can be simplified as:

$$\boldsymbol{\Omega}(n) = \boldsymbol{\Omega}(n-1) - \vartheta(n)\boldsymbol{F}(n)\boldsymbol{x}^H(n)e(n) \tag{6.5.24}$$

where

$$\vartheta(n) = \frac{1}{\frac{\lambda}{\psi(n,n)} + \boldsymbol{x}^H(n)\boldsymbol{F}(n-1)\boldsymbol{x}(n)}$$

When the exponential term is expanded with a second-order Taylor series, we have

$$\vartheta(n) = \frac{1}{\lambda[2\sigma^2 - |e(n)|^2]^{-1}/(2\sigma^2)^{-1} + \boldsymbol{x}^H(n)\boldsymbol{F}(n-1)\boldsymbol{x}(n)}$$

In the steady-state, the *a posteriori* error is defined as:

$$e_p(n) = \boldsymbol{x}^H(n)\boldsymbol{\Omega}(n)$$

Equation (6.5.24) is multiplied by $\boldsymbol{x}(n)$, and the relationship between the *a priori* and the *a posteriori* estimation errors can be written as:

$$e_p(n) = e_a(n) - ||\boldsymbol{x}(n)||^2_{\boldsymbol{F}(n)\vartheta(n)} e(n) \tag{6.5.25}$$

where $||\boldsymbol{x}(n)||^2_{\boldsymbol{\Lambda}} = \boldsymbol{x}^*\boldsymbol{\Lambda}\boldsymbol{x}$ represents the squared-weighted Euclidean norm of a vector $\boldsymbol{x}(n)$. When $\boldsymbol{x}(n) \neq 0$, the energy conservation relation (ECR) expression for the Volterra–CMCC algorithm can be given as:

$$\boldsymbol{\Omega}(n) + \frac{\vartheta(n)\boldsymbol{F}(n)\boldsymbol{x}^H(n)}{||\boldsymbol{x}(n)||^2_{\vartheta(n)\boldsymbol{F}(n)}} e_a(n) = \boldsymbol{\Omega}(n-1) + \frac{\vartheta(n)\boldsymbol{F}(n)\boldsymbol{x}^H(n)}{||\boldsymbol{x}(n)||^2_{\vartheta(n)\boldsymbol{F}(n)}} e_p(n) \tag{6.5.26}$$

Combining Equations (6.6.24) and (6.6.25), and using $\vartheta^{-1}(n)$ $\boldsymbol{F}^{-1}(n)$ as a weighting matrix for squared-weighted Euclidean norm, we have

$$||\boldsymbol{\Omega}(n)||^2_{\vartheta^{-1}(n)\boldsymbol{F}^{-1}(n)} + \frac{|e_a(n)|^2}{||\boldsymbol{x}(n)||^2_{\vartheta(n)\boldsymbol{F}(n)}} = ||\boldsymbol{\Omega}(n-1)||^2_{\vartheta^{-1}(n)\boldsymbol{F}^{-1}(n)} + \frac{|e_p(n)|^2}{||\boldsymbol{x}(n)||^2_{\vartheta(n)\boldsymbol{F}(n)}} \tag{6.5.27}$$

Suppose that the Volterra–CMCC algorithm converges, that is, when $n \to \infty$, the optimum EMSE can be obtained. Under this assumption, the following approximation can be made:

$$E\{\|\boldsymbol{\Omega}(n)\|^2_{\vartheta^{-1}(n)\boldsymbol{F}^{-1}(n)}\} \approx E\{\|\boldsymbol{\Omega}(n-1)\|^2_{\vartheta^{-1}(n)\boldsymbol{F}^{-1}(n)}\} \quad (6.5.28)$$

Taking expectation of both sides of Equation (6.5.27), substituting Equation (6.5.27) into Equation (6.5.27) results in

$$E\left\{\frac{|e_a(n)|^2}{\|\boldsymbol{x}(n)\|^2_{\vartheta(n)\boldsymbol{F}(n)}}\right\} \approx E\left\{\frac{|e_p(n)|^2}{\|\boldsymbol{x}(n)\|^2_{\vartheta(n)\boldsymbol{F}(n)}}\right\} \quad (6.5.29)$$

Now, substituting Equation (6.5.25) into Equation (6.5.29), we obtain that at steady state

$$E\{\|\mathbf{x}(n)\|^2_{\vartheta(\infty)\boldsymbol{F}(\infty)}|e(\infty)|^2\} = 2\mathrm{Re}(E\{e_a^*(\infty)e(\infty)\}) \quad (6.5.30)$$

where $\mathrm{Re}\{x\}$ represents the real part of x, taking into account $e(n) = e_a(n) + w(n)$, Equation (6.5.22) can be expressed as:

$$\sigma_w^2 E\{|\boldsymbol{x}(n)\|^2_{\vartheta(\infty)\boldsymbol{F}(\infty)}\} + E\{\|\boldsymbol{x}(n)\|^2_{\vartheta(\infty)\boldsymbol{F}(\infty)}\,|e_a(\infty)|^2\}$$
$$= 2E\{|e_a(\infty)|^2\} \quad (6.5.31)$$

Assume that $\|\boldsymbol{x}(n)\|^2_{\vartheta(\infty)\boldsymbol{F}(\infty)}$ and $|e_a(\infty)|^2$ are mutually independent at steady state, we have

$$\sigma_w^2 E\{\|\boldsymbol{x}(n)\|^2_{\vartheta(\infty)\boldsymbol{F}(\infty)}\} + E\{\|\boldsymbol{x}(n)\|^2_{\vartheta(\infty)\boldsymbol{F}(\infty)}\} \cdot E\{|e_a(\infty)|^2\}$$
$$= 2E\{|e_a(\infty)|^2\} \quad (6.5.32)$$

Substituting Equation (6.5.30) into Equation (6.5.32), we obtain

$$\zeta = \frac{\theta\sigma_w^2}{2-\theta} \quad (6.5.33)$$

where $\zeta = E\{|e_a(\infty)|^2\}$ and

$$\theta \triangleq E\{\|\boldsymbol{x}(n)\|^2_{\vartheta(\infty)\boldsymbol{F}(\infty)}\}$$
$$= E\left\{\frac{(2\sigma^2)^{-1}\mathrm{Tr}(\boldsymbol{x}(n)\boldsymbol{x}^H(n)\boldsymbol{F}(\infty))}{(2\sigma^2 - |e(\infty)|^2)^{-1}\lambda + (2\sigma^2)^{-1}\mathrm{Tr}(\boldsymbol{x}(n)\boldsymbol{x}^H(n)\boldsymbol{F}(\infty))}\right\}$$

According to Equation (6.5.23), the steady-state value of $\boldsymbol{F}^{-1}(n)$ is obtained as:

$$\boldsymbol{F}^{-1} \triangleq \lim_{n\to\infty} E\{\boldsymbol{F}^{-1}(n)\} = \frac{E\{\psi(n,n)\}\boldsymbol{Q}(n)}{1-\lambda}$$

where $\boldsymbol{Q}(n) = E\{\boldsymbol{x}(n)\boldsymbol{x}^H(n)\}$ is the covariance matrix. According to the approximation estimate in [41] and using the second-order Taylor series, the following approximation estimate can be made:

$$\begin{aligned} E\{\boldsymbol{F}(\infty)\} &\approx (E\{\boldsymbol{F}^{-1}(\infty)\})^{-1} = \frac{(1-\lambda)\boldsymbol{Q}^{-1}(\infty)}{E\{\exp(-\frac{|e(\infty)|^2}{2\sigma^2})\}} \\ &\approx \frac{E\{(2\sigma^2 - |e(\infty)|^2)^{-1}\}(1-\lambda)\boldsymbol{Q}^{-1}(\infty)}{(2\sigma^2)^{-1}} \end{aligned} \tag{6.5.34}$$

When $0 \leq \lambda \leq 1$, there is $|e_a(n)| \leq |w(n)|$ in the steady state. Therefore, substituting Equation (6.5.34) into Equation (6.5.33), one gets

$$\zeta = \frac{\sigma_w^2(1-\lambda)L\eta}{2-(1-\lambda)L\eta} \tag{6.5.35}$$

where

$$\eta = E\left[\frac{1}{\lambda E((2\sigma^2 - |w(n)|^2)^{-1})(2\sigma^2 - |w(n)|^2) + (1-\lambda)L}\right]$$

Equation (6.5.35) is difficult to further simplify. The theoretical result contains a random variable $w(n)$, but after the expect operation, an exact value can be obtained. In addition, Equation (6.5.35) is also applicable to the analysis of linear CMCC algorithms. The computation complexity of the Volterra–RLS algorithm and the Volterra–CMCC algorithm is shown in Table 6.5. For complex

Table 6.5 Comparison of computational complexity.

Operation	**Volterra–RLS**	**Volterra–CMCC**
Real addition	$5M^2 + 13M - 1$	$5M^2 + 17M + 1$
Real multiplication	$6M^2 + 18M + 1$	$6M^2 + 26M + 7$
Other operations	0	1 index operation

signals, the RLS algorithm requires $4M^2 + 12M - 1$ real additions and $4M^2 + 16M + 1$ real multiplications. By using the SOV expansion, the Volterra–RLS algorithm requires $M^2 + M$ real additions and $2M^2 + 2M$ real multiplications. Therefore, the Volterra–RLS algorithm requires a total of $5M^2 + 13M - 1$ real additions and $6M^2 + 16M + 1$ real multiplications. The Volterra–CMCC algorithm moderately increases the computational complexity.

6.5.4 *Simulation verification*

In order to evaluate the performance of the Volterra–CMCC algorithm, the Volterra–RLS algorithm was set as the comparison object. All simulation results in this section were achieved through 100 independent runs.

6.5.4.1 *Verification of EMSE*

The input signal was a complex Gaussian white noise signal (CWGN) with zero-mean and unit variance. Figure 6.22 illustrates a comparison between the theory and simulation results for Volterra–CMCC algorithm with SNR= [10, 20, 30] dB.

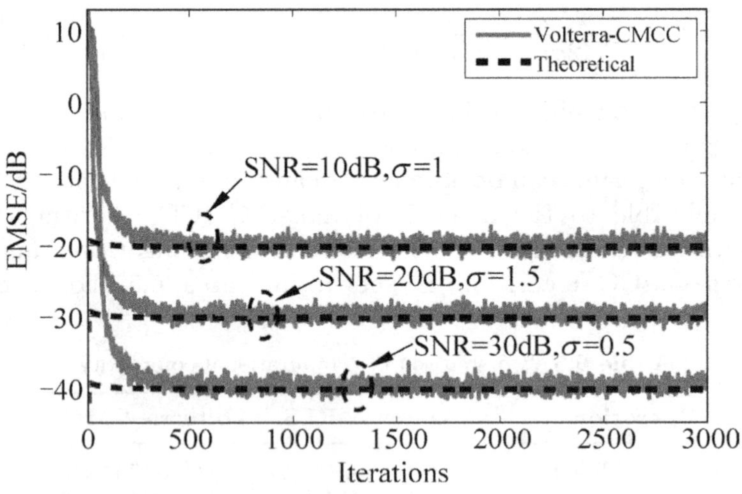

Figure 6.22 Theoretical and simulation learning curves for Volterra–CMCC algorithm ($\lambda = 0.99$).

Table 6.6 Equalization error of Volterra–CMCC algorithm.

Noise Power	Kernel Size	Filter Order	MMSE/dB	
			Theory	Simulation
0.01	1	14[18]	−31.46	−31.77
0.1	0.5	14[18]	−21.97	−22.26
0.001	1.5	14[15]	−41.30	−41.72
0.01	1	14[15]	−31.48	−31.77

Figure 6.22 shows that the theoretical results were in good agreement with the simulation results. In addition, Table 6.6 gives the theoretical EMSE values for different Volterra systems and noise environments. The number of iterations used is 3000. As can be seen, the simulation results agree quite well with the theoretical results.

6.5.4.2 *Application to nonlinear channel equalization*

In order to study the performance of the Volterra–CMCC algorithm for nonlinear channel equalization, the SOV filter [38] was used. To compare the BER performance of the Volterra–RLS and Volterra–CMCC algorithms fairly, the forgetting factor was changed to make them have similar convergence rates.

The SNR is defined as [38]:

$$\mathrm{SNR} = \frac{\sigma_x^2}{\gamma} \tag{6.5.36}$$

where σ_x^2 represents the power of the input vector.

The relationship between the transmitted signal and the channel output is written as:

$$\begin{aligned} x(n) = a(n) - 0.5a(n-1) + 0.001a^2(n) + 0.002a^2(n-1) \\ + 0.007a(n)a(n-1) \end{aligned} \tag{6.5.37}$$

The transmitted signal was a 2-PAM signal. The characteristic index α of the noise was set to 1.8. The length of the SOV filter was set to $L = 14$ with the memory length of linear kernels set to $M = 4$, and the memory length of the quadratic kernel was 10. In α-stable noise, the BER performance of the algorithms was shown

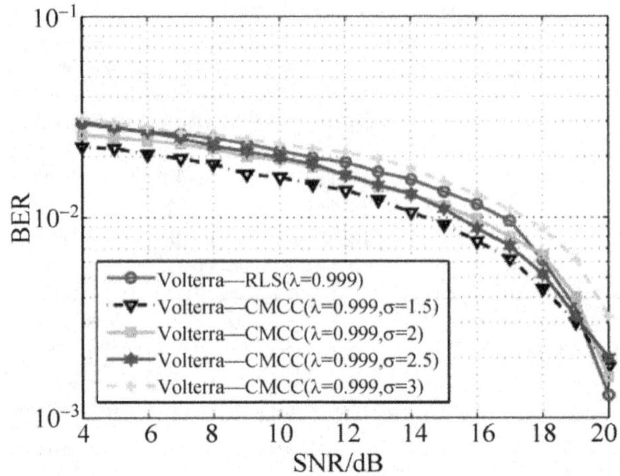

Figure 6.23 BERs performance of the equalizer.

in Figure 6.23. Figure 6.23 shows that the BER of all equalizers decreases with SNR for the nonlinear models. Again, the Volterra–CMCC algorithm outperforms the Volterra–RLS equalizer.

6.6 Complex Neural Network Polynomial Volterra Channel Blind Equalization Algorithm Based on Fuzzy Neural Network Controller

6.6.1 *Fuzzy neural network algorithm*

In 1965, the concept of fuzzy systems proposed by Professor Zadeh first, which is a rule-based design method, was mainly used to deal with some uncertainties and inaccuracies. In recent years, fuzzy neural networks (FNNs), which combines neural networks with fuzzy systems, have been proposed. In fuzzy neural networks, the numerical control parameters and connection matrix of neurons can be replaced by fuzzy parameters, which have stronger expressive ability, faster training speed, and more robust performance than conventional neural networks. In this case, ambiguity means more flexibility in the definition of the system, that is, the boundaries of the system are more ambiguous, and the system itself can be reconstructed for optimum performance.

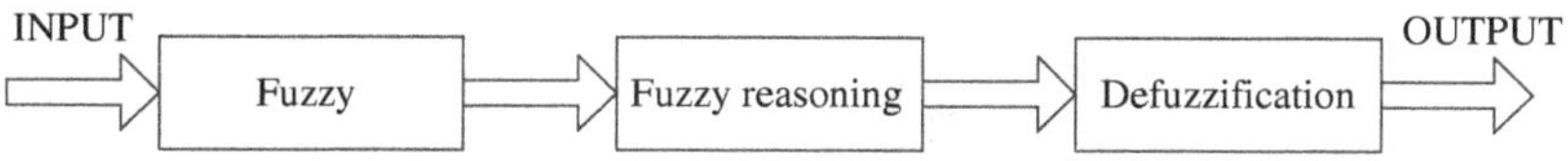

Figure 6.24 Structure of fuzzy neural network topology.

6.6.1.1 *Topological structure of fuzzy neural networks*

Different fuzzy neural network models have different network control structures, but they have the same topological structure, that is, fuzzy neural networks are basically three-layer structures including fuzzy layer, fuzzy reasoning layer, and defuzzification layer, its structure is shown in Figure 6.24.

The first layer is called the fuzzification layer. The fuzzification layer can implement the fuzzification of the input variables to compute a membership function. The fuzzification layer is an essential part of every fuzzy neural network.

The second layer is called the fuzzy reasoning layer. The fuzzy reasoning layer is the most important part of the network structure and is related to the premise and conclusion of fuzzy reasoning and realizes the fuzzy mapping of the network. The fuzzy reasoning layer with diversified structures may be a back propagation network, a radial basis function network, or other forms of neural networks. Different algorithms correspond to different structures, but the main difference of various different structures of the FNN lies in the fuzzy reasoning layer.

The third layer is called the defuzzification layer. This layer transforms the fuzzy state of the inference conclusion into a determined state and then gives a determined output to perform on the system. In some specific neural networks, it is not a necessary structure, which depends on the specific conditions.

In order to enhance the adaptability of the fuzzy neural network, the fuzzification layer, the fuzzy inference layer, and the defuzzification layer are usually composed of multiple networks, which need to be determined according to specific needs. In this way, the automatic adjustment of both the membership function and the fuzzy rule in the inference model can be realized through network learning.

6.6.1.2 *Fuzzy neural network control structure*

In the traditional constant modulus blind equalization algorithm, the convergence speed is improved and steady-state mean square

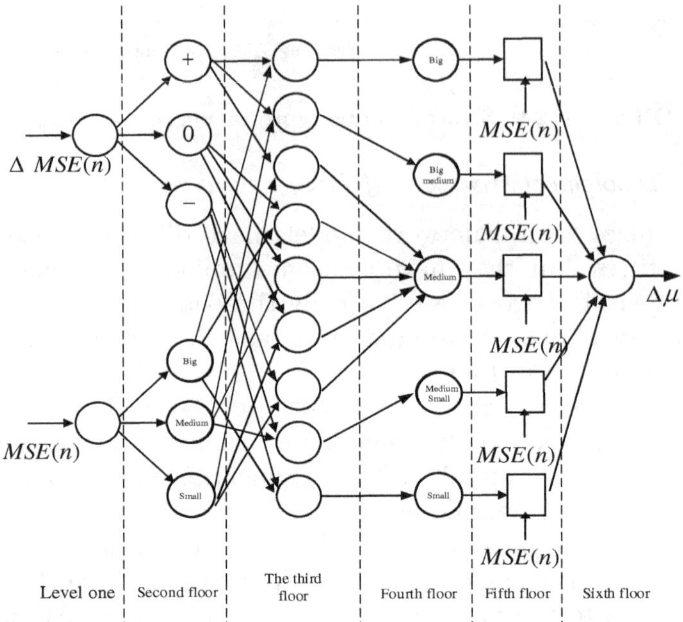

Figure 6.25 Structure of the fuzzy neural network controller.

error (MSE) is reduced by defining the step-size parameters, but it is difficult to improve the convergence speed and reduce the steady-state error at the same time. The fuzzy neural network controller uses human experience knowledge to construct fuzzy rules and extracts information from the MSE signals to control the parameters of the constant modulus blind equalization algorithm. To solve the contradiction between convergence speed and convergence accuracy, at the initial stage of running the algorithm, a large step-size is used to accelerate the convergence speed. As the algorithm continues to iterate, the step-size is decreased to reduce the steady-state residual error. The structure of the FNN controller is shown in Figure 6.25.

6.6.1.3 *Fuzzy neural network control process*

The first layer is called the input layer. ΔMSE and MSE are the input variables of the controller, i.e.,

$$\mu_1(1) = \Delta\text{MSE}, \quad \mu_1(2) = \text{MSE} \tag{6.6.1}$$

$$O_1(i,j) = \mu_1(i), \quad i = 1,2; \; j = 1,2,3 \tag{6.6.2}$$

where μ_1 is the input of the first layer and O_1 is the output of the first layer.

The second layer is called the fuzzification layer. ΔMSE and MSE are divided into three fuzzy levels, respectively.

$$\mu_2(i,j) = O_1(i,j) \quad i = 1, 2; \;\; j = 1, 2, 3 \tag{6.6.3}$$

$$O_2(i,j) = \exp\left[-\frac{\mu_2(i,j) - m(i,j)^2}{q(i,j)}\right] \tag{6.6.4}$$

where $m(i,j)$ and $q(i,j)$ represent the corresponding Gaussian function centers and bandwidth, respectively. For convenience, fixed center and bandwidth are used in this section.

The third layer is called the rule layer. The number of nodes in this layer is the product of the number of fuzzy sets of each input variable, and each node of the layer represents the front part of a rule, where

$$\begin{cases} O_3(1) = O_2(1,1) * O_2(2,1) \\ O_3(1) = O_2(1,1) * O_2(2,2) \\ O_3(3) = O_2(1,1) * O_2(2,3) \\ O_3(4) = O_2(1,2) * O_2(2,1) \\ O_3(5) = O_2(1,2) * O_2(2,2) \\ O_3(6) = O_2(1,2) * O_2(2,3) \\ O_3(7) = O_2(1,3) * O_2(2,1) \\ O_3(8) = O_2(1,3) * O_2(2,2) \\ O_3(9) = O_2(1,3) * O_2(2,3) \end{cases} \tag{6.6.5}$$

The fourth layer is called the fuzzy "OR" operation layer. The activation rules for input to the same neuron (they have the same rule post) are synthesized, i.e.,

$$\begin{cases} O_4(1) = O_3(1) \\ O_4(1) = O_3(1) \\ O_4(3) = \sum\limits_{k=3}^{7} O_3(k) \\ O_4(4) = O_3(4) \\ O_4(5) = O_3(5) \end{cases} \tag{6.6.6}$$

The fifth layer is called the layer of introducing $MSE(n)$.

$$O_5 = \alpha(i) * O_4 * \mathrm{MSE}(n). \tag{6.6.7}$$

where $\alpha(i)$ stands for the control quantity, which is mainly used to adjust the output of the layer and complete the latter part of the fuzzy rule.

The sixth layer is called the output layer of defuzzification.

$$\Delta\mu = O_6 = \sum_{h=1}^{5} w_6(n) O_5 \tag{6.6.8}$$

where $\boldsymbol{f}_6(n)$ is the weight vector of the sixth layer and iteratively updated. To do this, the cost function is defined as:

$$J_1 = \frac{1}{2}[|\Delta\mu(n)|^2 - R_1]^2 \tag{6.6.9}$$

where

$$R_1 = \frac{E[|\mathrm{MSE}(n)|^4]}{E[|\mathrm{MSE}(n)|^2]} \tag{6.6.10}$$

$\boldsymbol{f}_6(n)$ is updated as follows by the steepest gradient descent method.

$$\boldsymbol{f}_6(n+1) = \boldsymbol{f}_6(n) - \eta\frac{\partial J_1(n)}{\partial \boldsymbol{f}_6(n)} = 2[|\Delta\mu(n)|^2 - R_1]\Delta\mu(n)\frac{\partial \Delta\mu(n)}{\partial \boldsymbol{f}_6(n)} \tag{6.6.11}$$

where η is the iteration step-size.

6.6.2 *Complex neural polynomial network algorithm*

6.6.2.1 *Complex neural polynomial network structure*

The complex neural polynomial network structure is shown in Figure 6.26. This structure consists of two simple parts: a single-layer neural network and a nonlinear memoryless processor [47]. Among them, the input of the nonlinear processor is limited by a nonlinear constraint function in a certain area, which ensures the convergence and stability of the nonlinear processor.

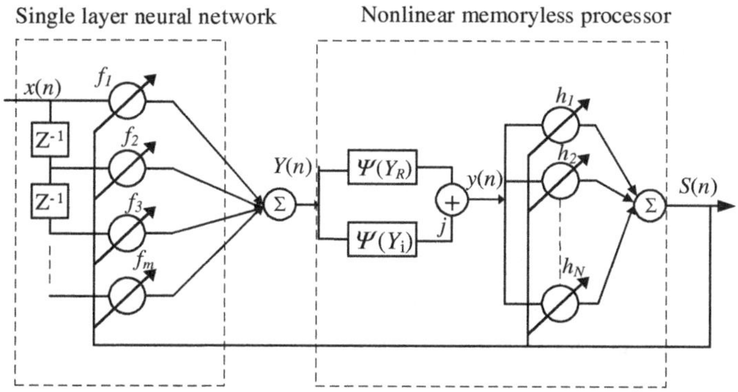

Figure 6.26 Structure of the complex neural polynomial network.

6.6.2.2 *Complex neural polynomial network algorithm*

The input of a single layer neural network can be expressed as:

$$\boldsymbol{Y}(n) = \boldsymbol{F}(n)\boldsymbol{X}^T(n) \tag{6.6.12}$$

where $\boldsymbol{X}(n)$ and $\boldsymbol{X}(n)$ are the input vector and weight coefficient vector of the tapped delay structure, respectively, and defined as:

$$\boldsymbol{X}(n) = [x(n), x(n-1), \dots, x(n-m+1)] \tag{6.6.13}$$

$$\boldsymbol{F}(n) = [f_1(n), f_2(n), \dots, f_m(n)] \tag{6.6.14}$$

The output of the single layer neural network filter is defined as:

$$y(n) = \psi(\boldsymbol{Y}(n)) = \psi(\boldsymbol{Y}_R(n)) + j\psi(\boldsymbol{Y}_I(n)) \tag{6.6.15}$$

where $\psi(\cdot)$ is an activation constraint function, i.e.,

$$\psi(x) = \frac{1-e^{-2x}}{1+e^{-2x}}, \quad \psi'(x) = \frac{4}{e^{2x}+e^{-2x}+2} \tag{6.6.16}$$

$$\begin{aligned} S(n) &= g(y(n)) \\ &= h_1(n) + h_2(n)y(n) + h_3(n)y^2(n) + \cdots \\ &= h_1(n) + h_2(n)\boldsymbol{X}(n)^T\boldsymbol{F}(n) \\ &\quad + h_3(n)\boldsymbol{F}(n)\boldsymbol{X}^T(n)\boldsymbol{F}(n)\boldsymbol{X}^T(n) + \cdots \\ &= \boldsymbol{H}(n)\boldsymbol{Z}^T(n) \end{aligned} \tag{6.6.17}$$

where $h_i(n)$ represents the weight coefficient, $\boldsymbol{Z}(n)$ is the input vector of the nonlinear processor, and $\boldsymbol{H}(n)$ is the weight vector.

$$\boldsymbol{Z}(n) = [1, y(n), y^2(n), y^3(n)] \tag{6.6.18}$$

$$\boldsymbol{H}(n) = [h_1(n), h_2(n), h_3(n), h_4(n)] \tag{6.6.19}$$

6.6.3 *Fuzzy neural network controlled complex neural polynomial Volterra channel blind equalization algorithm*

6.6.3.1 *64APSK signal*

With the increase of digital broadband satellite service and the expansion of system capacity, the spectrum bandwidth resources are becoming more and more scarce, and the use cost is more and more expensive. The signal transmission will consider the use of a combination of amplitude and phase of high-order modulation to improve the bandwidth efficiency. However, the traditional rectangular QAM has many amplitudes. When it passes through the satellite nonlinear transponder, some points are far away from the saturation point, and the power efficiency is not high. Moreover, the nonlinear distortion of the signal points close to the saturation state is very serious, which increases the complexity of the pre-distortion correction.

In the design of modulation constellation which adapts to high spectral efficiency of satellite channel, it is a trend to adopt a high-order scheme and reduce the fluctuation of signal amplitude as much as possible. In this trend, a star-shaped APSK signals with a circular constellation shape and a small number of circles will be the first choice [49, 50].

An APSK signal is equivalent to a special QAM signal, whose constellation diagram is composed of several concentric circles, and these signal points are, respectively, distributed on the concentric circles of the outer ring and the inner ring. On the one hand, the APSK signals have the characteristics of high band utilization of multi-modulus signals, on the other hand, its amplitude component is far less complicated than that of the QAM signal. Figure 6.27 shows the constellations of 64QAM and 64APSK signals with $n_1+n_2+n_3+n_4-$ APSK type. n_1 denotes the number of constellation points at the first inner circle of the $n_1+n_2+n_3+n_4-$ APSK type from the inside to the outside, n_2 stands for the number of constellation

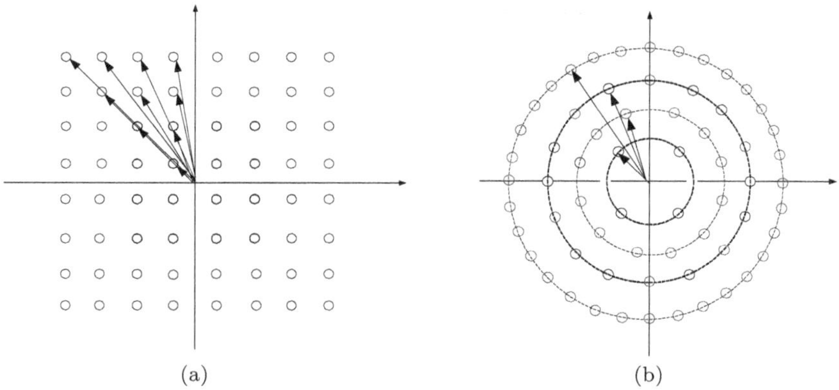

Figure 6.27 Constellations of 64QAM and 65APSK: (a) constellations of 64QAM and (b) constellations of $4+12+16+32$-APSK.

points at the second circle of the $n_1+n_2+n_3+n_4-$ APSK type, n_3 indicates the number of signal constellation points at the third circle, and n_4 represents that the number of signal constellation points at the fourth circle.

Figure 6.27 shows that 64QAM has 10 amplitude components, while $4+12+16+32$-APSK signal has only 4 amplitude components. In addition to the $4+12+16+32$-APSK signal, there is a $4+12+20+28$-APSK signal, and its constellations are also composed of four concentric circles and four different amplitude components. n_1, n_2, and n_3 are the number of constellation points of each circle of the $n_1+n_2+n_3+n_4-$APSK signal from the inner ring to the outer ring. For $n_1+n_2+n_3+n_4-$APSK signal, there are $4+12+16+32$-APSK signal, $4+12+20+28$-APSK signal, $4+8+24+28-$ APSK signal, and $4+8+20+32-$ APSK signal, and so on.

This section mainly studies the equalization problem of $4+12+20+28$-APSK and $4+12+16+32$-APSK signal passing through the Volterra channel. For the constellations of these two signals, the MATLAB simulation results are shown in Figure 6.28.

The constellation point set of the 64APSK signals is expressed as:

$$X_{\text{64-APSK}} = \begin{cases} r_1 \exp[j((2\pi/n_1 i) + \theta_1), i = 0,1,\ldots,n_1-1 \\ r_2 \exp[j((2\pi/n_2 i) + \theta_2), i = 0,1,\ldots,n_2-1 \\ r_3 \exp[j((2\pi/n_3 i) + \theta_3), i = 0,1,\ldots,n_3-1 \\ r_4 \exp[j((2\pi/n_4 i) + \theta_4), i = 0,1,\ldots,n_4-1 \end{cases} \tag{6.6.20}$$

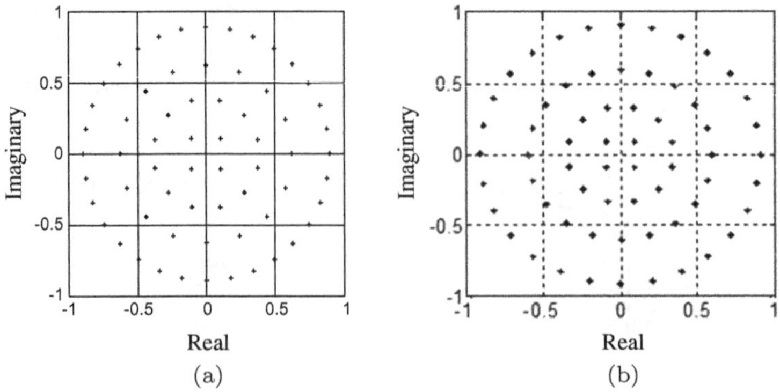

Figure 6.28 Constellations of 64APSK signals: (a) Constellations of 4 + 12 + 16 + 32-APSK; (b) Constellations of 4 + 8 + 20 + 32-APSK.

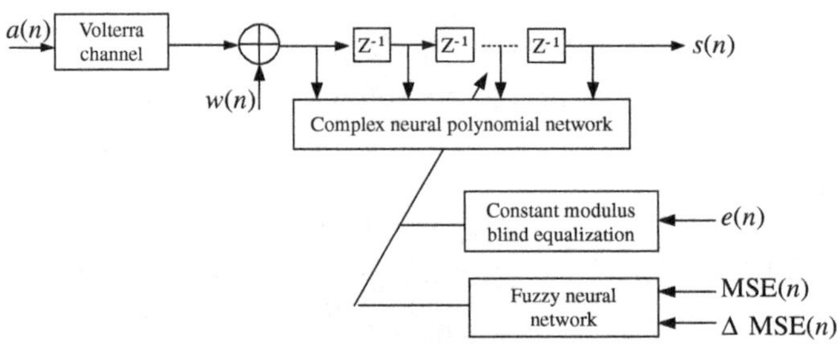

Figure 6.29 System model.

where n_1, n_2, n_3, and n_4 are the number of constellation points of the first, second, third, and fourth circle, respectively. r_1, r_2, r_3, and r_4 are the radius of each circle, and θ_1, θ_2, θ_3, and θ_4 are the initial phase of each circle.

6.6.3.2 *System block diagram and algorithm description*

The system model of complex neural polynomial constant modulus blind equalization algorithm based on the fuzzy neural network controller is shown in Figure 6.29.

In order to obtain the complex neural polynomial constant modulus blind equalization algorithm controlled by fuzzy neural network (FNN-CNP-CMA), we define the error function as:

$$e(n) = |S(n)|^2 - R^2 \tag{6.6.21}$$

where R^2 denotes the modulus value and is defined as:

$$R^2 = \frac{E[|a(n)|^4]}{E[|a(n)|^2]} \tag{6.6.22}$$

The cost function is defined as:

$$J(n) = -\frac{1}{2}|e(n)|^2 = -\frac{1}{2}e(n)e^*(n) \tag{6.6.23}$$

According to the principle of complex value, the weight vector $\boldsymbol{F}(n)$ can be expressed as:

$$\boldsymbol{F}(n) = \boldsymbol{F}_R(n) + j\boldsymbol{F}_I(n) \tag{6.6.24}$$

where $\boldsymbol{F}_R(n)$ and $\boldsymbol{F}_I(n)$ are the real part and the imaginary part of $\boldsymbol{F}(n)$, respectively.

According to Equations (6.6.17), (6.6.21), and (6.6.23), the following equations can be obtained.

$$\begin{aligned}
\frac{\partial J(n)}{\partial \boldsymbol{F}_R(n)} &= \frac{\partial[e(n)e^*(n)]}{\partial \boldsymbol{F}_R(n)} \\
&= \frac{\partial e(n)}{\partial \boldsymbol{F}_R(n)}e^*(n) + \frac{\partial e^*(n)}{\partial \boldsymbol{F}_R(n)}e(n) \\
&= \frac{\partial S(n)S^*(n)}{\partial \boldsymbol{F}_R(n)}e^*(n) + \frac{\partial S^*(n)S(n)}{\partial \boldsymbol{F}_R(n)}e(n) \\
&= \left[\frac{\partial S(n)}{\partial \boldsymbol{F}_R(n)}S^*(n) + \frac{\partial S^*(n)}{\partial \boldsymbol{F}_R(n)}S(n)\right]e^*(n) \\
&\quad + \left[\frac{\partial S^*(n)}{\partial \boldsymbol{F}_R(n)}S(n) + \frac{\partial S(n)}{\partial \boldsymbol{F}_R(n)}S^*(n)\right]e(n) \\
&= \frac{\partial S(n)}{\partial \boldsymbol{Z}(n)}\frac{\partial \boldsymbol{Z}(n)}{\partial y(n)}\frac{\partial y(n)}{\partial \boldsymbol{F}_R(n)}S^*(n)e^*(n) \\
&\quad + \frac{\partial S^*(n)}{\partial \boldsymbol{Z}^*(n)}\frac{\partial \boldsymbol{Z}^*(n)}{\partial y^*(n)}\frac{\partial y^*(n)}{\partial \boldsymbol{F}_R(n)}S(n)e^*(n) \\
&\quad + \frac{\partial S^*(n)}{\partial \boldsymbol{Z}^*(n)}\frac{\partial \boldsymbol{Z}^*(n)}{\partial y^*(n)}\frac{\partial y^*(n)}{\partial \boldsymbol{F}_R(n)}S(n)e(n) \\
&\quad + \frac{\partial S(n)}{\partial \boldsymbol{Z}(n)}\frac{\partial \boldsymbol{Z}(n)}{\partial y(n)}\frac{\partial y(n)}{\partial \boldsymbol{F}_R(n)}S^*(n)e(n)
\end{aligned}$$

$$
\begin{aligned}
&= \boldsymbol{H}(n)\boldsymbol{Z}'(n)\frac{\partial y(n)}{\partial \boldsymbol{F}_R(n)}S^*(n)e^*(n) \\
&\quad + \boldsymbol{H}^*(n)(\boldsymbol{Z}'(n))^*\frac{\partial y^*(n)}{\partial \boldsymbol{F}_R(n)}S(n)e^*(n) \\
&\quad + \boldsymbol{H}^*(n)(\boldsymbol{Z}'(n))^*\frac{\partial y^*(n)}{\partial \boldsymbol{F}_R(n)}S(n)e(n) \\
&\quad + \mathbf{H}(n)\boldsymbol{Z}'(n)\frac{\partial y(n)}{\partial \boldsymbol{F}_R(n)}S^*(n)e(n) \\
&= \boldsymbol{H}(n)\boldsymbol{Z}'(n)\psi'(\boldsymbol{Y}(n))\boldsymbol{X}(n)S^*(n)e^*(n) \\
&\quad + \boldsymbol{H}^*(n)(\boldsymbol{Z}'(n))^*(\psi'(\boldsymbol{Y}(n)))^*\boldsymbol{X}^*(n)S(n)e^*(n) \qquad (6.6.25)
\end{aligned}
$$

For the same reason, one gets

$$
\begin{aligned}
\frac{\partial J(n)}{\partial F_I(n)} &= \frac{\partial e(n)}{\partial F_I(n)}e^*(n) + \frac{\partial e^*(n)}{\partial \boldsymbol{F}_I(n)}e(n) \\
&= j\boldsymbol{H}(n)\boldsymbol{Z}'(n)\psi'(\boldsymbol{Y}(n))\boldsymbol{X}(n)S^*(n)e^*(n) \\
&\quad + jH^*(n)(Z'(n))^*(\psi'(Y(n)))^*X^*(n)S(n)e^*(n) \\
&\quad + j\boldsymbol{H}^*(n)(\boldsymbol{Z}'(n))^*(\psi'(\boldsymbol{Y}(n)))^*\boldsymbol{X}^*(n)S(n)e(n) \\
&\quad + j\boldsymbol{H}(n)\boldsymbol{Z}'(n)\psi'(\boldsymbol{Y}(n))\boldsymbol{X}(n)S^*(n)e(n) \qquad (6.6.26)
\end{aligned}
$$

So, $\boldsymbol{F}_R(n)$ and $\boldsymbol{F}_I(n)$ can be updated as follows:

$$
\begin{aligned}
\boldsymbol{F}_R(n+1) &= \boldsymbol{F}_R(n) - \mu_1\frac{\partial J(n)}{\partial \boldsymbol{F}_R(n)} \\
&= \boldsymbol{F}_R(n) - \mu_1\boldsymbol{H}(n)\boldsymbol{Z}'(n)\psi'(\boldsymbol{Y}(n))\boldsymbol{X}(n)S^*(n)e^*(n) \\
&\quad - \mu_1\boldsymbol{H}(n)\boldsymbol{Z}'(n)\psi'(\boldsymbol{Y}(n))\boldsymbol{X}(n)S^*(n)e(n) \\
&\quad - \mu_1\boldsymbol{H}^*(n)(\boldsymbol{Z}'(n))^*(\psi'(\boldsymbol{Y}(n)))^*\boldsymbol{X}^*(n)S(n)e^*(n) \\
&\quad - \mu_1\boldsymbol{H}^*(n)(\boldsymbol{Z}'(n))^*(\psi'(\boldsymbol{Y}(n)))^*\boldsymbol{X}^*(n)S(n)e(n) \\
&\qquad (6.6.27)
\end{aligned}
$$

$$
\begin{aligned}
\boldsymbol{F}_I(n+1) &= \boldsymbol{F}_I(n) - \mu_1\frac{\partial J(n)}{\partial \boldsymbol{F}_I(n)} \\
&= \boldsymbol{F}_I(n) - j\mu_1\boldsymbol{H}(n)\boldsymbol{Z}'(n)\psi'(\boldsymbol{Y}(n))\boldsymbol{X}(n)S^*(n)e^*(n)
\end{aligned}
$$

$$
\begin{aligned}
&- j\mu_1 \boldsymbol{H}(n)\boldsymbol{Z}'(n)\psi'(\boldsymbol{Y}(n))\boldsymbol{X}(n)S^*(n)e(n)\\
&- j\mu_1 \boldsymbol{H}^*(n)(\boldsymbol{Z}'(n))^*(\psi'(\boldsymbol{Y}(n)))^*\boldsymbol{X}^*(n)S(n)e^*(n)\\
&- j\mu_1 \boldsymbol{H}^*(n)(\boldsymbol{Z}'(n))^*(\psi'(\boldsymbol{Y}(n)))^*\boldsymbol{X}^*(n)S(n)e(n)
\end{aligned}
\tag{6.6.28}
$$

Therefore, the weight vector of a single-layer neural network can be written as:

$$
\boldsymbol{F}(n+1) = \boldsymbol{F}(n) - \mu_1 \left[\frac{\partial J(n)}{\partial \boldsymbol{F}_R(n)} + \frac{\partial J(n)}{\partial \boldsymbol{F}_I(n)}\right] \tag{6.6.29}
$$

where $\boldsymbol{Z}'(n)$ is the derivative of $\boldsymbol{Z}(n)$ and

$$
\boldsymbol{Z}'(n) = [0, 1, 2y(n), 3y(n)^2] \tag{6.6.30}
$$

In the same way, the partial derivative of $\boldsymbol{H}(n)$ can be obtained by:

$$
\frac{\partial J(n)}{\partial \boldsymbol{H}(n)} = e(n)\boldsymbol{Z}^*(n)S(n) \tag{6.6.31}
$$

Therefore, the weight vector of the nonlinear processor is updated by:

$$
\boldsymbol{H}(n+1) = \boldsymbol{H}(n) + \mu_2 e(n)\boldsymbol{Z}^*(n)S(n) \tag{6.6.32}
$$

where μ_1 and μ_2 are the iterative step-size of a single-layer neural network and a nonlinear processor, respectively. The iterative step-size μ_1 and μ_2 can become $\mu_1+\Delta\mu$, $\mu_2+\Delta\mu$ after being controlled by the fuzzy neural network controller. This constitutes a blind equalization algorithm which uses FNN controller to control the iterative step-size of linear filter and nonlinear memoryless processor.

6.6.4 *Simulation verification*

In order to verify the performance of the FNN-CNP-CMA, it is compared with a conventional constant modulus blind equilibrium algorithm (CMA) and complex neural polynomial network constant modulus blind equalization algorithm (CNP-CMA). The test channel used in this section was shown in Equation (6.6.33) and the channel

Table 6.7 Channel Volterra kernel.

First-order Kernel	Third-order Kernel
$h(0) = 0.8529 + j0.4502$	$h(0,0,1) = 0.0979 - j0.0979$
$h(1) = 0.0881 - j0.0014$	$h(0,0,2) = 0.1091 - j0.0615$
$h(2) = -0.0336 - j0.0196$	$h(0,0,3) = -0.1119 - j0.0252$
$h(3) = 0.0503 + j0.0433$	$h(1,1,0) = -0.028 - j0.0475$
	$h(3,3,0) = 0.0503 - j0.0503$

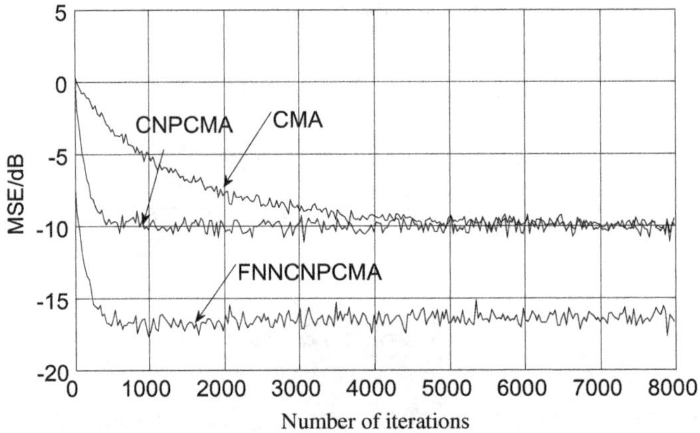

Figure 6.30 Convergence curves of all algorithms.

parameters were shown in Table 6.7. The transmitted signals were $4+12+16+36-\text{APSK}$ and $4+12+20+28-\text{APSK}$, respectively. SNR was set to 25. 50 Monte Carlo simulations and its results are shown in Figure 6.30.

$$\begin{aligned} x(n) = w(n) + \sum_{i=o}^{L_0} (h(i)a(n-i) + h(0,0,2)a^2(n)a^*(n-2) \\ + h(3,3,0)a^2(n-3)a^*(n) + h(0,0,1)a^2(n)a^*(n-1) \\ + h(0,0,3)a^2(n)a^*(n-3) + h(1,1,0)a^2(n-1)a^*(n)) \end{aligned} \tag{6.6.33}$$

Figure 6.30 shows that the convergence speed of the FNN-CNP-CMA has an improvement of about 5000 steps and 800 steps in

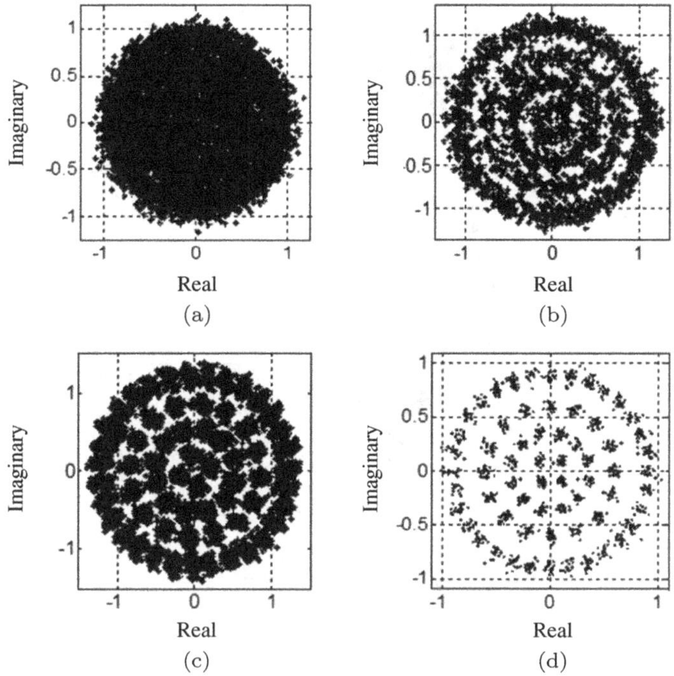

Figure 6.31 Simulation results of $(4+12+16+32)$ — APSK signal: (a) input constellations of equalizer; (b) constellations for CMA; (c) constellations for CNP-CMA and (d) constellations for FNN-CNP-CMA.

comparison with that of the CMA and the CNP-CMA, respectively; the steady-state error of FNN-CNP-CMA has a drop of about 11 dB and 8 dB in comparison with that of the CMA and the CNP-CMA, respectively.

Figure 6.31 shows the performance of the CNP-CMA and the FNN-CNP-CMA in equalizing $(4+12+16+32)$-APSK signals, and Figure 6.32 shows the equalization effect of the CNP-CMA and the FNN-CNP-CMA for $(4+12+20+28)$-APSK signals. It can be seen from the two figures that the output constellations of the FNN-CNP-CMA are clearer and more compact than those of the CMA and the CNP-CMA, its eye opening is clearer than that of the CMA and the CNP-CMA, and the phase shift is effectively overcome. The output signals of the equalizer are closer to the transmitted signals, so the equalization performance of the FNN-CNP-CMA outperforms that of the CNP-CMA.

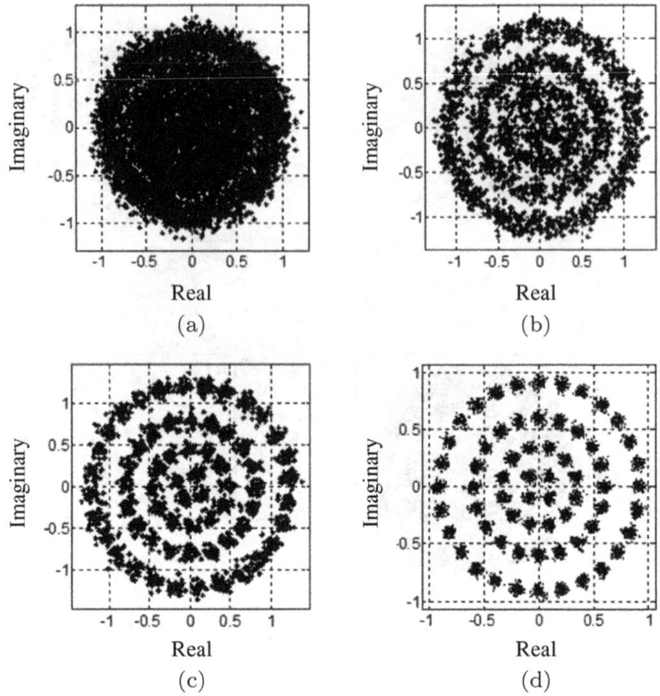

Figure 6.32 Simulation results of (4 + 12 + 20 + 28) — APSK signal: (a) input constellations of equalizer; (b) constellations for CMA; (c) constellations for CNP-CMA and (d) constellations for FNN-CNP-CMA.

References

[1] Zhuochen L. The current status and developmental trends of space travelling wave tube amplifier. *Space Electronic Technology*, 2012, 41(4): 28–34.
[2] Yecai G. *Blind Equalization Theory, Algorithm and Implementation of Fuzzy Wavelet Neural Network*. Beijing: Science Press, 2011.
[3] Zhenya H. *Adaptive Signal Processing*. Beijing: Science Press, 2002.
[4] Yecai G. *Blind Equalization Algorithm and Simulation of Communication Channel*. Beijing: Meteorological Publishing House, 2009.
[5] Benedetto S., Biglieri E. and Daffara R. Modeling and performance evaluation of nonlinear satellite links-a Volterra series approach. *IEEE Transactions on Aerospace and Electronic Systems*, 1979, (4): 494–507.
[6] He L., Malkemes R., Reed C., *et al.* A comparison of two blind equalization algorithms for broadband indoor wireless communication. *Proceedings of the Sixth International Symposium on Signal Processing and its Applications (Cat. No. 01EX467)*, 2001, 2: 505–508.

[7] Giugno L., Luise M. and Lottici V. Adaptive pre- and post-compensation of nonlinear distortions for high-level data modulations. *IEEE Transactions on Wireless Communications*, 2004, 3(5): 1490–1495.

[8] Piazza R., Shankar B., Grotz J., *et al.* Sensitivity analysis of multicarrier digital pre-distortion equalization techniques for non-linear satellite channels. *31st AIAA International Communications Satellite Systems Conference (ICSSC)*, IEEE Press, Williamsburg, VA, USA, 2013, p. 5692.

[9] Xiaojuan Z. and Changqi W. Adaptive Volterra equalizer based on variable step-size frequency domain LMS. *Journal of Circuits and Systems*, 2012, 17(4): 99–102.

[10] Guo J., Guo Y., Xu F., *et al.* Wavelet frequency domain weighted multi-modulus blind equalization algorithm based on lower order statistics. *Informatics and Management Science III*, 2013, 20(6): 765–773.

[11] Xiumei Z. Study on the algorithms of adaptive Volterra filter. Master's thesis, Hangzhou: Hangzhou University of Electronic Science and Technology, 2009.

[12] Benammar B., Thomas N., Poulliat C., *et al.* On linear MMSE based Turbo-equalization of nonlinear Volterra channels. *2013 IEEE International Conference on Acoustics, Speech and Signal Processing (ICASSP)*, IEEE Press, Vancouver, BC, Canada, 2013, pp. 4703–4707.

[13] Wang D. and Wang D. Comparison of nonlinear CMA-criterion-based blind equalizers. *2010 6th International Conference on Wireless Communications Networking and Mobile Computing (WiCOM)*, IEEE Press, Chengdu, China, 2010, pp. 1–4.

[14] Shan Z. Blind equalization algorithm of Volterra satellite channel. Master's thesis, Nanjing: Nanjing University of Information Science and Technology, 2016.

[15] Yecai G., Weiei M., Shan Z., *et al.* On linear MMSE based Turbo-equalization of nonlinear Volterra channels. *Journal of System Simulation*, 2016, 28(11): 2742–2746.

[16] Liu D. N. and Fitz M. P. Iterative equalization in nonlinear satellite channels. *The 7th International Symposium on Turbo Codes and Iterative Information Processing (ISTC)*, IEEE Press, Gothenburg, Sweden, 2012, pp. 220–224.

[17] Qiang L., Xia L. and Xianpin L. *Iterative Equalization Technique in Wireless Communication.* Beijing: National Defense Industry Press, 2010.

[18] Colavolpe G. and Piemontese A. Novel SISO detection algorithms for nonlinear satellite channels. *Wireless Communications Letters*, IEEE, 2012, 1(1): 22–25.

[19] Fang L., Guo Q. and Huang D., *et al.* A low cost soft mapper for turbo equalization with high order modulation. *International SoC Design Conference (ISOCC)*, IEEE Press, Jeju, Korea (South), 2012, pp. 305–308.
[20] Delamotte T., Bauch G. and Dantona V. Transmit precoding for MIMO-BICM-ID satellite system with nonlinear power amplifiers. *WSA, Proceedings of the 17th International ITG Workshop on Smart Antennas*, Piscataway: IEEE Press, 2013, pp. 1–8.
[21] Ryan W. E. and Gutierrez A. Performance of adaptive Volterra equalizers on nonlinear magnetic recording channels. *IEEE Transactions on Magnetics*, 1995, 31(6): 3054–3056.
[22] Burnet C. E. and Cowley W. G. Performance analysis of turbo equalization for nonlinear channels. *International Symposium on Information Theory, 2005.* ISIT 2005. Proceedings, IEEE Press, Adelaide, SA, Australia, 2005, pp. 2026–2030.
[23] Ngo N. H., Pietrobon S. S., Barbulescu S. A., *et al.* Optimization of 16APSK constellation with a new iterative decoder on the nonlinear channel. *Communications Theory Workshop, 2006. Proceedings, 7th Australian*, IEEE Press, Perth, WA, Australia, 2006, pp. 71–75.
[24] Benedetto S. and Biglieri E. Nonlinear equalization of digital satellite channels. *IEEE Journal on Selected Areas in Communications*, 1983, 1(1): 57–62.
[25] Hermann R. Volterra modeling of digital magnetic saturation recording channels. *IEEE Transactions on Magnetics*, 1990, 26(5): 2125–2127.
[26] Witrisal K., Leus G., Pausini M., *et al.* Equivalent system model and equalization of differential impulse radio UWB systems. *IEEE Journal on Selected Areas in Communications*, 2005, 23(9): 1851–1862.
[27] Gutierrez A. and Ryan W. E. Performance of adaptive Volterra equalizers on nonlinear satellite channels. *1995 IEEE International Conference on Communications. ICC'95 Seattle, 'Gateway to Globalization'*, IEEE, 1995, 1: 488–492.
[28] Tuchler M. and Singer A. C. Turbo equalization: An overview. *IEEE Transactions on Information Theory,* 2011, 57(2): 920–952.
[29] Douillard C., Jézéquel M., Berrou C., *et al.* Iterative correction of intersymbol interference: Turbo- equalization. *Transactions on Emerging Telecommunications Technologies*, 1995, 6(5): 507–511.
[30] Ampeliotis D., Rontogiannis A. A., Berberidis K., *et al.* Turbo equalization of nonlinear satellite channels using soft interference cancellation. *Advanced Satellite Mobile Systems, 2008. ASMS 2008. 4th. IEEE*, IEEE Press, Bologna, Italy, 2008, pp. 289–292.
[31] Benammar B., Thomas N., Poulliat C., *et al.* On linear frequency domain Turbo-equalization of nonlinear Volterra channels. *2014*

8th International Symposium on Turbo Codes and Iterative Information Processing (ISTC), IEEE Press, Bremen, Germany, 2014, pp. 183–187.

[32] Im S. Adaptive equalization of nonlinear digital satellite channels using a frequency-domain Volterra filter. *Military Communications Conference*, 1996, 3, 843–848.

[33] Yangwang F., Licheng J. and Jin P. Volterra filter equalization: A frequency domain approach. *5th International Conference on Signal Processing Proceedings, 2000*. IEEE Press, Beijing, China, WCCC-ICSP 2000, Vol. 1, pp. 302–305.

[34] Liu Z. and Li C. Recursive least-squares for censored regression. *IEEE Transactions on Signal Processing*, 2017, 65(6): 1565–1579.

[35] Hong X., Gao J. and Chen S. Zero-attracting recursive least squares algorithms. *IEEE Transactions on Vehicular Technology*, 2017, 66(1): 213–221.

[36] Navia-Vazquez A. and Arenas-Garcia J. Combination of recursive least P-norm algorithms for robust adaptive filtering in alpha-stable noise. *IEEE Transactions on Signal Processing*, 2012, 60(3): 1478–1482.

[37] Shao M. and Nikias C. L. Signal processing with fractional lower order moments: Stable processes and their applications. *Proceedings of the IEEE*, 1993, 81(7): 986–1010.

[38] Weng B. and Barner K. E. Nonlinear system identification in impulsive environments. *IEEE Transactions on Signal Processing*, 2005, 53(7): 2588–2594.

[39] Chen B. and Príncipe J. C. Maximum correntropy estimation is a smoothed MAP estimation. *IEEE Signal Processing Letters*, 2012, 19(8): 491–494.

[40] Belge M. and Miller E. L. A sliding window RLS-link adaptive algorithm for filtering alpha-stable noise. *IEEE Signal Process Letters*, 2000, 7(4): 86–89.

[41] Sayed A. H. *Fundamental of Adaptive Filtering*. John Wiley & Sons, USA, 2003.

[42] Lee J. and Mathews V. J. A fast recursive least squares adaptive second order Volterra filter and its performance analysis. *IEEE Transactions on Signal Processing*, 1993, 41(3): 1087–1102.

[43] Yecai G. and Lihua W. A hybrid wavelet neural network blind equalization algorithm based on Fuzzy controlling. *Acta Electronica Sinica*, 2011, 39(4): 975–980 (in Chinese).

[44] Zhenxing L., Yecai G., Min G., *et al.* Multi-wavelet fuzzy neural network blind equalization algorithm in underwater acoustic channel. *Acta Armamentarii*, 2010, 31(9): 1137–1144 (in Chinese).

[45] Yecai G., Shan Z., Lu Z., *et al.* Satellite channel complex neural polynomial network blind equalization system and method. Nanjing University of Information Science and Technology, China Patent: CN105162738A, 2015-07-30 (in Chinese).

[46] Yecai G. and Zhenxin L. Fuzzy neural network blind equalization method based on balanced orthogonal multiwavelet transform. Nanjing University of Information Science and Technology, China Patent: CN101958860A, 2010-08-30.

[47] Yecai G. and Lihua W. Hybrid wavelet neural network controlled blind equalization method based on fuzzy neural network. Nanjing University of Information Science and Technology, China Patent: CN 101924718A, 2010-08-30.

[48] Yecai G., Menghan Z., Shan Z., *et al.* Multi-wavelet fuzzy neural network blind equalization algorithm in underwater acoustic channel. *Journal of Data Acquisition and Processing*, 2017, 32(6): 1082–1088.

[49] Nakazawa M., Yoshida M., Hirooka T., *et al.* QAM quantum stream cipher using digital coherent optical transmission. *Optics Express*, 2014, 22(4): 4098–4107.

[50] Chow Y. C., Nix A. R., McGeehan J. P. Analysis of 16-APSK modulation in AWGN and Rayleigh fading channel. *Electronics Letters*, 1992, 28(17): 1608–1610.

[51] Yangwang F. and Licheng J. Identification of MIMO nonlinear system: An approach to Volterra series. *Journal of Circuits and Systems*, 2000, 5(4): 38–42.

Chapter 7

Satellite and Molecular MIMO Channel Markov Chain Model Based on Machine Learning

Summary

This chapter takes multiple input and multiple output (MIMO) system as the research object. Based on the analysis of single input single output (SISO) channel model characteristics, MIMO channel enhanced two-state Markov chain model and improved enhanced two-state Markov chain model modeling and test analysis method are presented. Aiming at the land mobile satellite (LMS) network as an important part of the future mobile network and broadcasting systems, a simple empirical–stochastic dual-circular polarization LMS-MIMO narrowband channel model was proposed and verified. Aiming at the low transmission rate of molecular communication, an efficient modeling and simulation method of molecular MIMO channel based on machine learning was analyzed by using molecular MIMO technology and machine learning method.

7.1 Single Input Single Output (SISO) and Multiple Input and Multiple Output (MIMO) Channel Enhanced Two-State Markov Chain Model

The design and evaluation of the air interface for modern land mobile satellite (LMS) systems between 1 and 4 GHz requires the use of reference generative propagation channel models. In a sense, it can generate baseband complex series of channel time evolution. Among the possible generative modeling methods, the empirical stochastic model has been widely used in the past two decades. In fact, it usually requires few input parameters and reduces computation time. For example, the SISO channel three-state Markov chain model was used by the TM-SSP DVB group [1, 2]. In some cases, the available input parameters were affected by measurement biases and uncertainties due to the limited length of the original experimental dataset. In recent years, the SISO channel enhanced two-state Markov chain model [3], three-state Markov chain model, and improved enhanced two-state Markov chain model [4–6] have been developed. Better dynamic ranges and longer distances of the improved enhanced two-state Markov chain model have been demonstrated by performing L-band, S-band, and C-band experiments [4, 7]. In 2011 and 2012, the French Space Agency (CNES) performed experiments at 2.2 GHz and 3.8 GHz for various elevation angles, environments, and antenna configurations. These experimental datasets have been used to verify the improved enhanced two-state model [4, 5], the dual polarization MIMO configurations, and the MIMO configurations extended by SISO model [4]. In the verification, fading factor, Rice factor, level crossing rate (LCR), and average fading were used as evaluation indicators [8, 9].

7.1.1 *Two improved enhanced two-state Markov chain models*

The enhanced two-state semi-Markov model described in [4, 5] is an extension of the reference generative model for LMS propagation channels. It concerns statistical modeling (introduction of confidence intervals), Doppler spectrum modeling, MIMO extension, etc.

7.1.1.1 *Experimental datasets*

The Auch 2012 dataset includes SISO, single input multiple output (SIMO), and MIMO experimental data collected at 2.2 and 3.8 GHz at various conditions:

(1) Elevation angles: A helicopter emulating a satellite at 20°, 30°, 45°, 60°, and 70°.
(2) Environments: small village, woodland, suburban, urban, residential areas, commercial districts, and industrial area.

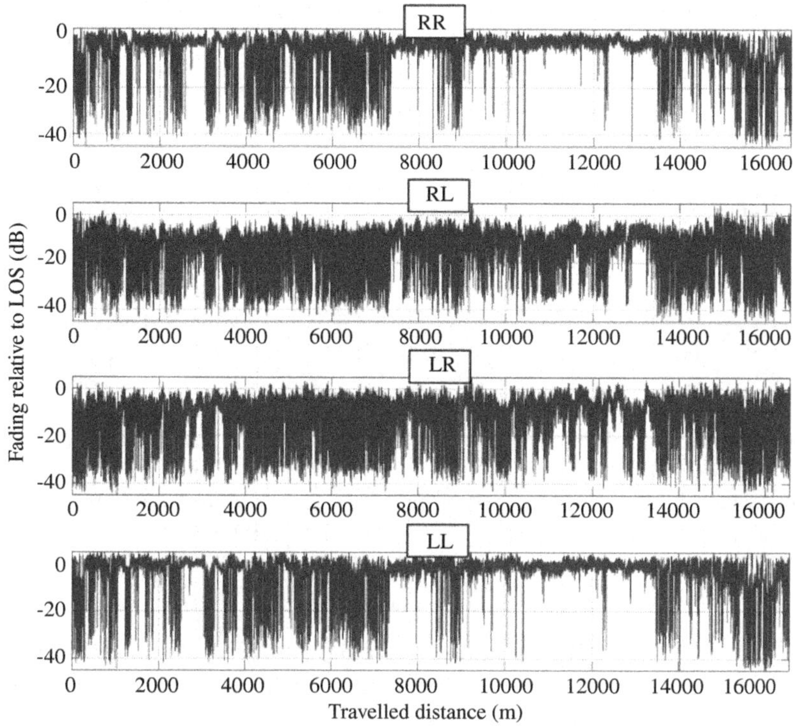

Figure 7.1 Dual-polarization 2 × 2 MIMO fading channel series collected in Auch at S-band, 30° elevation.

(3) Antenna configurations: SISO (RHCP/LHCP/dipole), MIMO (RHCP + LHCP), SIMO (two dipoles, RHCP + dipole).
(4) Total length of measurement route: 16.5 km.

These measurements and post-processing results are given in [9]. An example of the MIMO series measurement is shown in Figure 7.1.

This section addresses the results obtained using dataset collected at S-band with circular polarization antennas (SISO and MIMO configurations).

7.1.1.2 *SISO channel two-state semi-Markov input parameters*

The input parameters of the SISO enhanced two-state semi-Markov model (see [4]) are described in Table 7.1

Table 7.1 Semi-Markov two-state model parameters.

Domain	Parameter	Description
Two-state semi-Markov chain	$(m, \sigma)_{G,B}$	Log-normal distribution parameters for states duration
C. Loo distribution	$(M, \sum)_{G,B}$	Normal distribution parameters for <A> (dB)
	Std(A) = $g_{G,B}$ (<A>)	Expression linking the <A> in one state to the standard deviation of A.
	<K> = h(<A>)	Expression linking state average values of A and K (carrier-to-multipath ratio)
Direct path amplitude correlation	L_{corr}	Correlation length for both good and bad states
Transitions	Sp = Sp(ΔA)	Slope in the transitions between states

The "G" and "B" indices respectively refer to good and bad states. C. Loo distribution of the fading amplitude within one channel event (either good or bad) is assumed. <A> and std(A) respectively refer to the mean and standard deviation of the channel direct path amplitude A (in dB). K refers to the Rice factor (in dB) given by K = A–MP, where MP is the multipath power assumed to be constant within one channel event. These input parameters have been inferred in suburban and wooded environments at 30° and 60° elevation angles.

7.1.1.3 *Confidence intervals*

Since the input parameters for the enhanced two-state semi-Markov chain model are obtained from experimental statistics, they can be affected by statistical uncertainties. In particular, the model relies on the following assumptions:

(1) Good- and bad-state durations belong to a log-normal distribution.
(2) Average direct path amplitude in good and bad states belongs to a Gaussian distribution.

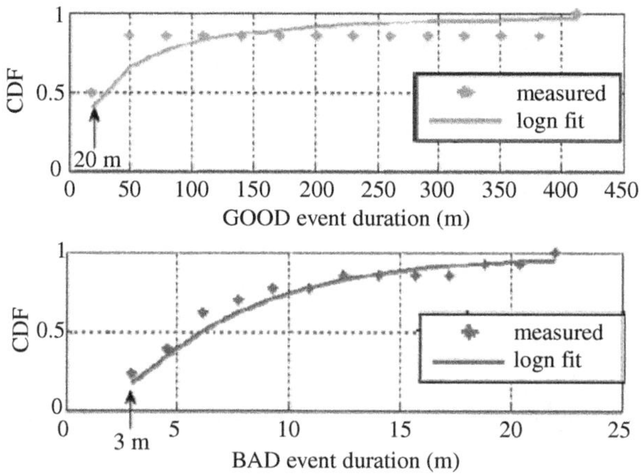

Figure 7.2 State duration distributions and minimum state length (top: good, bottom: bad).

In order to avoid the extreme instantaneous state durations and the mean direct path amplitude values far from the experimental values, the confidence intervals of the model have been investigated, and the following conditions are introduced in the model:

(1) *State duration*: the state duration is no lower than the minimum measurement state duration.
(2) *Mean direct path amplitude*: the optimum confidence intervals are 80% and 90% for the bad and good states, respectively.

Figure 7.2 shows the relationship between the cumulative distribution function (CDF) and the duration of the good and bad states, and Figure 7.3 shows the relationship between the CDF and the channel direct path for the example of suburban environment at 60° elevation angle at 2.2 GHz.

7.1.1.4 *Doppler spectrum*

Few results in the existing literatures can describe the realistic Doppler spectrum of LMSs. From a conservative perspective, the classic Jakes spectrum is widely used, though it relies on theoretical assumptions suitable for terrestrial links only. Butterworth-shaped filters are also commonly used. An empirical parameterization of such a Doppler spectrum is provided in [3]:

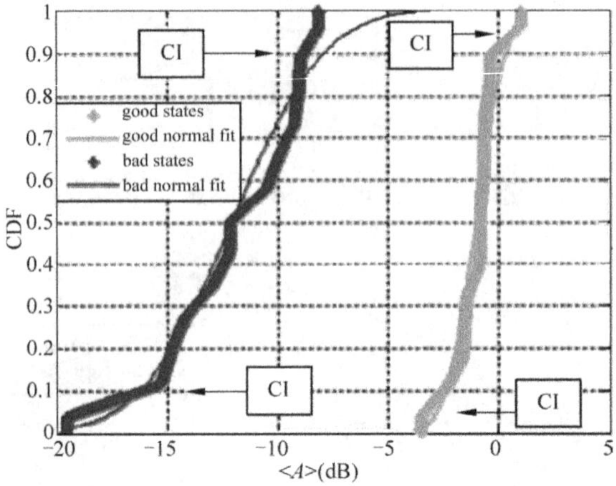

Figure 7.3 Mean direct path amplitude distributions and confidence intervals (CIs).

(1) maximum ripple of 3 dBs up to $0.9 \times v_M/\lambda(v_M$ is the mobile speed);
(2) attenuation of 100 dBs at $3 \times v_M/\lambda$.

In [4], an approach based on a sum-of-sinusoids (SoS) method is proposed depending on two parameters (Table 7.2):

(1) K_0: restriction parameter of the classical spectrum to its lowest frequency components;
(2) χ: multiplicative parameter of the maximum Doppler frequency.

In this section, a non-analytical method based on experimental LCR statistics is proposed to derive the optimum input parameters for the SoS model. This indirect method is particularly convenient for experimental datasets with the measured channel amplitude only. In the framework of Auch 2012 post-processing, a third parameter was added to the SoS model: α_d, the Doppler shift of the direct path normalized by the maximum Doppler frequency v_M/λ. For a non-shadowed direct path and a σ_0^2 multipath power, the Doppler spectrum is given as

$$S(f) = \delta(f - \alpha_d f_{\max}) + \begin{cases} \dfrac{\sigma_0^2}{\pi f_{\max}\sqrt{1 - \left(\frac{f}{\chi \times f_{\max}}\right)^2}} & \text{if } |f| < \chi \cdot K_0 f_{\max} \\ 0 & \text{if } |f| \geq \chi \cdot K_0 f_{\max}, \end{cases} \tag{7.1.1}$$

Table 7.2 Auch 2012 experimental input parameters.

Parameters	Wooded/30°	Wooded/60°	Suburban/30°	Suburban/60°
$(m,\sigma)_G$	(2.56,1.24)	(2.93,1.32)	(3.01,1.42)	(3.41,1.43)
$(m,\sigma)_B$	(2.38,1.18)	(2.20,1.11)	(2.45,0.76)	(1.99,0.71)
$(M,\sum)_G$	(−1.32,1.66)	(−0.62,1.17)	(−0.70,1.21)	(−0.78,0.80)
$(M,\sum)_B$	(−10.42,2.44)	(−9.96,2.22)	(−11.98,3.47)	(−12.14,3.18)
g_G	(−0.17,0.72)	(−0.25,0.54)	(−0.13,0.90)	(−0.23,0.28)
g_B	(−0.05,2.23)	(−0.09,1.47)	(−0.16,0.28)	(−0.29,−0.60)
h	(1.13,12.89)	(1.43,16.15)	(0.86,13.67)	(1.31,17.92)
L_{corr}	1.09	1.76	1.71	2.02
Sp	(0.03,3.85)	(0.11,4.22)	(0.11,3.30)	(−0.41,8.74)

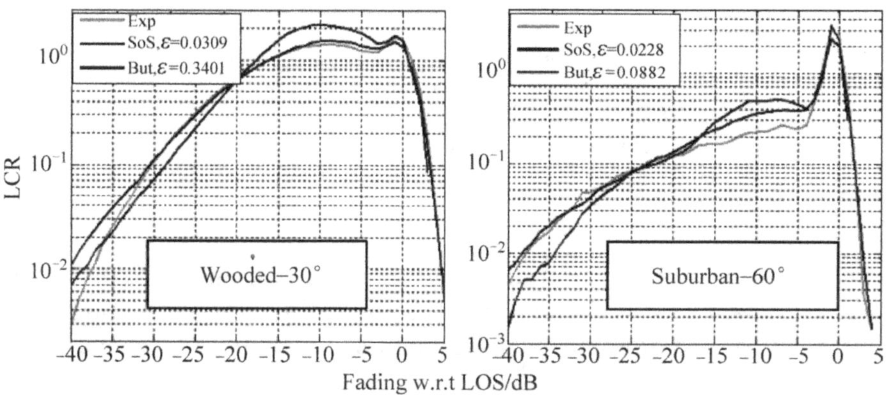

Figure 7.4 LCR statistics for experimental and modeled Doppler spectra (left: wooded at 30°; right: suburban at 60°).

where f is the spatial frequency and $f_{\max}$ is the maximum Doppler frequency. The χ, K_0, and α_d parameters have been optimized for each environment, elevation angle, and link frequency of Auch 2012 database following the test-and-trial approach described in [4]. Figure 7.4 gives sample comparisons of experimental and synthetic LCR statistics obtained with both SoS and Butterworth models when the wooded elevation angle is 30° and the suburban elevation angle is 60° at 2.2 GHz. Table 7.3 shows the root mean square error (RMSE) between the experimental and synthetic LCR and AFD statistics for the SoS and Butterworth models. From these sample results, it can be observed that the LCR and AFD RMSEs are reduced by

Table 7.3 AFD and LCR RMSEs for the Butterworth and SoS Doppler spectrum models (S-band).

	SoS		Butterworth	
	$\varepsilon_{\mathbf{AFD}}$	$\varepsilon_{\mathbf{LCR}}$	$\varepsilon_{\mathbf{AFD}}$	$\varepsilon_{\mathbf{LCR}}$
Wooded (30°)	0.012	0.03	0.083	0.34
Suburban (30°)	0.03	0.10	0.02	0.09
Suburban (60°)	0.02	0.12	0.02	0.25

Table 7.4 S-band experimental parameters of the sum-of sinusoids model.

	χ	K_0	α
Wooded (30°)	2.5	0.21	−0.51
Wooded (60°)	2.6	0.22	−0.31
Suburban (30°)	1.5	0.39	−0.56
Suburban (60°)	0.9	0.189	−0.30

about one order of magnitude when using the optimum SoS model. Table 7.4 shows the input parameters of the SoS model assessed for wooded and suburban environments at 30° elevation angles at S-band.

7.1.1.5 *MIMO extension*

A MIMO extension of the SISO semi-Markov chain model is proposed in [4]. This MIMO model takes into account the correlation of the direct path amplitudes experienced on co-polarized and cross-polarized channels. For a dual-circular polarization MIMO link within a channel state (either good or bad), the direct path amplitude SH is modeled as follows:

$$\begin{pmatrix} A_{RR} \\ A_{RL} \\ A_{LR} \\ A_{LL} \end{pmatrix} = \begin{pmatrix} m_{RR} \\ m_{RL} \\ m_{LR} \\ m_{LL} \end{pmatrix} + \begin{pmatrix} \sigma_{RR} \\ \sigma_{RL} \\ \sigma_{LR} \\ \sigma_{LL} \end{pmatrix} \cdot \boldsymbol{C}_A^{\frac{1}{2}} \begin{pmatrix} X_1 \\ X_2 \\ X_3 \\ X_4 \end{pmatrix}, \tag{7.1.2}$$

where

(1) $Y \cdot Z$ is the Hadamard product of the Y and Z matrices,
(2) R and L indices respectively refer to RHCP and LHCP (first letter: transmit antenna, second letter: receive antenna),
(3) $(X_i)_{1<i<4}$ are uncorrelated Gaussian processes with zero mean and unit variance,
(4) m and σ are respectively the mean and standard deviation of the direct path amplitude within the channel event, and
(5) $\boldsymbol{C}_A = (c_{ij})_{i<i,j<4}$ is the correlation matrix of the direct path amplitudes converted into zero-mean unit-variance Gaussian processes [4]. For example, for the Auch 2012 S-band dataset, in suburban environment at 60° elevation, $\boldsymbol{C}_A$ is obtained as follows:

$$\begin{array}{cc} & \begin{array}{cccc} RR & LR & LL & RL \end{array} \\ \boldsymbol{C}_A = & \begin{pmatrix} 1 & 0.30 & 0.73 & 0.17 \\ 0.30 & 1 & 0.14 & 0.30 \\ 0.73 & 0.14 & 1 & 0.31 \\ 0.17 & 0.30 & 0.31 & 1 \end{pmatrix} \begin{array}{c} RR \\ LR \\ LL \\ RL \end{array}. \end{array}$$

In [4,10], it is assumed that all channels are simultaneously in the same channel state, and $m_{RR} = m_{LL} = m_{co}$ and $\sigma_{RR} = \sigma_{LL} = \sigma_{co}$. Here, these assumptions are considered to be valid so that m_{co} can be randomly extracted from the $(\mathrm{M}, \sum)_{\mathrm{B,G}}$ Gaussian distributions. Then, σ_{co} can be solved using $\sigma_{co} = g_{G,B}(m_{co})$. However, [4] does not explicitly address the way to get $m_{LR} = m_{RL} = m_{\text{cross}}$ and $\sigma_{LR} = \sigma_{RL} = \sigma_{\text{cross}}$. In order to get a full MIMO analytical model, a polynomial relation between m_{cross} and m_{co} is assumed. From the Auch 2012 experiment datasets, the relationship between co-polarized and cross-polarized parameters can be well established using a second-order polynomial function:

$$m_{\text{cross}} = f(m_{co}) = f_2 \times m_{co}^2 + f_1 \times m_{co} + f_0, \tag{7.1.3}$$

where σ_{cross} can be solved using $\sigma_{\text{cross}} = g_{G,B}(m_{\text{cross}})$. The relationship between the co-polarized and cross-polarized mean direct paths in the suburban environment at S-band is shown in Figure 7.5. The blue circles represent all direct path amplitude samples, whereas the red circles represent only mean values

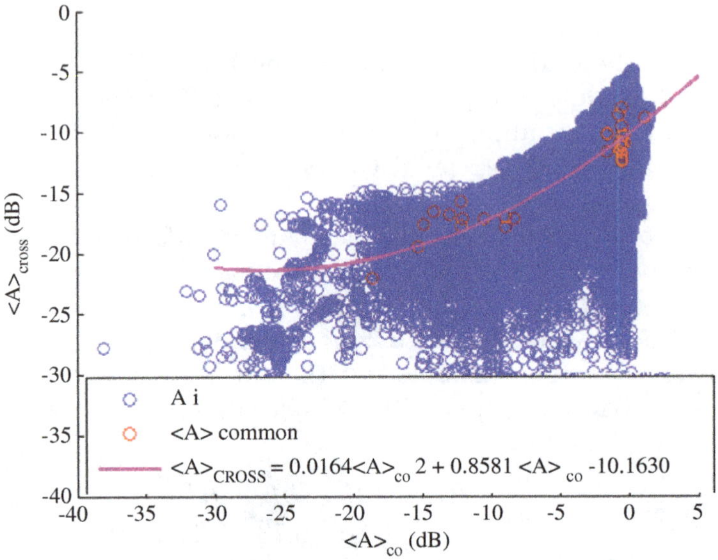

Figure 7.5 Connection between co-polarized and cross-polarized mean direct path amplitude.

obtained for good-state and bad-state events. The curve is the second-order polynomial function fitting the red circles. The asymptotic behavior of this function may be due to the antennas CPD.

7.1.2 *Testing analysis*

An exhaustive testing of the enhanced two-state semi-Markov model has been performed over the whole Auch 2012 database. Three types of experimental and synthetic statistics have been compared:

(1) Fading margin (in dB) complementary cumulative distribution function (CCDF);
(2) Rice factor (in dB) CCDF;
(3) AFD (in m) statistics.

The fading margin and Rice factor distributions have been computed for the following percentages (based on values proposed by ITU-R Study Group 3 Databank Tables): $[0.1, 1, 5, 10, 30, 50, 90, 99, 99.9]\%$.

The AFD has been computed for fading thresholds from −20 to 0 dB (with 1 dB steps).

The testing variable ε is the root mean square error (RMSE) difference between fading margin values (error in dB), Rice factor values (error in dB), and AFD values (error in m).

Sample results for suburban environment at 60° elevation angle at 2.2 GHz are shown in Figures 7.6 and 7.7, where "*exp*", "*IC all*", "*IC loo*", and "*no IC*" respectively refer to experimental statistics, and model results with the following options: all (state-duration and C. Loo) confidence intervals, C. Loo confidence intervals only, and no confidence intervals. Table 7.5 shows the results for the two tested environments and elevation angles.

In the model validation, low RMSEs are observed. In addition, for these specific cases, all confidence intervals outperformed slightly better at 60° elevation and slightly worse at 30° elevation. This can be explained by the fact that at higher elevation angles, few bad-channel events have been experienced in Auch 2012. Therefore, the experimental state duration and fading distributions in the bad state

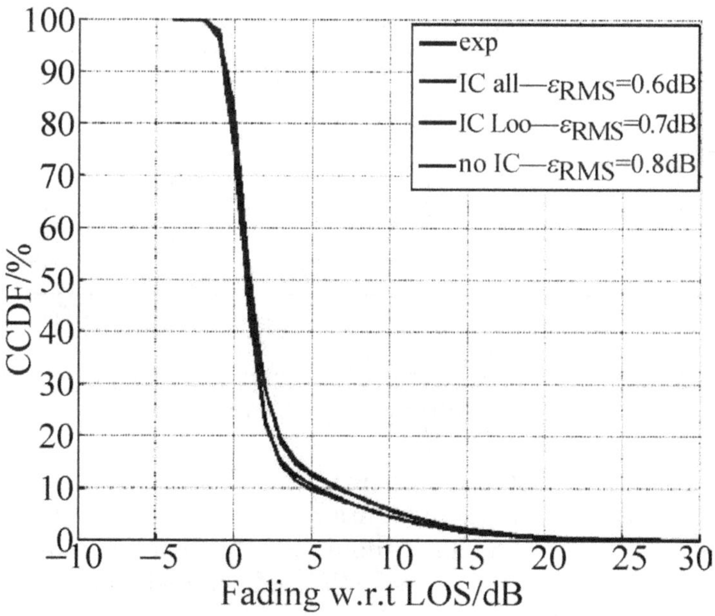

Figure 7.6 Testing of fading margin statistics (suburban, 60°, S-band).

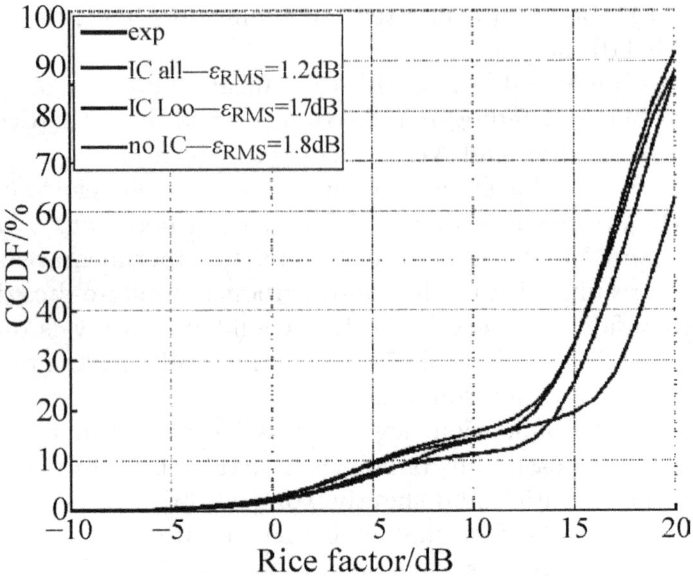

Figure 7.7 Testing of Rice factor statistics (suburban, 60°, S-band).

Table 7.5 Testing analysis results.

	Fade Margin (dB)		**Rice Factor (dB)**		**AFD (m)**	
	No IC	**IC all**	**No IC**	**IC all**	**No IC**	**IC all**
Wooded/30°	0.7	1.1	1.2	0.8	0.08	0.09
Wooded/60°	0.5	0.2	1.3	1.3	0.04	0.04
Suburban/30°	0.6	1.7	1.6	2.3	0.03	0.05
Suburban/60°	0.8	0.6	1.8	1.2	0.06	0.06

are less reliable. However, the confidence intervals can prevent from drawing unrealistic state durations or C. Loo parameters values, and it should be more likely to be used in the final model.

As far as the MIMO extension to the SISO model is concerned, the performance of the fading margin statistic has been experimentally verified. Figure 7.8 shows simulated ("S") and measured ("M") values for co-polarized and cross-polarized channel statistics for suburban environment at 60° elevation. For this specific case, the MIMO model seems to reproduce the experimental fading margin statistics

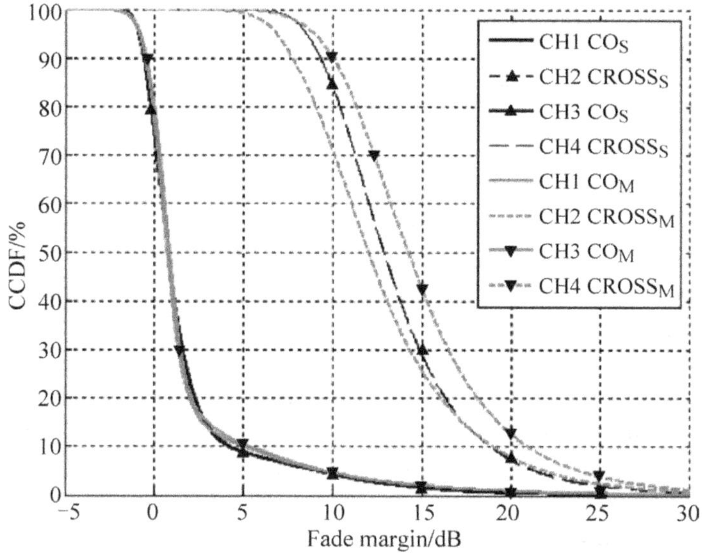

Figure 7.8 Testing of MIMO fading margin statistics (suburban, 60°, S-band).

quite well. Similar results have been observed for wooded and suburban environments at 30° and 60° elevation angles.

7.2 LMS-MIMO Channel Empirical–Stochastic Markov Model

In terms of increasing capacity and diversity gains, it is beneficial to apply MIMO techniques to LMSs [11]. In recent years, the use of MIMO channel model in terrestrial wireless systems including the next-generation wireless networks, IEEE 802.11n [12] as well as wide area networks, IEEE 802.16m [13] and long-term evolution of third-generation mobile (3GPP LTE) [14] has become widespread. Recently, the use of MIMO for LMS has gained interest with regards to the satellite based on digital video broadcasting standards DVB-SH and DVB-NGH [15]. However, it is necessary to establish a suitable and simple channel model of the radio wireless environment before the LMS-MIMO systems are widely used. This section defines an empirical–stochastic channel model for such use.

As a typical LMS channel model, as well as the model discussed in this section, large-scale fading includes Markov chains to represent the "on/off" nature of the channel and filtered log-normal simulation to represent the shadowing effects, and Rician simulation to represent the small-scale fading effects [16]. However, this model improves the capabilities to the MIMO case, where its stochastic properties are derived from an S-band tree-lined road measurement campaign using dual-circular polarizations at low elevations. In addition, this model is in particular compared to existing models to date [17] considers the interdependence of small-scale fading. Further, the most appropriate means to accommodate this interdependence is to use a Rician fading model, where the co-polar and cross-polar components are suitably correlated.

This section first describes the typical factors unique to a dual-circular polarization LMS-MIMO channel modeling [18], including the need for a four-state Markov chain to form simultaneous dual polar shadowing models as well as small-scale fading models. Second, the Markov chain behavior in the LMS-MIMO channel is verified. A step-by-step guide is then provided to generate the proposed empirical–stochastic model informed by measurement data. Finally, validation tests are carried out to clarify the application of this model in both line-of-sight (LOS) and non-line-of-sight (NLOS) regions.

7.2.1 *LMS-MIMO channel model*

The LMS-MIMO channel model is a 2×2 MIMO system, wherein the two antennas at each end are dual-circular polarized antennas with right-hand and left-hand circular polarizations (RHCP and LHCP), as illustrated in Figure 7.9, which is suitable for the L-band and S-band frequency ranges, so that the ionospheric and tropospheric effects are negligible. In the ionosphere, the largest impact would be the Faraday rotation, which would be overcome using circular polarization. If the effects of the ionosphere and troposphere could be ignored, the main factor affecting the satellite channel MIMO model would be the multipath effects caused by local scattering near to the mobile. The scattering of this region will create some depolarization from RHCP to LHCP and from LHCP to RHCP, which are represented in a 2×2 MIMO channel matrix $\boldsymbol{H}$, where there are two co-polar (RHCP to RHCP and LCHP to LHCP) and two cross-polar circularly polarized channels (RHCP to LHCP and LCHP to RHCP). These four channels can be conveniently distinguished by

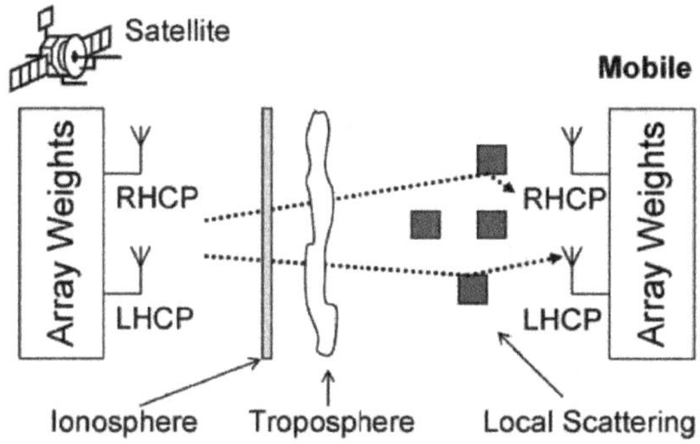

Figure 7.9 Satellite MIMO channel structure.

subscripts R and L:

$$\boldsymbol{H} = \begin{pmatrix} h_{RR} & h_{LR} \\ h_{RL} & h_{LR} \end{pmatrix}. \tag{7.2.1}$$

Accordingly, the output vector received at the mobile $\boldsymbol{y}(t)$ relates to the input vector $\boldsymbol{x}(t)$ by the following equation:

$$\boldsymbol{y}(t) = \boldsymbol{H}(t)\boldsymbol{x}(t) + \boldsymbol{w}(t), \tag{7.2.2}$$

where $\boldsymbol{w}(t)$ is a vector to represent additive white Gaussian noise at the receiver. Therefore, it is meaningful to model the channel state at time t, $H(t)$. The simplest method to model the LMS–MIMO channel is to use a stochastic process, which will define the distribution of the channel states (or first-order statistics), but it is also necessary to define how the channel evolves by considering what is known as the second-order statistics.

For satellite MIMO, it is best to break the channel $\boldsymbol{H}$ into three components:

(i) *Free space path loss*: This is the well-known Friis formula [15]. In order to analyze the channel conveniently and consider it to be an offset, it must be normalized.
(ii) *Shadowing or large-scale fading*: When the mobile station is on the ground, it may either have a direct LOS link to the satellite or there may be a building, tree, or other large scattering objects blocking the direct path. Therefore, the mobile is within the shadow of the scatterer and will be subject to extra

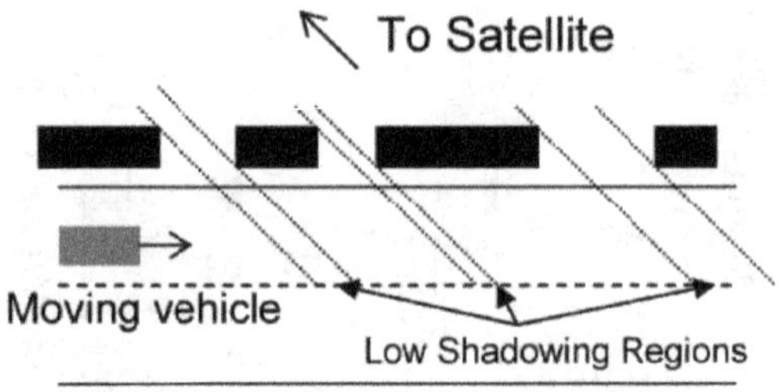

Figure 7.10 Example of heavy and light shadowing regions.

path loss. In the case of the satellite communications, the mobile is constantly moving in and out of the shadow area, as illustrated in Figure 7.10. Here, the mobile is a vehicle moving along the road; it will enter regions between scatterers on its left-hand side where it will have an LOS link with the satellite. Therefore, the statistical modeling of both heavy and light shadowing scenes is required. Further, it is necessary to use the Markov chain for modeling when the mobile is switching between heavy and light shadowing [19].

(iii) *Small-scale fading*: There are several scattering objects in the local area around the mobile, which will produce reflection, refraction, and diffraction signals to the input. These reflections, refractions, and diffractions change as the mobile moves. Therefore, the received signal is constantly changing. In some cases, the reflected signals will add up constructively in phase, while in other cases, they will add up destructively out of phase and the received signal will go into a deep fade.

Therefore, it is necessary to separately model the small-scale and large-scale fading channels as it changes over time and with movement of the mobile.

For LMS-MIMO systems, many factors, which are not considered in the existing LMS channel modeling, are considered as follows:

(1) MIMO branches are in a circular polarization domain rather than spatial domain. Therefore, channel multiplexing occurs in this domain, which will bring different eigenvalues to the eigenvalue decomposition of the channel. Therefore, it is necessary for the traditional channel with different eigenvalues to be separately modeled [20].

(2) When switching from heavy to light shadowing, the large-scale fading changes constantly. This produces different characteristics for the co-polar and cross-polar channel branches, and likewise should be modeled to be consistent with simultaneous measurements of these channels. In addition, the shadowing characteristics of the two polarizations should be compared in order to maintain their interdependence.
(3) In the small-scale fading channel, especially when there is an LOS link, each of the four MIMO paths is not necessarily independent. Therefore, in such circumstances, the interdependence must be reasonably modeled so that the model and the real measured data have comparable eigen decomposition.

Before developing the models, it is necessary to identify the Markov chain, polarization, and interdependence characteristics from real measurement data. After describing the measurements, the methods for modeling the large-scale fading channels, Markov chain models, and small-scale fading channels must be analyzed.

7.2.2 *Measurement setup*

Extensive measurements were carried out on the edge of the town of Guildford, UK, where dense forests in the suburbs/rural areas of the region are suitable for LMS-MIMO channel measurements. An artificial terrestrially based platform (acting as a satellite) was placed on top of the hill, as illustrated in Figure 7.11, containing directional RHCP and LHCP antennas, spaced just under one wavelength apart. Each antenna had a gain of 12 dBi and a 3-dB beamwidth of 30°. The mobile vehicle contained a receiver and its roof was fitted with omnidirectional RHCP and LHCP antennas spaced four wavelengths apart. These two antennas had a beamwidth of 70° in elevation. As the mobile end moves along a tree-lined road, the satellite elevation angle ranged from 7° to 18°. Although many operational satellite elevations exceed these, some geostationary and low-Earth orbiting mobile satellite services are required to work at low elevations. Therefore, the presented experiment regards these as the worst cases, where such elevations enable the highest degrees of multipath. The Rice factor variation and channel correlation defined by the Rice distribution provide a usefully wide range of empirical results.

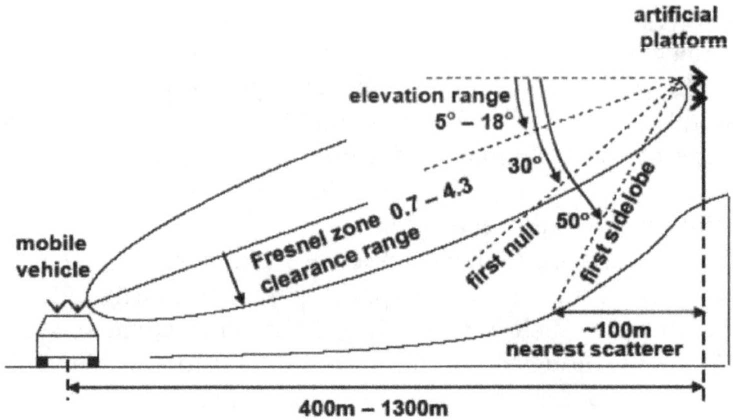

Figure 7.11 Measurement setup of artificial platform.

A wideband MIMO channel sounder was configured for a carrier frequency of 2.45 GHz and a bandwidth of 200 MHz. Each MIMO channel was obtained sequentially by using fast switching at a rate of 152.7 Hz, which is over twice the maximum Doppler shift of 73 Hz for the vehicle speed, thus meeting Nyquist criteria. It is assumed that the experiment was carried out on a tree-lined road with few moving vehicles. From the data, suitable results were found whereby the channel could be captured within the channel sounder's sensitivity giving a SNR that enabled the multipath to be measured without error due to receiver noise. The interference contained in the data from the local wireless area networks that had to be filtered out from measurement data and replaced with interpolated data. Since the data sampling rate was over the required Nyquist criteria, this meant it was possible to achieve this. Figure 7.12 shows the Markov chain behavior of the heavy and light shadowing for all four MIMO branches using the measurement data. It can be seen from Figure 7.12 that there are different small-scale fading characteristics in heavy and light shadowing regions, as well as Markov chain characteristics that are related for co-polar and cross-polar channels, though they are required to be generated separately. In the next section, how the large-scale fading, Markov chains, and small-scale fading in three separate steps and then integrated to form a working channel matrix are explained in detail.

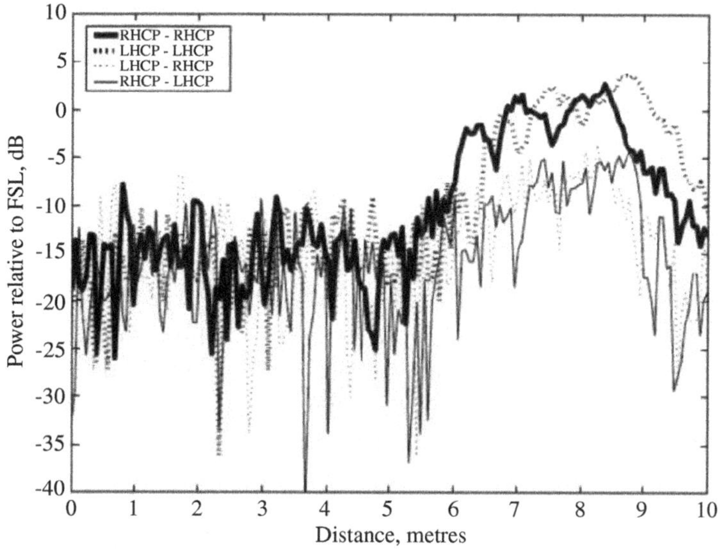

Figure 7.12 Sample measurement showing the Markov chain characteristics of the LMS-MIMO channel.

7.2.3 *Model generation*

Step 1: Generate large-scale fading for heavy and light shadowing.

For 2 × 2 MIMO systems, four simultaneous models of heavy shadowing that vary over distance (i.e., in the NLOS region) are required to produce vector $\boldsymbol{s}_{\text{heavy}}$ and four simultaneous models of light shadowing (i.e., in the LOS region) are required to produce vector $\boldsymbol{s}_{\text{light}}$, in dB by using zero-mean, unit-standard-deviation Gaussian random noise signals. In order to make the two sets of four shadowing models be interdependent, a 4 × 4 correlation matrix for large-scale fading, $\boldsymbol{C}_{\text{large}}$, is then applied to the heavy and light shadowing as analysis of measurement data has shown that both shadowing cases follow the Gaussian distribution. The interdependence between the four MIMO branches is accommodated so that the correlated shadowing $\boldsymbol{s}_{\text{heavy}|\text{c}}$ and $\boldsymbol{s}_{\text{light}|\text{c}}$ is expressed as follows:

$$\boldsymbol{s}_{\text{heavy}|\text{c}} = \boldsymbol{C}_{\text{large}}^{\frac{1}{2}} \boldsymbol{s}_{\text{heavy}}, \ \boldsymbol{s}_{\text{light}|\text{c}} = \boldsymbol{C}_{\text{large}}^{\frac{1}{2}} \boldsymbol{s}_{\text{light}}, \tag{7.2.3}$$

where the 1/2 denotes the Cholesky factorization. All 16 possible permutations of this notation are shown in the matrix as follows:

$$\boldsymbol{C}_{\text{large}} = \begin{pmatrix} c_{\text{large}|\text{RR,RR}} & c_{\text{large}|\text{LL,RR}} & c_{\text{large}|\text{RL,RR}} & c_{\text{large}|\text{LR,RR}} \\ c_{\text{large}|\text{RR,LL}} & c_{\text{large}|\text{LL,LL}} & c_{\text{large}|\text{RL,LL}} & c_{\text{large}|\text{LR,LL}} \\ c_{\text{large}|\text{RR,RL}} & c_{\text{large}|\text{LL,RL}} & c_{\text{large}|\text{RL,RL}} & c_{\text{large}|\text{LR,RL}} \\ c_{\text{large}|\text{RR,LR}} & c_{\text{large}|\text{LL,LR}} & c_{\text{large}|\text{RL,LR}} & c_{\text{large}|\text{LR,LR}} \end{pmatrix}, \tag{7.2.4}$$

where $\boldsymbol{C}_{\text{large}|\text{RR,LL}}$ denotes the correlation of the shadowing between the right to right-hand branch and the left to left-hand branch.

An appropriate set of values taken from the measurement data representing a typical tree-lined road in a suburban environment are obtained as follows:

$$\boldsymbol{C}_{\text{large}} = \begin{pmatrix} 1 & 0.86 & 0.85 & 0.9 \\ 0.86 & 1 & 0.91 & 0.9 \\ 0.85 & 0.91 & 1 & 0.88 \\ 0.9 & 0.87 & 0.88 & 1 \end{pmatrix}. \tag{7.2.5}$$

Due to the close proximity of the two transmit antennas and the two receive antennas, the correlation values are high, and other measurements have shown that high correlation is maintained in different channel scenarios. Using these correlations, eight time-synchronized simulations are created that are defined by the following process to form the second-order statistics [21]:

$$\begin{aligned} \boldsymbol{s}_{\text{heavy}|\text{cf}}(n) &= \boldsymbol{s}_{\text{heavy}|\text{c}}(n) + e^{-\frac{v_m \Delta t}{r_c}} \boldsymbol{s}_{\text{heavy}|\text{cf}}(n-1), \\ \boldsymbol{s}_{\text{light}|\text{cf}}(n) &= \boldsymbol{s}_{\text{light}|\text{c}}(n) + e^{-\frac{v_m \Delta t}{r_c}} \boldsymbol{s}_{\text{light}|\text{cf}}(n-1), \end{aligned} \tag{7.2.6}$$

where the time variation is determined by a coherence distance, r_c, for a given mobile speed, v_m, with sample time, Δt, and $\boldsymbol{s}_{\text{heavy}|\text{c}}$ and $\boldsymbol{s}_{\text{light}|\text{c}}$ are the correlated Gaussian distributed random variables with zero mean and unit standard deviation. Measurements have shown r_c to be 25 m on average for a tree-lined road environment. The range of r_c values recorded spanned from 23 m through to 29 m. It is assumed in this case that each sample, n, is taken for every meter. The shadowing then requires normalization, where a set of standard deviations to the shadowing represented by vectors $\boldsymbol{\sigma}_{\text{heavy}}$

and $\boldsymbol{\sigma}_{\text{light}}$ and mean values, represented by vectors $\boldsymbol{\mu}_{\text{heavy}}$ and $\boldsymbol{\mu}_{\text{light}}$, all in dB, are applied, where $\circ$ denotes element-wise multiplication and the normalized, filtered, and correlated shadowing, $\boldsymbol{s}_{\text{heavy|cfn}}$ and $\boldsymbol{s}_{\text{light|cfn}}$ are given as

$$\boldsymbol{s}_{\text{heavy|cfn}} = \left(\boldsymbol{s}_{\text{heavy|cf}} \circ \boldsymbol{\sigma}_{\text{heavy}} \circ \sqrt{1 - e^{-\frac{2v_m \Delta t}{r_c}}}\right) + \boldsymbol{\mu}_{\text{heavy}},$$

$$\boldsymbol{s}_{\text{light|cfn}} = \left(\boldsymbol{s}_{\text{light|cf}} \circ \boldsymbol{\sigma}_{\text{light}} \circ \sqrt{1 - e^{-\frac{2v_m \Delta t}{r_c}}}\right) + \boldsymbol{\mu}_{\text{light}}. \quad (7.2.7)$$

Empirical values of standard deviations $\boldsymbol{\sigma}_{\text{heavy}}$ and $\boldsymbol{\sigma}_{\text{light}}$ and mean values $\boldsymbol{\mu}_{\text{heavy}}$ and $\boldsymbol{\mu}_{\text{light}}$ for co-polar and cross-polar channels in the tree-lined road environment in dB, as shown in Table 7.6, are derived generically from all measurement data taken as the highest and lowest values. Finally, the data must be reshaped to create two separate 2×2 channel matrices $\boldsymbol{H}_{\text{heavy}}$ and $\boldsymbol{H}_{\text{light}}$.

Table 7.6 Mean and standard deviation of shadowing model.

	Heavy Shadowing (dB)		**Light Shadowing (dB)**	
Polarization	μ_{heavy}	σ_{heavy}	μ_{light}	σ_{light}
Co-polar	−20.5	6.5	−1.5	4.0
Cross-polar	−21.5	6.0	−4.5	3.0

Table 7.7 Markov state and state transition table.

State Transitions					**State**
0.6822	0.1579	0.0561	0.1037	0.0764	CP light XP light
0.2887	0.2474	0.0447	0.4192	0.0416	CP light XP heavy
0.1682	0.0996	0.1745	0.5607	0.0229	CP heavy XP light
0.0098	0.0199	0.0150	0.9554	0.8591	CP heavy XP heavy
CP light XP light	CP light XP heavy	CP heavy XP light	CP heavy XP heavy		

Step 2: Generate Markov chain.
After the data for heavy and light shadowing sequences are generated, the conversion between the heavy and light shadowing regions for co-polar and the cross-polar channels can be described by the Markov chain [17]. As shown in Figure 7.10, this sharp transition occurs as the mobile moves past buildings and can be modeled appropriately. Therefore, if two polarizations are considered, there are four possible Markov states, as shown in Figure 7.13. It is assumed that the behavior is the same whichever of RHCP or LHCP is being transmitted. Once the polarization is defined at the transmit end, these four possible states therefore consider whether the co-polar or cross-polar channel at the receiving end are both in a heavy or light state or in opposite states, as can happen in certain instances. There are 16 possible state transitions for the four possible states, as shown by the arrows. By analyzing the Markov chain properties of all measured samples, the Markov chain statistics extracted from the measured data (once a threshold is selected for heavy and light shadowing in the measured data) are shown in Table 7.7. The Markov chain derived from the measurement data using the chosen threshold was analysed to ensure that where state transitions did occur, they were true cases of real transitions. There can be cases with extreme light probability, where a heavy state shadowing falls below the threshold while also a light state threshold rises above the threshold. The Markov chain test eliminated these possible situations. The columns of state transition represent the probability of one state moving to another

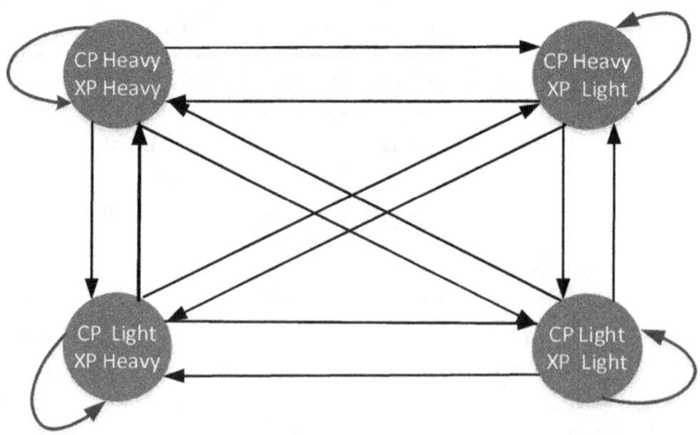

Figure 7.13 Illustration of the four Markov states for the LMS–MIMO channel.

Table 7.8 Comparison of mean and standard deviation values of complex correlation magnitude against Rice factor.

Rice Factor	Mean r_{CP}	Standard Deviation r_{CP}	Mean r_{XP}	Standard Deviation r_{XP}
0 to 2	0.41	0.24	0.28	0.177
2 to 4	0.73	0.19	0.18	0.13
4 to 10	0.87	0.0052	0.42	0.24

state listed in the right-hand column, whereas each row represents the probability of moving to the state shown on the right-hand column from the previous state shown on the bottom row. Thus, the top right-hand state transition of 0.1037 is the probability of moving from the state "CP heavy, XP heavy" to "CP light, XP light", where CP is a co-polar channel and XP is a cross-polar channel. In most cases, both CP and XP are in a heavy-shadowing state, which reflects the fact that the measurement environment is a road lined with houses and trees with foliage, thus providing dense multipath and the satellite at a light elevation. By applying the Markov chain analysis, sampled every meter, to the large-scale shadowing, a 2×2 channel matrix, $\boldsymbol{H}_{\text{large}}$ can be finally created.

After generating the Markov chain, the data must be up-sampled to match small-scale fading that is generated next. Up-sampling is necessary because the rate of change of the large-scale fading is quite light compared to that of the small-scale fading. Therefore, in the first case, such a large sampling size (based on the size of r_c) is not required. However, for the large-scale fading to be integrated with the small-scale fading at a later stage, the large-scale fading must match the sampling rate for the small-scale fading.

Step 3: Generate small-scale fast fading.
In this case, the small-scale fading is modeled by a Rice distribution, where the light shadowing region will have a higher Rice factor than that of the heavy shadowing region, which is subject to denser multipath. The usage of values of Rice factors based on measurements is shown in Table 7.8. Using the Rician factors [21], Rician fading with appropriate second-order statistics can be generated for each MIMO branch by using a ring scatterer model (of course, small-scale

fading can also be generated by Doppler filtering and autocorrelation matrices [22]). For each sample n, the small-scale fading elements, denoted by subscripts $_{xx}$ of $\boldsymbol{H}_{\text{small}}$, are derived in this case as follows:

$$h_{\text{small}xx}(n) = \frac{\sqrt{k_{xx}}e^{j\frac{n2\pi f_m}{N_f}} + \sum_{i=1}^{N_S} M_{\text{norm}} e^{j(\frac{2\pi n \sin\phi_i}{N_f}+\theta_i)}}{\sqrt{k_{xx}+1}}, \qquad (7.2.8)$$

where N_f is the sampling factor equal to the sampling frequency divided by the maximum Doppler shift, f_m, due to mobile movement. The Rayleigh (or scattered) part of the small-scale fading is normalized by M_{norm} so that its mean is unity. The components of the Rayleigh part will arrive at angle ϕ_i and have a random phase θ_i. The four $h_{\text{small}xx}$ elements are arranged into the 2×2 matrix $\boldsymbol{H}_{\text{small}}$.

In the heavy shadowing regions, there is generally an NLOS condition and the Rice factor is low. In the case of the LMS, the scattering rich environment local to the mobile terminal provides low correlation between the antenna branches, while at the satellite, there are directional antennas with highly orthogonal circular polarizations, thus their correlation is also low. Therefore, at each end, the correlation is independent and remains low in an NLOS scenario, which therefore justifies that the channel is separable in order to allow the well-known Kronecker model to be applied [23] to model the correlated small-scale fading, as follows:

$$\text{vec}(\boldsymbol{H}_{\text{small}|\text{c}}) = \boldsymbol{R}^{\frac{1}{2}}_{\text{small}|\text{c}} \text{vec}(\boldsymbol{H}_{\text{small}|\text{NLOS}}), \qquad (7.2.9)$$

where vec is the vectorize function. The small-scale fading correlation matrix is written as

$$\boldsymbol{R}_{\text{small}} = \begin{pmatrix} r_{\text{small}|\text{RR,RR}} & r_{\text{small}|\text{LL,RR}} & r_{\text{small}|\text{RL,RR}} & r_{\text{small}|\text{LR,RR}} \\ r_{\text{small}|\text{RR,LL}} & r_{\text{small}|\text{LL,LL}} & r_{\text{small}|\text{RL,LL}} & r_{\text{small}|\text{LR,LL}} \\ r_{\text{small}|\text{RR,RL}} & r_{\text{small}|\text{LL,RL}} & r_{\text{small}|\text{RL,RL}} & r_{\text{small}|\text{LR,RL}} \\ r_{\text{small}|\text{RR,LR}} & r_{\text{small}|\text{LL,LR}} & r_{\text{small}|\text{RL,LR}} & r_{\text{small}|\text{LR,LR}} \end{pmatrix}.$$

The values of $\boldsymbol{R}_{\text{small}}$ are derived as an average of correlation values evaluated over the whole sampled measurement data, so

$$\boldsymbol{R}_{\text{small}} = \begin{pmatrix} 1 & 0.03 & 0.02 & 0.09 \\ 0.03 & 1 & 0.12 & 0.01 \\ 0.02 & 0.12 & 1 & 0.01 \\ 0.09 & 0.01 & 0.02 & 1 \end{pmatrix}. \qquad (7.2.10)$$

In the case where there is light shadowing and an LOS, the channel cannot be considered a separable, and Kronecker's assumption does not hold. Therefore, the Kronecker model has been extended here such that it is suited to a 2 × 2 satellite MIMO system in an LOS environment with dual-circular polarization, which enables polarization multiplexing.

First, the co-polar correlation matrix R_{CP} is written as

$$\boldsymbol{R}_{CP} = \begin{pmatrix} 1 & r^*_{CP} \\ r_{CP} & 1 \end{pmatrix}, \tag{7.2.11}$$

where r_{CP} is the correlation between h_{RR} and h_{RR}. The phase information in the matrix is not necessary when generating a model.

Second, R_{XP} is also defined in a similar way. r_{XP} is the correlation between h_{LR} and h_{LL}.

Again, the phase information is not essential. In terms of the gain patterns and polarization purity, the two antennas at the transmit end have similar characteristics. Although the transmit antenna does not have to be the same as the receive end, it must be the same at the receiving end. The 2 × 2 correlation matrix for the cross-polar is defined as follows:

$$\boldsymbol{R}_{CP} = \begin{pmatrix} 1 & r^*_{XP} \\ r_{XP} & 1 \end{pmatrix}. \tag{7.2.12}$$

The two matrices in Equations (7.2.11) and (7.2.12) will affect the orthogonality of the right-hand and left-hand polarizations, which will be key to the MIMO channel. A 1 × 2 co-polarized channel vector $\boldsymbol{h}_{\text{CP|small}}$ can be obtained by the Rice factor determined by Equation (7.2.7), and then the two co-polarizations are correlated to obtain $\boldsymbol{h}_{\text{CP|small|C}}$ as follows:

$$\boldsymbol{h}_{\text{CP|small|C}} = \boldsymbol{R}_{\text{CP}}^{\frac{1}{2}} \boldsymbol{h}_{\text{CP|small}}. \tag{7.2.13}$$

Likewise, for the cross-polarizations, $\boldsymbol{h}_{\text{CP|small|C}}$ is given as follows:

$$\boldsymbol{h}_{\text{XP|small|C}} = \boldsymbol{R}_{\text{XP}}^{\frac{1}{2}} \boldsymbol{h}_{\text{XP|small}}. \tag{7.2.14}$$

The four elements once generated can easily be inserted into a 2×2 matrix to generate the small-scale fading, $\boldsymbol{H}_{\text{small}}$. After applying the correlation, it is necessary to re-normalize the mean values of all four MIMO branches. For the cross-polar components, $\boldsymbol{h}_{\text{smallLR}}$ and $\boldsymbol{h}_{\text{smallRL}}$, the mean values must also be divided by the square root of the cross-polar ratio, $\sqrt{\text{XPD}}$. The XPD is defined as the ratio of the mean co-polar power to the mean cross-polar power. It is assumed that XPD is the same whether the co-polar component is left-hand or right-hand circularly polarized. The measured data shows that for co-polarization, the Rice factor ranges from 0 to 10. The cross-polar Rice factors were also found to be wide ranging, though as a rule they are always less than the co-polar Rice factor for a set of samples, thus any Rice factors can be selected for a model that adhere to the rule, though corresponding XPD and correlations have to be used alongside given Rice factors. Tables 7.8 and 7.9 give suitable corresponding values of correlation and XPD, respectively, based on the analysis of the measured data. In the case of a high Rice factor, the co-polar correlation is inherently high, while for a low Rice factor, the correlation is lower and has a larger variance. For a high Rice factor, XPD can be as high as 15 dB, but on average, it is closer to 10 dB. For lower Ricean factor, where scattering results in significant de-polarization, the average XPD is closer to 0 dB.

Step 4: Integrate steps 1, 2, and 3.
After the large and small scales of fading channels are created, they can be simply multiplied together to form the final channel model such that

$$\boldsymbol{H} = \boldsymbol{H}_{\text{large}} \circ \boldsymbol{H}_{\text{small}}, \tag{7.2.15}$$

where $\circ$ denotes an element-wise multiplication of the two matrices. It should be noted that the resultant large-scale fading $\boldsymbol{H}_{\text{large}}$ is already normalized to the bulk mean free space path loss and any

Table 7.9 Comparison of maximum and minimum XPDs against Rice factors.

Rice factor	Max. XPD (dB)	Min. XPD (dB)
0 to 2	6	0
2 to 4	6	0
4 to 10	15	4

other losses in the ionosphere or troposphere. However, the resultant small-scale fading $\boldsymbol{H}_{\text{small}}$ must be normalized to unit mean power.

7.2.4 *LMS-MIMO channel model validation of small-scale fading*

An important validation for the MIMO channel model is to ensure that the eigenvalue cumulative distributions produced by the model are in good agreement with the measured data. This will not only ensure that the first- and second-order statistics of the physical channel are suitably modeled but also that the interdependence between them is suitably accounted for to demonstrate the diversity and multiplexing capabilities of the model. In order to demonstrate the validation of the model, two LOS and NLOS cases were chosen, which had values of Rice factor, XPD, and correlation shown in Table 7.10. The Rice factor values are in linear form. The following three sections compare the first-order, second-order, and eigen analysis as a validation of the model proposed using appropriate sections of the measurement data. The sampling rate of the measured data for testing model is more than twice the maximum Doppler shift frequency to meet the Nyquist criterion.

7.2.4.1 *First-order statistics*

Figures 7.14 and 7.15 illustrate the cumulative distribution of the small-scale fading for all four branches of the NLOS and LOS regions. In both cases, there is good agreement between the measurement data and model, where in the LOS scenario, a wider gap can be identified between co-polar and cross-polar branches. It is interesting to note that compared with RHCP, the mean value of the cross-polar component for LHCP transmission is shifted by over 6 dB, where the multipath was found to combine better constructively within the

Table 7.10 Correlation, XPD, and Rice factor values used in the NLOS and LOS cases for validation.

Scenario	XPD (dB)	k_{CP}	k_{XP}	r_{CP}	r_{XP}
NLOS	5.9	2.43	0.97	0.65	0.34
LOS	8.1	6.01	2.04	0.92	0.61

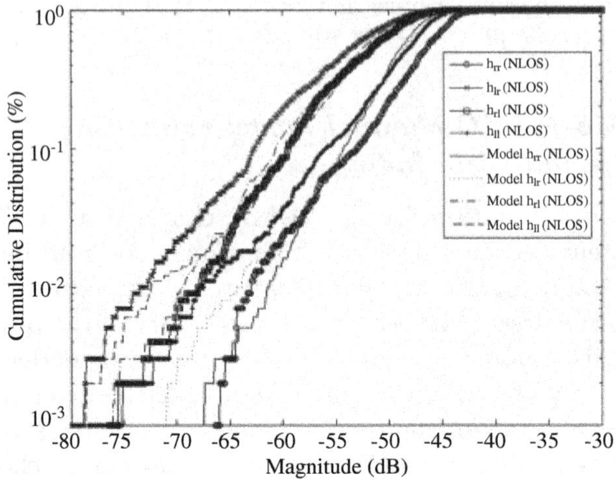

Figure 7.14 Cumulative distribution plot of the modeled and measured channels for heavy shadowing (NLOS).

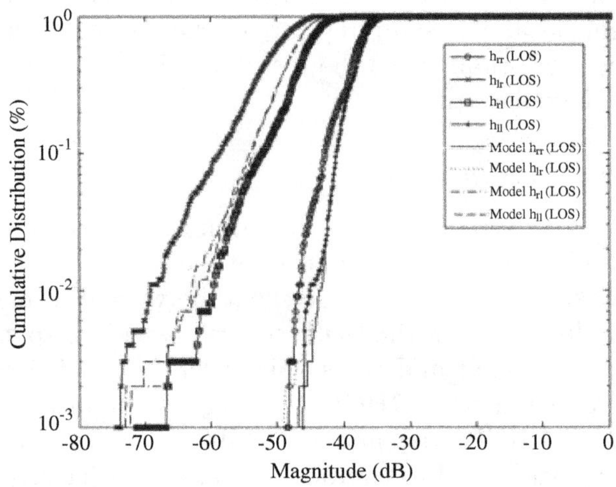

Figure 7.15 Cumulative distribution plot of the modeled and measured channels for light shadowing (LOS).

sampled time window. However, the model has not accommodated this difference since the negligible contribution of the cross-polar element to the MIMO channel has no real effect on the capacity or eigen analysis.

7.2.4.2 *Second-order statistics*

Figure 7.16 shows the Doppler spread in both left-hand to left-hand h_{LL} and right-hand to right-hand h_{RR} polarized cases, where the polarization has little difference in LOS or NLOS. The Doppler spread shown verifies the suitability of the Rician distribution for small-scale modeling based on a classical bath tub model with the addition of a delta function for the Rice component [16]. A high Rice component is identified in Figure 7.16 for LOS, while the NLOS case still has a small Rice component due to the non-uniform angle of arrival. Figure 7.17 shows further validation of the second-order statistic, which shows a good agreement between LCR and AFD when comparing the measured and modeled narrowband data. Free space loss and the maximum Doppler frequency shown in Figure 7.17 are labeled as FSL and f_m, respectively.

7.2.4.3 *Eigen analysis*

Figure 7.18 presents results of the eigen analysis of the model and measurement data as well as the modeled data for the NLOS case. Figure 7.18 shows that the modeled and measured data have good consistency, which verifies that the Kronecker model method is suitable for the NLOS case. In the graph, the notation s_n denotes an eigenvalue, where n is 1 or 2 for a 2×2 MIMO system.

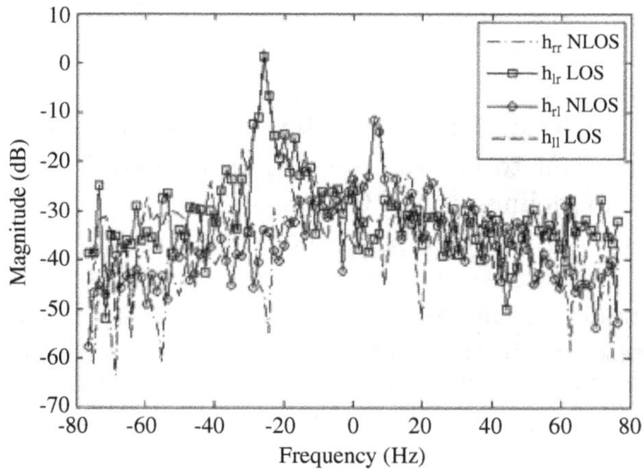

Figure 7.16 Comparison of the right to right-hand and left to left-hand polarized Doppler spread in NLOS and LOS.

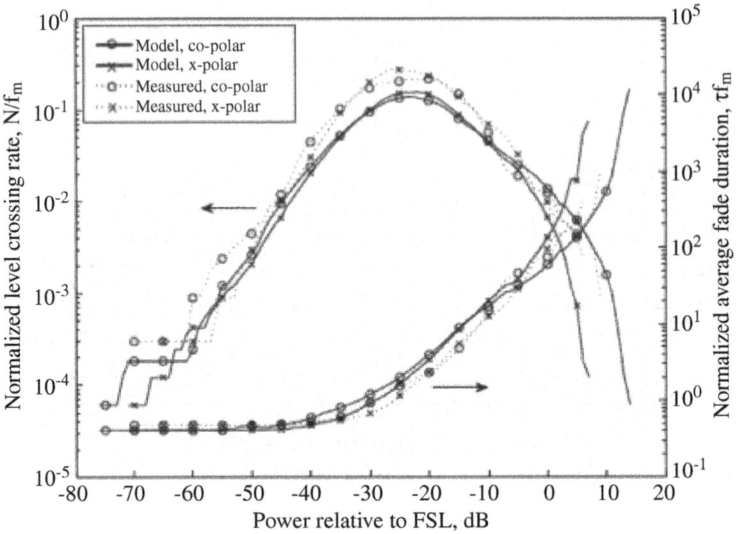

Figure 7.17 Validation of the second-order statistics based on LCR and AFD.

On the other hand, Figure 7.19 compares measured and modeled eigenvalues in the LOS case. Compared with the NLOS case, because the eigenvalue distribution is closer, the LOS channel has rich polarization multiplexing rather than diversity, which may be caused by less scatterers. Although the experimental results are consistent, it should be noted that in this validation, the phase information was applied in the correlation matrix. Were the phase information not applied, the second eigenvalue would marginally change its gradient, moving away from the measured data by less than 2 dB. Since the effect of this deviation on the modeling channel capacity is negligible, it is not important to use the phase information. A similar scenario occurs when modeling other LOS regions.

Based on the switching between heavy and light shadowing regions with different multipath conditions, a simple empirical–stochastic model for the dual-circular polar 2×2 LMS-MIMO channel has been presented in this section, along with results to validate the model at low elevation. The validation of the model at such elevations will also be suited to higher elevation angles, where the multipath is reduced and the opportunity to implement polarization multiplexing is increased. It is well known that the Kronecker model is suitable for the NLOS case, and the model presented in this section is suitable for

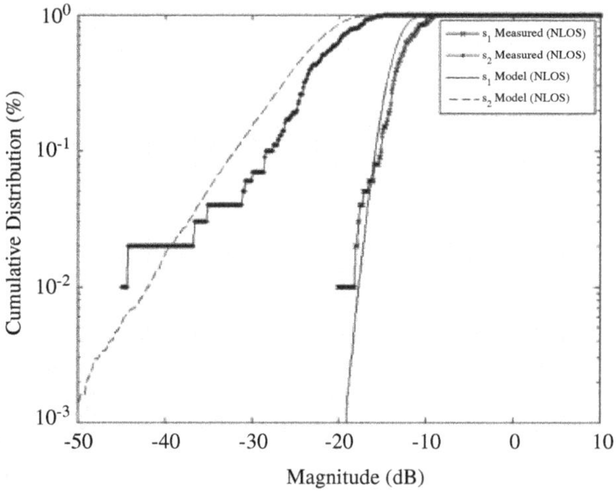

Figure 7.18 Comparison of the model and measured eigenvalues for the heavy shadowing (NLOS) region.

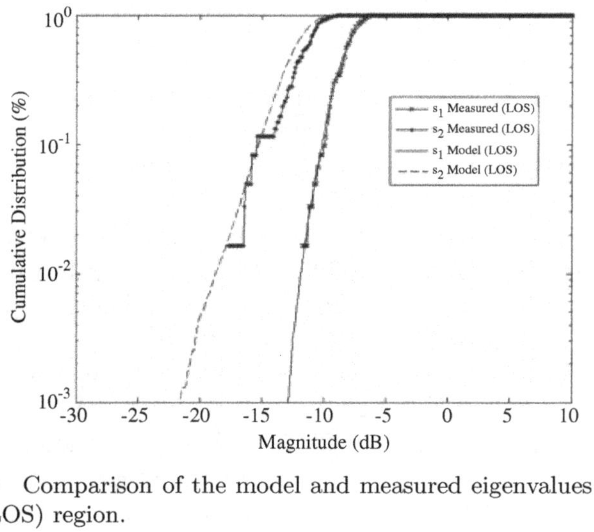

Figure 7.19 Comparison of the model and measured eigenvalues for the light shadowing (LOS) region.

a polarization multiplexing rich scenario in the LOS case. The comparisons in both cases show good accuracy. Given the simplicity of generating a Markov chain and correlated small-scale and large-scale fading, it is highly appropriate for conformance testing of satellite MIMO applications by controlling the simplicity of the channel by changing the Rice factor, correlation, and XPD values according to the proposed guidelines.

7.3 Molecular MIMO Channel Model Based on Machine Learning

Molecular communication (MC) is quite different from traditional electromagnetic wave communication in that it conveys information by utilizing molecules. Such small-scale communication is a challenging task [24]. Recently, researchers have been paying attention to MC because it has the potential for enabling complex applications of nanotechnology that requires collaboration of very small entities. For example, MC can be applied in the nanorobots' communication and health monitoring issues [24–26].

MC has three advantages in nanoscale communication compared to traditional electromagnetic wave communication. First, low energy consumption is one of the important requirements for nanoscale communication, and MC shows better performance (in terms of energy consumption) than traditional methods. Second, MC can reduce the size of the antenna, which is validated by many examples in nature. Finally, as long as harmless molecules are used, the MC *in vivo* system has greater biocompatibility characteristics than traditional communication.

In a diffusion-based MC, the received signal is determined by the received molecules, and the information particles move via a diffusion process, which shows random movement. One of the main challenges in MC is to develop valid models for representing received signals in different environments. Studies on MC, such as modulation, detection, and receiver design, utilize channel models that rely on the environmental conditions [28–32]. Previous studies have mostly focused on the molecular SISO communication systems, whereas few studies have focused on molecular MIMO communication systems. In [33], the authors proposed an analytical channel model for absorbing receivers in a one-dimensional environment. In [34], the authors enhanced the channel model by considering the flow effects. In [35], the authors proposed a theoretical molecular channel model for a three-dimensional (3-D) environment. The model is based on a point transmitter and an absorbing spherical receiver, where the expected cumulative number of received molecules is formulated with respect to time. In [36], the channel model described in [35] has been enhanced by adding the receptor effect. When a molecular MIMO channel with two absorbing antennas is considered, analytical derivation becomes difficult to handle. Therefore, semi-analytical methods are used in which channel taps are taken from the simulations and plugged into the analysis derivations. In [37], the authors presented a molecular MIMO channel model using simulation data. A known challenge is to calculate an accurate channel model for molecular MIMO. Therefore, a more practical method is needed to acquire channel taps without simulations.

Modeling molecular MIMO channels with absorbing receiver antennas remains an open issue. In this section, a machine learning approach is used to model a 2×2 molecular MIMO channel with point transmitter antennas and absorbing receiver antennas. Artificial neural network (ANN)-based channel modeling techniques are used to develop two different techniques that utilize the ANN-based *one-machine* and *two-machines* techniques. The differences between the two is studied here. Then, the model of the proposed technique is used to calculate the bit error rate (BER) [38].

7.3.1 *System model*

Consider a 2×2 molecular MIMO system, as depicted in Figure 7.20. It shows two point transmitters and two spherical receivers connected to a cuboid body. In this model, the antennas are assumed to be perfectly aligned. The transmitter antennas are separated by a distance h, which also holds for the receiver antennas. The radius of a spherical receiver antenna is denoted as R, and the distance between corresponding transmitter and receiver antennas is denoted by d.

When molecules are emitted from the point transmitter antennas, they propagate by the diffusion process in the 3-D environment. The diffusion process is characterized by the diffusion coefficient D. The receiver antennas absorb molecules that impinge on the surface of the receiver surface, and each receiver antenna calculates the number of received molecules. The absorbed molecules are counted only once (i.e., the molecule contributes to the molecular signal only once). Here, assume that if a molecule hits the cuboid body, the molecule is reflected to the medium.

In this section, a molecular MIMO channel model based on the SISO channel model is developed. In [35], the expectation of the

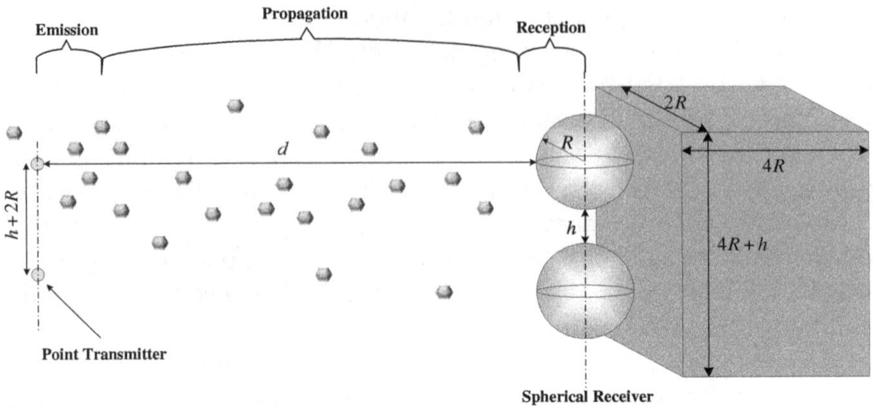

Figure 7.20 Schematic of 2 × 2 molecular MIMO communication model with two point transmitters and two spherical receivers.

SISO channel response is given and analyzed from the perspective of channel characteristics. The authors proposed a molecular fractional formula that hits the receiver before time t, as follows [39]:

$$F_{\text{hit}}^{\text{3D}} = \frac{R}{d+R}\text{erfc}\left(\frac{d}{\sqrt{4Dt}}\right), \tag{7.3.1}$$

where d and erfc(·) represent the distance and the complementary error function, respectively. When there is only one spherical receiver, there is a circular symmetry, all the points at the same radius are equivalent and the solution for the system of differential equations is enabled. However, this model cannot define the molecular MIMO channel accurately. Therefore, the above SISO channel model with additional adjustment parameters for each receiver antenna may be utilized.

7.3.2 *Molecular MIMO channel model*

Finding model parameters for a 2 × 2 molecular MIMO channel defines all the required data for evaluating the channel impulse response. In this communication system, there are four emission scenarios: only either transmitter 1 (Tx_1) or 2 (Tx_2) emits, two transmitters emit simultaneously, and none of them emit molecules. Due to the rectangular symmetry of the topology, the molecular MIMO channel by considering the emission of a single emitter can be modeled. In this case, all four cases can be formulated by using the channel response of a single emitter case. Therefore, by using a molecular MIMO communication simulator, the data for the channel responses of receivers 1 (Rx_1) and 2 (Rx_2) are collected when a single antenna emitted molecules.

The proposed technique involves two processes: fitting the channel model parameters and learning the input–output dataset patterns by ANNs. After the training phase, the trained ANN can effectively predict channel model parameters without simulation. A representative scheme of the proposed technique is depicted in Figure 7.21.

7.3.2.1 *Channel model and fitting*

In this section, two different model functions are used to fit the simulation data. Because Rx_1's and Rx_2's responses exhibit different schemes due to the distance difference, the corresponding distances in model functions are used. When considering the case that only the Tx_1 emits molecules for analyzing the cumulative channel functions at Rx_1 (i.e., $F_{11}(\cdot)$) and Rx_2 (i.e., $F_{21}(\cdot)$), the model function with three scaling factors is given as follows:

$$F_{11}(t, b_1, b_2, b_3) = b_1 \frac{R}{d+R} \text{erfc}\left(\frac{d}{(4D)^{b_2} t^{b_3}}\right), \tag{7.3.2}$$

where b_1, b_2, and b_3 represent the model-fitting parameters.

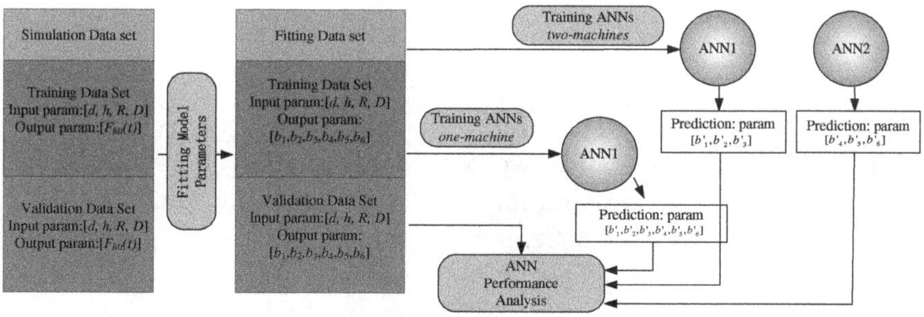

Figure 7.21 Flowchart of the proposed techniques: *one machine* and *two machines*. From left to right, process 1 deals with fitting model parameters by utilizing the dataset from the simulator; process 2 deals with training ANNs on the training dataset that is obtained from the process 1. The output from process 1 consists of input–output pairs, where the input is (d, h, R, D) and the output is the model parameters.

These model-fitting parameters are introduced so as to compensate for the discrepancy between the SISO and MIMO models. Similarly, the response at the Rx_2 (due to the cross-link interference) is defined as follows:

$$F_{21} = (t, b_4, b_5, b_6) = b_4 \frac{R}{\sqrt{d^2 + h^2} + R} \operatorname{erfc}\left(\frac{\sqrt{d^2 + h^2}}{(4D)^{b_5} t^{b_6}} \right), \tag{7.3.3}$$

where b_4, b_5, and b_6 represent the model-fitting parameters.

To find the model parameters, the nonlinear least squares curve-fitting technique is adopted. Assuming that there are N observations during the simulation, the parameter estimation problem with m parameters is given as follows:

$$\arg\min_{b_n, \mathrm{L}, b_m} \sum_{k=1}^{N} \left(F_{\mathrm{ij}}(t_k, b_n, \mathrm{L}, b_m) - S_{ij}^{\mathrm{3D}}(t_k) \right)^2, \tag{7.3.4}$$

where $S_{ij}^{\mathrm{3D}}(t)$ corresponds to the mean simulation data that represent the ratio of hitting molecules until time t. Note that for each receiver, the corresponding model function from Equations (7.3.2) or (7.3.3) may be used.

The output of the curve-fitting process consists of model parameters. Therefore, the model parameters for each simulation case, which form the dataset of the next process, are obtained. This dataset structure is shown in Figure 7.21 (i.e., the model parameters in the fitting dataset blocks).

7.3.2.2 *Training ANN*

The channel response in the diffusion-based 2×2 molecular MIMO communication system with two point transmitters and two spherical receivers is modeled by using machine learning techniques. ANN, one of the popular machine learning technologies, has simple, neuron-like nodes with thresholds and connections with weights. Basically, the thresholds and weights are adjusted by backpropagation (which evaluates the gradient and takes a step accordingly to minimize the loss function) for learning input–output pattern.

The dataset is divided into two disjointed subsets, namely training (TDS) and validation (VDS) datasets. The TDS is used for training the ANN to obtain the desired output for a given input. For training the ANN, the backpropagation with Bayesian regularization is used to update the weights and bias values according to Levenberg–Marquardt optimization. Bayesian regularization helps to minimize a combination of squared errors and weights together. Therefore, it helps to determine the ANN parameters that generalize the pattern in the input–output pairs without overlearning. After the learning phase, the trained ANN is used to estimate the channel parameters for different cases. Note that the trained ANN does not require any simulation data, i.e., the required inputs are the system parameters, namely d, h, R, and D.

Furthermore, in order to consider the effect of the number of ANNs for the modeling purpose, a machine learning technique is adopted. There are two different methods: *one machine* and *two machines* (Figure 7.21). First, the input parameters of d, h, R, and D are used to train a single ANN, and all the output parameters are regarded as model parameters (i.e., $b_1, b_2, b_3, b_4, b_5, b_6$). Second, for two ANNs, the parameters of d, h, R, and D in a similar way are used as inputs but the output parameters are separated. The first ANN's output parameters are b_1, b_2, b_3 and the second ANN's are b_4, b_5, b_6. The first ANN predicts the channel model parameters of Rx_1, which is aligned with the emitting transmitter, and the second ANN predicts the channel model parameters of Rx_2. For both techniques, it is necessary to use one fully connected hidden layer approach while keeping the total number of nodes same, i.e., one-machine technique's ANN consists of 30 nodes, so each ANN of the two-machines technique uses 15 nodes.

7.3.2.3 *Using ANN output for theoretical BER evaluation*

For a case study, the ANN results are used to evaluate the BER of a selected case. In the molecular MIMO communication system, information is transmitted using a sequence of symbols that are spread over sequential time slots (t_s). Binary concentration shift keying (BCSK) modulation at each antenna is utilized, i.e., N molecules are emitted for a bit-1 and no emission is done for a bit-0 [40–42]. Demodulation takes place at the end of each symbol slot, and the BCSK symbols are demodulated by thresholding the number of received molecules at Rx_1 (N^{Rx_1}) and Rx_2 (N^{Rx_2}) in a given symbol slot. The demodulated symbol at the mth symbol slot of Rx_i is denoted by $s_m^{\mathrm{Rx}_i}$. In MC, the number of received molecules is highly affected by previous emissions, which leads to inter-symbol interference (ISI). For tractability, a finite number of symbol slots for ISI is considered and the insignificant part of ISI is ignored. With a given ISI window (η), the number of received molecules at the mth symbol slot is the sum of binomial-distributed random variables, where each summand represents the number of hitting molecules due to a previous emission. For tractability, by using Gaussian approximation for binomial random variables and considering an ISI window of η, the general case is as follows:

$$\begin{aligned}
N^{\mathrm{Rx}_i}(m) &\sim \mathcal{N}(m_G, \sigma^2),\\
m_G &= \sum_{k=0}^{\eta} N s_{m-k}^{\mathrm{Tx}_i} F_{\mathrm{ii}}(k) + \sum_{k=0}^{\eta} N s_{m-k}^{\mathrm{Tx}_j} F_{\mathrm{ij}}(k),\\
\sigma_G^2 &= \sum_{k=0}^{\eta} N s_{m-k}^{\mathrm{Tx}_i} F_{\mathrm{ii}}(k)(1 - F_{\mathrm{ii}}(k))\\
&\quad + \sum_{k=0}^{\eta} N s_{m-k}^{\mathrm{Tx}_j} F_{\mathrm{ij}}(k)(1 - F_{\mathrm{ij}}(k)),
\end{aligned} \tag{7.3.5}$$

where $\mathcal{N}(m_G, \sigma_G^2)$, $N^{\mathrm{Rx}_i}(m)$, $s_k^{\mathrm{Tx}_i}$, and $F_{\mathrm{ij}}(k)$ denote a Gaussian random variable with mean m_G and variance σ_G^2, the number of received molecules for Rx_i at the mth symbol slot, the bit value for Tx_i to send at the kth symbol slot, and the probability of hitting to Rx_i for the molecules emitted from Tx_j at the kth following symbol slot, respectively. Note that $F_{\mathrm{ij}}(0)$ corresponds to the hitting probability at the current symbol slot. By using the output of the ANN technique, the evaluation of BER analytically can be modeled as follows:

$$\boldsymbol{P}_e = \sum_{s_{m-\eta:m}^{\mathrm{Tx}_1}, s_{m-\eta:m}^{\mathrm{Tx}_2}} \boldsymbol{P}_{e|s_{m-\eta:m}^{\mathrm{Tx}_1}, s_{m-\eta:m}^{\mathrm{Tx}_2}} \boldsymbol{P}(s_{m-\eta:m}^{\mathrm{Tx}_1}) \boldsymbol{P}(s_{m-\eta:m}^{\mathrm{Tx}_2}), \tag{7.3.6}$$

where

$$\begin{aligned}
\boldsymbol{P}_{e|s_{m-\eta:m}^{\mathrm{Tx}_1}, s_{m-\eta:m}^{\mathrm{Tx}_2}} &= \frac{1}{2}\boldsymbol{P}(s_m^{\mathrm{Rx}_1} \neq s_m^{\mathrm{Tx}_1} | s_{m-\eta:m}^{\mathrm{Tx}_1}, s_{m-\eta:m}^{\mathrm{Tx}_2})\\
&\quad + \frac{1}{2}\boldsymbol{P}(s_m^{\mathrm{Rx}_2} \neq s_m^{\mathrm{Tx}_2} | s_{m-\eta:m}^{\mathrm{Tx}_1}, s_{m-\eta:m}^{\mathrm{Tx}_2}).
\end{aligned} \tag{7.3.7}$$

Table 7.11 Range of parameters used in the experiments.

Parameter	**Value**
Number of emitted molecules	3000
Duration of channel simulations (t_{end})	1.5 sv
Replication	500
TDS distances (d)	$\{3, 5, 7, 9, 11\}$ μm
VDS distances (d)	$\{2, 4, 6, 8, 10\}$ μm
Distance between antennas (h)	$\{0, 1, 2\}$ μm
Diffusion coefficients (D)	$\{50, 100\}$ $\mu m^2/s$
Receiver radius (R)	$\{3, 5, 7\}$ μm

These probabilities are evaluated by the tail probabilities of the Gaussian random variable defined in Equation (7.3.5).

7.3.3 *Results and analysis*

Common system parameters for simulations, TDS, and VDS are presented in Table 7.11. From the given datasets, each of the VDS and TDS have $5 \times 3 \times 2 \times 3 = 90$ different cases, making a total of 180 cases. Each simulation case is replicated 500 times to estimate the mean channel model.

7.3.3.1 *Received signal analysis*

In Figures 7.22 and 7.23, the received signals are plotted for simulation data, curve-fitting, and ANN (*one-machine and two-machines*) techniques. Note that the ANN technique requires no simulation data while the curve-fitting method does. After training an ANN, channel model parameters for VDS by giving only the system parameters as input were estimated. It can be seen from Figures 7.22 and 7.23 that the trained ANN estimates the channel model parameters effectively for $d = 2\,\mu$m and $d = 8\,\mu$m cases. It can be found that the trained ANNs perform better at longer distances, a finding that will be detailed and quantified in the following section.

7.3.3.2 *RMSE analysis*

To analyze the performance of *one machine and two machines*, the RMSE of cases with respect to simulation data in terms of the

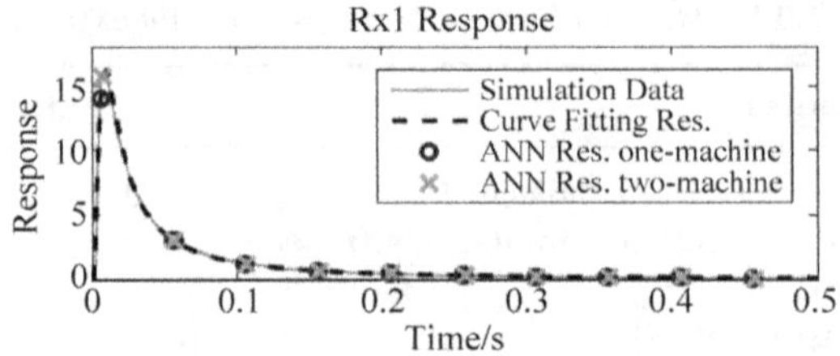

(a) F_{11}

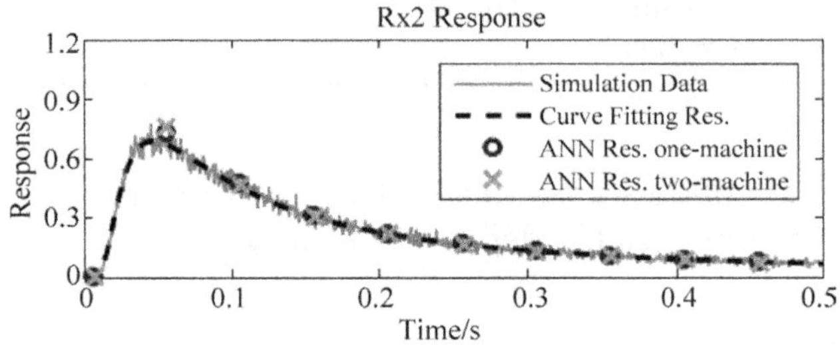

(b) F_{21}

Figure 7.22 Received signal plots for $d = 2\,\mu\text{m}$, $h = 1\,\mu\text{m}$, $R = 3\,\mu\text{m}$, $D = 100\,\mu\text{m}^2/s$ with the time resolution of 1 ms: (a) F_{11}; (b) F_{21}.

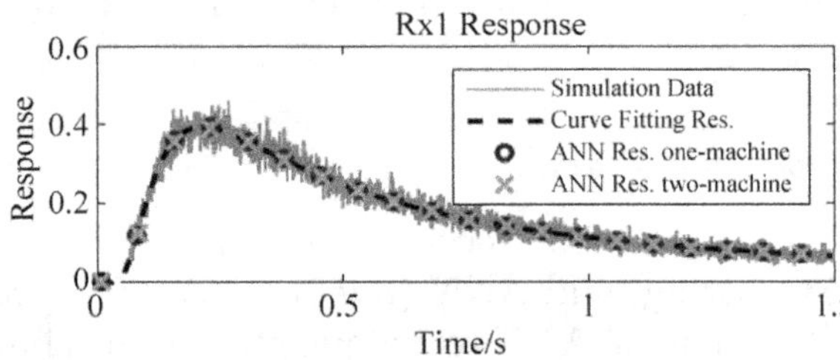

(a) F_{11}

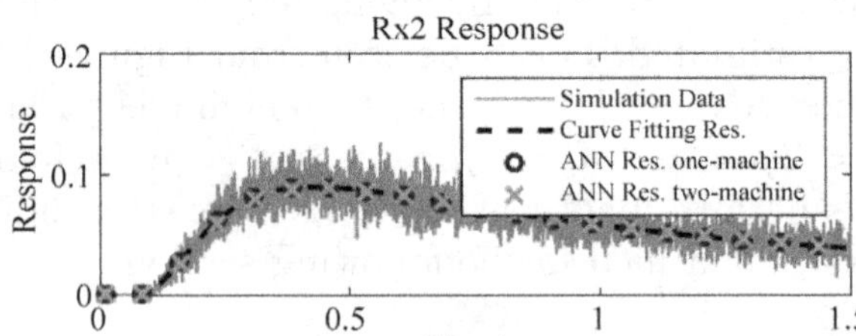

(b) F_{21}

Figure 7.23 Received signal plots for $d = 8\,\mu\text{m}$, $h = 1\,\mu\text{m}$, $R = 5\,\mu\text{m}$, $D = 50\,\mu\text{m}^2/s$ with the time resolution of 1 ms: (a) F_{11}; (b) F_{21}.

Table 7.12 Mean RMSE values for ANN techniques with full TDS.

		$D = 50\,\mu m^2/s$		$D = 100\,\mu m^2/s$	
		One ANN	Two ANN	One ANN	Two ANN
$d = 2\,\mu m$	$F_{11}(t)$	0.1561	0.1596	0.3043	0.2827
	$F_{21}(t)$	0.0437	0.0458	0.0537	0.0540
$d = 4\,\mu m$	$F_{11}(t)$	0.0574	0.0575	0.0922	0.0924
	$F_{21}(t)$	0.0195	0.0201	0.0299	0.0234
$d = 6\,\mu m$	$F_{11}(t)$	0.0395	0.0395	0.0538	0.0537
	$F_{21}(t)$	0.0176	0.0177	0.0217	0.0218
$d = 8\,\mu m$	$F_{11}(t)$	0.0305	0.0305	0.0383	0.0383
	$F_{21}(t)$	0.0152	0.0153	0.0198	0.0199
$d = 10\,\mu m$	$F_{11}(t)$	0.0239	0.0239	0.0303	0.0304
	$F_{21}(t)$	0.0133	0.0133	0.0182	0.0183

Table 7.13 Mean RMSE values for ANN techniques with full TDS.

		$D = 50\,\mu m^2/s$		$D = 100\,\mu m^2/s$	
		One ANN	Two ANN	One ANN	Two ANN
$d = 2\,\mu m$	$F_{11}(t)$	0.1875	0.1688	0.3155	0.3222
	$F_{21}(t)$	0.0628	0.0553	0.0799	0.0729
$d = 4\,\mu m$	$F_{11}(t)$	0.0592	0.0590	0.0922	0.0926
	$F_{21}(t)$	0.0231	0.0219	0.0275	0.0270
$d = 6\,\mu m$	$F_{11}(t)$	0.0404	0.0401	0.0543	0.0540
	$F_{21}(t)$	0.0188	0.0184	0.0237	0.0229
$d = 8\,\mu m$	$F_{11}(t)$	0.0314	0.0312	0.0390	0.0388
	$F_{21}(t)$	0.0165	0.0156	0.0207	0.0205
$d = 10\,\mu m$	$F_{11}(t)$	0.0247	0.0246	0.0309	0.0306
	$F_{21}(t)$	0.0147	0.0137	0.0182	0.0188

channel response was evaluated. Tables 7.12 and 7.13 show the following: (1) The *one-machine* method learns the channel model parameters slightly better than the *two-machines* method. (2) The proposed techniques perform better when increasing the distance. For the short distances, the compensation parameters cannot work effectively enough to model the molecular MIMO channel parameters. The gain of the proposed technique is higher for the longer distances both in terms of RMSE (i.e., modeling performance) and the run time (simulations for the longer distances require much more run time).

In order to analyze the dataset requirements of the proposed methods, only half of the TDS was used to train (data for the half TDS was removed randomly). The different performance for ANNs was observed according to the amount of input data. The two-machines technique showed a better performance when the half of the TDS for the training phase (Table 7.13) is used. Hence, more training data was needed to make the one-machine technique perform better at modeling the channel.

7.3.3.3 *Theoretical BER analysis*

As a case study, the channel model from ANN was considered and the BER was evaluated by using Equation (7.3.6). To validate the proposed model, simulations with the given parameters were carried out and the empirical BER results were obtained. Simulations were replicated with consecutive 106 bits. It can be seen from Figure 7.24 that the simulation and model results are remarkably close for the selected case from VDS. In the zoomed inset plot, there is a small deviation from the simulation result.

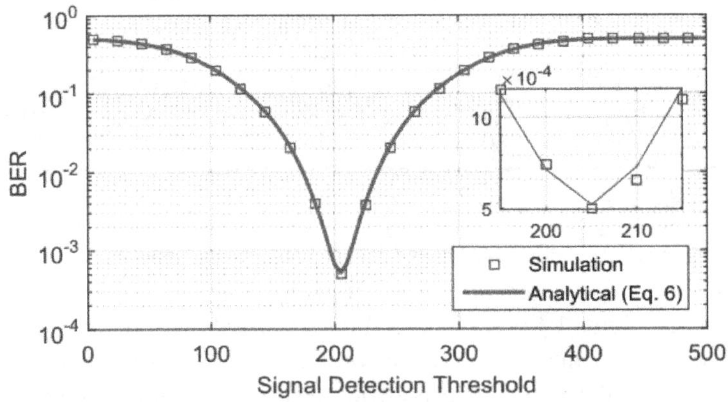

Figure 7.24 BER plot for $d = 6\,\mu\text{m}$, $h = 1\,\mu\text{m}$, $R = 5\,\mu\text{m}$, $D = 100\,\mu\text{m}^2/s$, $N = 1000$, $t_s = 0.5\,s$, $\eta = 5$.

References

[1] Perez Fontan F., Vazquez Castro M. A., Buonomo S., *et al.* S-band LMS propagation channel behavior for different environments, degrees of shadowing and elevation angles. *IEEE Transactions on Broadcasting*, 1998, 44(1): 40–76.

[2] Fontan F. P., Vázquez-Castro M., Cabado C. E., *et al.* Statistical modeling of the LMS channel. *IEEE Transactions on Vehicular Technology*, 2001, 50(6): 1549–1567.

[3] Prieto Cerdeira R., Perez Fontan F., Burzigotti P., *et al.* Versatile two — state land mobile satellite channel model with first application to DVB — SH analysis. *International Journal of Satellite Communications and Networking*, 2010, 28(5–6): 291–315.

[4] Carrie G., Pérez Fontán F., Lacoste F., *et al.* A generative MIMO channel model encompassing single satellite and satellite diversity cases. *European Conference on Antennas and Propagation(EUCAP)*, IEEE Press, Prague, Czech Republic, 2012, pp. 2164–3342.

[5] Arndt D., Heyn T., König J., Ihlow A., *et al.* Extended two-state narrowband LMS propagation model for S-Band. *IEEE International Symposium on Broadband Multimedia Systems and Broadcasting*, IEEE Press, Seoul, Korea (South), 2012, pp. 1–6.

[6] Seunghwan C., Jong-Seob B., Jong-Soo S., Analysis of dual-polarized MIMO transmission for digital terrestrial broadcasting systems over multipath Ricean fading channel. *IEEE International Symposium on Broadband Multimedia Systems and Broadcasting*, IEEE Press, Seoul, Korea (South), 2012: 1–6.

[7] Heyn T., Eberlein E., Arndt D., *et al.* Mobile satellite channel with angle diversity: The MiLADY project. *IEEE European Conference on Antennas & Propagation*, IEEE Press, Prague, Czech Republic, 2010, pp. 1–5.

[8] Lacoste F., Villacieros B. M., Prieto-Cerdeira R., *et al.* SISO and MIMO enhanced 2-state modelling of the Land Mobile satellite channel for various frequencies, environments and elevation angles. *Antennas and Propagation (EuCAP), 2014 8th European Conference on IEEE*, IEEE Press, The Hague, Netherlands, 2014, pp. 2277–2281.

[9] Lacoste F., Lemorton J., Casadebaig L., *et al.* SISO, MIMO and SIMO characterisation of the land mobile and nomadic satellite propagation channels. *IEEE European Conference on Antennas and Propagation*, IEEE Press, Gothenburg, Sweden, 2013, pp. 2291–2295.

[10] Liolis K. P., Gomez Vilardebo J., Casini E., *et al.* Statistical modeling of dual-polarized MIMO land mobile satellite channels. *IEEE Transactions on Communications*, 2010, 58(11): 3077–3083.

[11] King P. R. and Stavrou S. Capacity improvement for a land mobile single satellite MIMO system. *IEEE Antennas & Wireless Propagation Letters*, 2006, 5(1): 98–100.
[12] IEEE 802 LAN/MAN Standards Committee. Standard for information technology- telecommunications and information exchange between systems-local and metropolitan area networks-specific requirements [Online]. IEEE 802.11n, 2009. http://standards.ieee.org.
[13] IEEE 802.16 Working Group. Standard for local metropolitan area networks [Online]. IEEE 802.16, 2009. http://standards.ieee.org.
[14] Third generation partnership project [Online]. http://www.3gpp.org.
[15] Digital video broadcasting standards [Online]. http://www.dvb.org.
[16] Saunders S. R. and Aragón-Zavala A. A. *Antennas and Propagation for Wireless Communications.* 2nd edn., London: Wiley, 2007.
[17] Sellathurai M., Guinand P., and Lodge J. Space–Time coding in mobile satellite communications using dual-polarized channels. *IEEE Transactions on Vehicular Technology*, 2006, 55(1): 188–199.
[18] King P. R., Brown T. W. C., Kyrgiazos A., *et al.* Empirical-stochastic LMS-MIMO channel model implementation and validation. *IEEE Transactions on Antennas and Propagation*, 2012, 60(2): 606–614.
[19] Fontan F. P. and Espiñeira P. M. *Modelling the Wireless Propagation Channel: A Simulation Approach with MATLAB.* London: Wiley, 2008.
[20] Vaughan R. G. and Andersen J. B. *Channels, Propagation and Antennas for Mobile Communications.* London: IEEE Press, 2002.
[21] Gudmundson M. Correlation model for shadow fading in mobile radio systems. *Electronics Letters*, 1991, 27(23): 2145–2146.
[22] Pätzold M. *Mobile Fading Channels.* London: Wiley, 2002.
[23] Kermoal J. P., Schumacher L., Pedersen K. I., *et al.* A stochastic MIMO radio channel model with experimental validation. *IEEE Journal on Selected Areas in Communications*, 2002, 20(6): 1211–1226.
[24] Farsad N., Yilmaz H. B., Eckford A., *et al.* A comprehensive survey of recent advancements in molecular communication. *IEEE Communications Surveys & Tutorials*, 2016, 18(3): 1887–1919.
[25] Nakano T., Eckford A. W., and Haraguchi T. *Molecular Communication.* Cambridge: Cambridge University Press, 2013.
[26] Akyildiz I. F., Jornet J. M., and Pierobon M. Nanonetworks: A new frontier in communications. *International Conference on Wireless Information Networks and Systems (WINSYS)*, IEEE Press, Athens, Greece, 2010, pp. IS-5 – IS-5.
[27] Guo W., Mias C., Farsad N., *et al.* Molecular versus electromagnetic wave propagation loss in macro-scale environments. *IEEE Transactions on Molecular, Biological and Multi-Scale Communications*, 2015, 1(1): 18–25.

[28] Arjmandi H., Gohari A., Kenari M. N., *et al.* Diffusion-based nanonetworking: A new modulation technique and performance analysis. *IEEE Communications Letters*, 2013, 17(4): 645–648.

[29] Noel A., Cheung K. C., and Schober R. Optimal receiver design for diffusive molecular communication with flow and additive noise. *IEEE Transactions on Nano-Bioscience*, 2014, 13(3): 350–362.

[30] Jamali V., Ahmadzadeh A., and Schober R. On the design of matched filters for molecule counting receivers. *IEEE Communications Letters*, 2017, 21(8): 1711–1714.

[31] Kilinc D. and Akan O. B. Receiver design for molecular communication. *IEEE Journal on Selected Areas in Communications*, 2013, 31(12): 705–714.

[32] Azadi M. and Abouei J. A novel electrical model for advection–diffusion-based molecular communication in nanonetworks. *IEEE Transactions on Nano-Bioscience*, 2016, 15(3): 246–257.

[33] Srinivas K. V., Eckford A. W., and Adve R. S. Molecular communication in fluid media: The additive inverse gaussian noise channel. *IEEE Transactions on Information Theory*, 2012, 58(7): 4678–4692.

[34] Kadloor S., Adve R. S., and Eckford A. W. Molecular communication using Brownian motion with drift. *IEEE Transactions on Nano-Bioscience*, 2012, 11(2): 89–99.

[35] Yilmaz H. B., Heren A. C., Tugcu T., *et al.* Three-dimensional channel characteristics for molecular communications with an absorbing receiver. *IEEE Communications Letters*, 2014, 18(6): 929–932.

[36] Bilgin B. A. and Akan O. B. A fast algorithm for analysis of molecular communication in artificial synapse. *IEEE Transactions on Nano Bioscience*, 2017, 16(7): 408–417.

[37] Akkaya A., Yilmaz H. B., Chae C. B., *et al.* Effect of receptor density and size on signal reception in molecular communication via diffusion with an absorbing receiver. *IEEE Communications Letters*, 2015, 19(2): 155–158.

[38] Lee C., Yilmaz H. B., and Chae C.-B. Machine learning based channel modeling for molecular MIMO communications. *2017 IEEE 18th International Workshop on Signal Processing Advances in Wireless Communications (SPAWC)*, IEEE Press, Sapporo, Japan, 2017, pp. 1–5.

[39] Koo B. H., Lee C., Yilmaz H. B., *et al.* Molecular MIMO: From theory to prototype. *IEEE Journal on Selected Areas in Communications*, 2016, 34(3): 600–614.

[40] Rouzegar S. M. and Spagnolini U. Channel estimation for diffusive MIMO molecular communications. *2017 European Conference on Networks and Communications (EuCNC)*, IEEE Press, Oulu, Finland, 2017, pp. 1–5.

[41] Kuran M. S., Yilmaz H. B., Tugcu T., *et al.* Modulation techniques for communication via diffusion in nanonetworks. *Communications (ICC), 2011 IEEE International Conference*, IEEE Press, Kyoto, Japan, 2011, pp. 1–5.

[42] Kim N. R. and Chae C. B. Novel modulation techniques using isomers as messenger molecules for nano communication networks via diffusion. *IEEE Journal on Selected Areas in Communications*, 2013, 31(12): 847–856.

Index

S

V

W

CPSIA information can be obtained
at www.ICGtesting.com
Printed in the USA
JSHW031935080722
27874JS00003B/6

9 789811 249440